Pipeline Corrosion Control

Level Two

Trainee Guide
Third Edition

PEARSON

Boston Columbus Indianapolis New York San Francisco Amsterdam
Cape Town Dubai London Madrid Milan Munich Paris Montreal Toronto Delhi
Mexico City Sao Paulo Sydney Hong Kong Seoul Singapore Taipei Tokyo

NCCER

President: Don Whyte
Vice President: Steve Greene
Chief Operations Officer: Katrina Kersch
Director of Product Development: Daniele Dixon
Pipeline Mechanical Project Managers: Patty Bird, Elizabeth Schlaupitz, Justin Street, Chris Wilson
Senior Development Manager: Mark Thomas
Senior Production Manager: Tim Davis
Quality Assurance Coordinator: Karyn Payne
Desktop Publishing Coordinator: James McKay
Permissions Specialist: Adrienne Payne
Production Specialist: Adrienne Payne
Editors: Graham Hack, Karyn Payne, Debie Hicks, Jackie Strawbridge, Adrienne Payne

Writing and development services provided by Topaz Publications, Liverpool, NY

Lead Writers/Project Managers: Don Congdon, Veronica Westfall
Desktop Publisher: Joanne Hart
Art Director: Alison Richmond
Permissions Editors: Andrea LaBarge, Briana Morgan
Writers: Don Congdon, Tony Clay, Thomas Burke

Pearson

Director of Alliance/Partnership Management: Andrew Taylor
Editorial Assistant: Collin Lamothe
Program Manager: Alexandrina B. Wolf
Digital Studio Project Managers: Heather Darby, Tanika Henderson, Jose Carchi
Director of Marketing: Leigh Ann Simms
Senior Marketing Manager: Brian Hoehl
Composition: NCCER
Printer/Binder: LSC Communications
Cover Printer: LSC Communications
Text Fonts: Palatino and Univers

Credits and acknowledgments for content borrowed from other sources and reproduced, with permission, in this textbook appear at the end of each module.

PEARSON

ISBN-13: 978-0-13-471651-0
ISBN-10: 0-13-471651-5

ScoutAutomatedPrintCode

Preface

To the Trainee

NCCER is pleased to release the 3rd edition of Pipeline Training and Qualification with updates that will engage you and give you the best training in the industry. The curriculum has been redesigned to better align with the learning objectives, and also includes new end-of-section review questions to compliment the Module Review. Changes to the NCCER's covered task (CT) modules include updates to the latest technology and standards, as well as alignment to the American Petroleum Institute's *API Recommended Practice for Pipeline Operator Qualification (OQ) 1161*, 3rd edition (2014). Each CT module contains task-specific Abnormal Operating Conditions (AOCs) to which the trainee must be prepared to respond. General AOCs for both Field (Liquid and Gas) and Control Center settings are also available as single modules.

CT modules are available as traditional printed levels and as individual module eBooks. Modules are completed by successfully passing the proctored Module Examination, and completing the Performance Profile under the supervision of an NCCER-certified evaluator. These completions are tracked within the NCCER Registry for training delivered through an Accredited Training Sponsor. Covered task module completions in the Registry are also documented in ISNetworld.

The modules are designed to be used for Operational Qualification, and they are each a vital part of NCCER's pipeline workforce development curricula. They provide in-depth training on the fundamental knowledge and skills required for pipeline work, as well as technical training that increases the depth and breadth of a pipeline worker's capability within several specialty areas.

We invite you to visit the NCCER website at **www.nccer.org** for information on the latest product releases and training, as well as online versions of the *Cornerstone* magazine and Pearson's NCCER product catalog.

Your feedback is welcome. You may email your comments to **curriculum@nccer.org** or send general comments and inquiries to **info@nccer.org**.

NCCER Standardized Curricula

NCCER is a not-for-profit 501(c)(3) education foundation established in 1996 by the world's largest and most progressive construction companies and national construction associations. It was founded to address the severe workforce shortage facing the industry and to develop a standardized training process and curricula. Today, NCCER is supported by hundreds of leading construction and maintenance companies, manufacturers, and national associations. The NCCER Standardized Curricula was developed by NCCER in partnership with Pearson, the world's largest educational publisher.

Some features of the NCCER Standardized Curricula are as follows:

- An industry-proven record of success
- Curricula developed by the industry for the industry
- National standardization providing portability of learned job skills and educational credits
- Compliance with the Office of Apprenticeship requirements for related classroom training (*CFR 29:29*)
- Well-illustrated, up-to-date, and practical information

NCCER also maintains the NCCER Registry, which provides transcripts, certificates, and wallet cards to individuals who have successfully completed a level of training within a craft in NCCER's Curricula. *Training programs must be delivered by an NCCER Accredited Training Sponsor in order to receive these credentials.*

Additional Pipeline Resources

The following additional Pipeline Resources are available from NCCER's website at **www.nccer.org/pipeline**:

1. A complete copy of NCCER's Covered Task (CT) list
2. Access to:
    a. An online bookstore to order print and digital versions of NCCER's pipeline curricula.
    b. Instructor resources (Lesson Plans and PowerPoints)
    c. Exams for Operator Qualification
    d. A site to custom build curriculum with pipeline and/or other NCCER-developed modules

Pipeline Career Pathway Training

Entry-level Pipeline Craft Trainee → **Pipeline Fundamentals** → **Level 1**

The newly established pipeline career pathway represented on this page provides the recognition pipeline professionals deserve through standardized training and industry-recognized credentials.

Levels of training have been established by combining training modules into specific focus areas within the industry. This allows trainees to progress through a standardized program and earn a professional credential.

In addition, Levels Two and Three are comprised of covered task training modules that can lead to operator qualifications. *Pipeline Corrosion Control is the only exception which actually provides covered task training starting in Level One.*

The blue boxes denote the professional credential that can be earned with successful completion of each program.

A trained workforce is a safe and qualified workforce!

Pipeline Fundamentals:

Core Curriculum

Introduction to the Pipeline Industry

and

a.) Field Abnormal Operating Conditions
Module: AOCFG-17

or

b.) Control Center Abnormal Operating Conditions
Module: AOCCC_17

Level 1:

Field and Control Center Operations

Pipeline Maintenance and Mechanical

Electrical and Instrumentation

Pipeline Corrosion Control CT

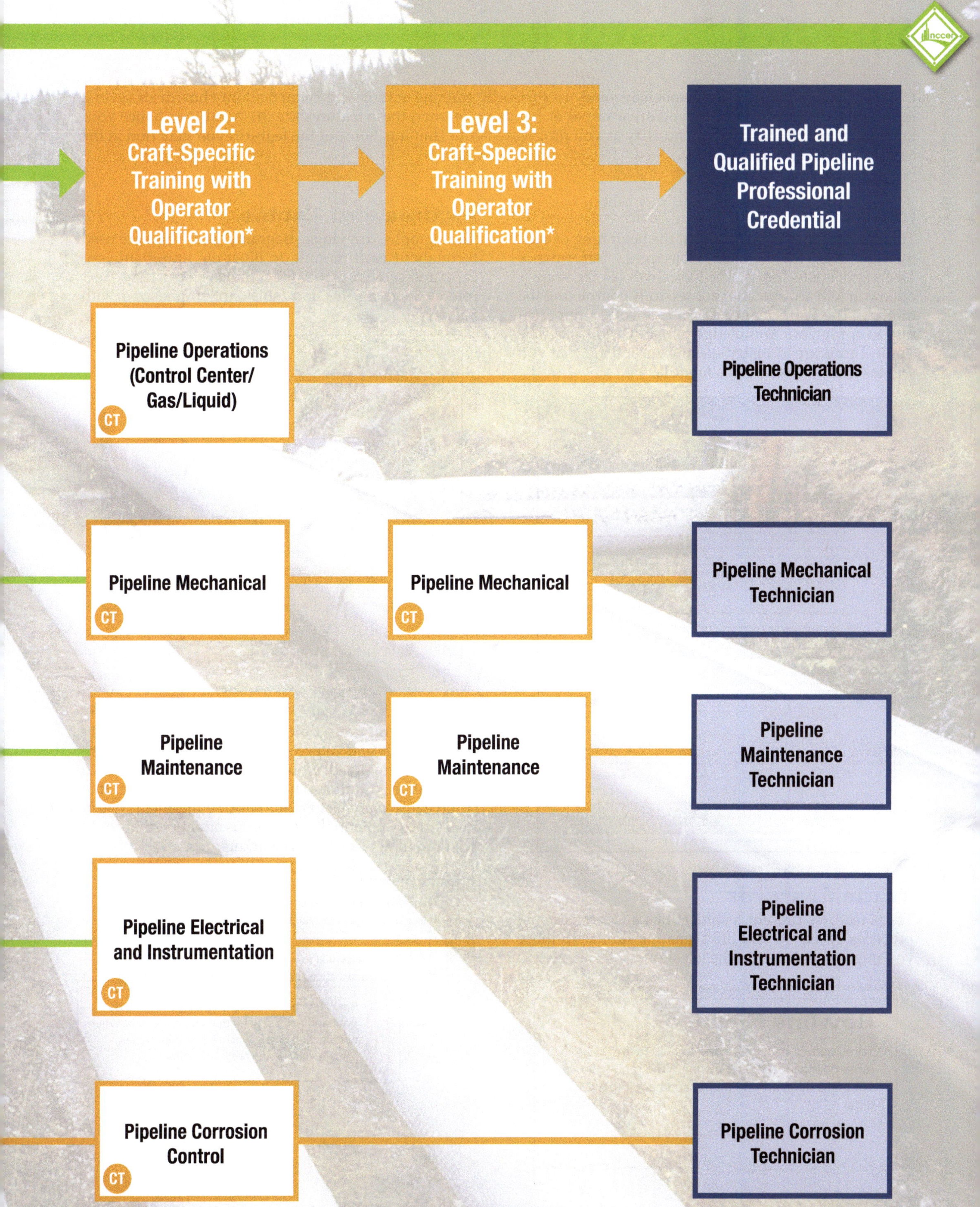

CT *Each of these titles have corresponding operator qualifications. The Covered Task list is available online at www.nccer.org/pipeline-program*

Special Features

In an effort to provide a comprehensive and user-friendly training resource, this curriculum showcases several informative features. Whether you are a visual or hands-on learner, these features are intended to enhance your knowledge of the construction industry as you progress in your training. Some of the features you may find in the curriculum are explained below.

Introduction

This introductory page, found at the beginning of each module, lists the module Objectives, Performance Tasks and Trade Terms. The Objectives list the knowledge you will acquire after successfully completing the module. The Performance Tasks give you an opportunity to apply your knowledge to real-world tasks. The Trade Terms are industry specific vocabulary that you will learn as you study this module.

00104-15
INTRODUCTION TO POWER TOOLS

Objectives

When you have completed this module, you will be able to do the following:

1. Identify and explain how to use various types of power drills and impact wrenches.
   a. Identify and explain how to use common power drills and bits.
   b. Identify and explain how to use a hammer drill.
   c. Identify and explain how to use pneumatic drills and impact wrenches.
2. Identify and explain how to use various types of power saws.
   a. Identify and explain how to use a circular saw.
   b. Identify and explain how to use saber and reciprocating saws.
   c. Identify and explain how to use a portable band saw.
   d. Identify and explain how to use miter and cutoff saws.
3. Identify and explain how to use various grinders and grinder attachments.
   a. Identify and explain how to use various types of grinders.
   b. Identify and explain how to use various grinder accessories and attachments.
4. Identify and explain how to use miscellaneous power tools.
   a. Identify and explain how to use pneumatic and powder-actuated fastening tools.
   b. Identify and explain how to use pavement breakers.
   c. Identify and explain the uses of hydraulic jacks.

Performance Tasks

Under supervision of your instructor, you should be able to do the following:

1. Safely and properly demonstrate the use of three of the following tools:
   - Electric drill
   - Hammer drill or rotary hammer
   - Circular saw
   - Reciprocating saw
   - Portable band saw
   - Miter or cutoff saw
   - Portable or bench grinder
   - Pneumatic nail gun
   - Pavement breaker

Trade Terms

Abrasive	Chuck key	Ground fault circuit interrupter (GFCI)	Revolutions per minute (rpm)
Alternating current (AC)	Countersink	Ground fault protection	Ring test
Arbor	Direct current (DC)	Kerf	Shank
Auger bit	Forstner bit	Masonry bit	Trigger lock
Carbide	Grit	Reciprocating	
Chuck			

Industry Recognized Credentials

If you are training through an NCCER-accredited sponsor, you may be eligible for credentials from NCCER's Registry. The ID number for this module is 00104-15. Note that this module may have been used in other NCCER curricula and may apply to other level completions. Contact NCCER's Registry at 888.622.3720 or go to **www.nccer.org** for more information.

Trade Features

Trade features present technical tips and professional practices based on real-life scenarios similar to those you might encounter on the job site.

Bowline Trivia

Some people use this saying to help them remember how to tie a bowline: "The rabbit comes out of his hole, around a tree, and back into the hole."

Figures and Tables

Photographs, drawings, diagrams, and tables are used throughout each module to illustrate important concepts and provide clarity for complex instructions. Text references to figures and tables are emphasized with *italic* type.

Notes, Cautions, and Warnings

Safety features are set off from the main text in highlighted boxes and categorized according to the potential danger involved. Notes simply provide additional information. Cautions flag a hazardous issue that could cause damage to materials or equipment. Warnings stress a potentially dangerous situation that could result in injury or death to workers.

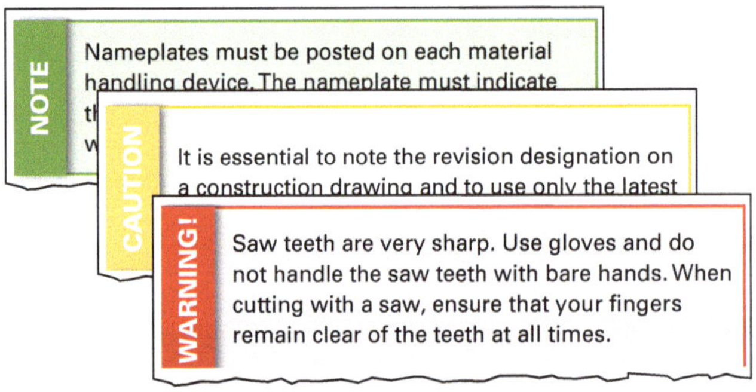

Case History

Case History features emphasize the importance of safety by citing examples of the costly (and often devastating) consequences of ignoring best practices or OSHA regulations.

> **Case History**
>
> **Requesting an Outage**
>
> An electrical contractor requested an outage when asked to install two bolt-in, 240V breakers in panels in a data processing room. It was denied due to the 24/7 worldwide information processing hosted by the facility. The contractor agreed to proceed only if the client would sign a letter agreeing not to hold them responsible if an event occurred that damaged computers or resulted in loss of data. No member of upper management would accept liability for this possibility, and the outage was scheduled.
>
> **The Bottom Line:** If you can communicate the liability associated with an electrical event, you can influence management's decision to work energized.

Going Green

Going Green features present steps being taken within the construction industry to protect the environment and save energy, emphasizing choices that can be made on the job to preserve the health of the planet.

> GOING GREEN
>
> **Reducing Your Carbon Footprint**
>
> Many companies are taking part in the paperless movement. They reduce their environmental impact by reducing the amount of paper they use. Using email helps to reduce the amount of paper used,

Did You Know

Did You Know features introduce historical tidbits or interesting and sometimes surprising facts about the trade.

> **Did You Know?**
>
> **Safety First**
>
> Safety training is required for all activities. Never operate tools, machinery, or equipment without prior training. Always refer to the manufacturer's instructions.

Step-by-Step Instructions

Step-by-step instructions are used throughout to guide you through technical procedures and tasks from start to finish. These steps show you how to perform a task safely and efficiently.

> Perform the following steps to erect this system area scaffold:
>
> *Step 1* Gather and inspect all scaffold equipment for the scaffold arrangement.
>
> *Step 2* Place appropriate mudsills in their approximate locations.
>
> *Step 3* Attach the screw jacks to the mudsills.

Trade Terms

Each module presents a list of Trade Terms that are discussed within the text and defined in the Glossary at the end of the module. These terms are presented in the text with **bold, blue** type upon their first occurrence. To make searches for key information easier, a comprehensive Glossary of Trade Terms from all modules is located at the back of this book.

> During a rigging operation, the **load** being lifted or moved must be connected to the apparatus, such as a crane, that will provide the power for movement. The connector—the link between the load and the apparatus—is often a sling made of synthetic, chain, or **wire rope** materials. This section focuses on three types of slings:

Section Review

Each section of the module wraps up with a list of Additional Resources for further study and Section Review questions designed to test your knowledge of the Objectives for that section.

> **Additional Resources**
>
> *Materials Handling Handbook*, The American Society of Mechanical Engineers (ASME) and The International Material Management Society (IMMS), Raymond A. Kulwiec, Editor-in-Chief. 1985. New York, NY: Wiley-Interscience.
>
> *Manufacturing Facilities Design & Material Handling*, Matthew P. Stevens, Fred E. Meyers. 2013. West Lafayette, IN: Purdue University Press.
>
> **1.0.0 Section Review**
>
> 1. For material handling tasks, it is just as important to be mentally fit as it is to be _____.
>   a. physically fit
>   b. physically aggressive
>   c. closely supervised
>   d. over 200 pounds
> 2. Which of the following is a type of knot that is often used to join the ends of two ropes in non-critical, low-strain applications?
>   a. Bowline
>   b. Clove hitch
>   c. Half hitch
>   d. Square knot

Review Questions

The end-of-module Review Questions can be used to measure and reinforce your knowledge of the module's content.

> **Review Questions**
>
> 1. Identification tags for slings must include the _____.
>   a. type of protective pads to use
>   b. type of damage sustained during use
>   c. color of the tattle-tail
>   d. manufacturer's name or trademark
> 2. The type of wire rope core that is susceptible to heat damage at relatively low temperatures is the _____.
>   a. fiber core
>   b. strand core
>   c. independent wire rope core
>   d. metallic link supporting core
> 3. Synthetic slings must be inspected _____.
>   a. once every month
>   b. visually at the start of each work week
>   c. before every use
>   d. once wear or damage becomes apparent
> 4. An alloy steel chain sling must be removed from service if there is evidence that _____.
>   a. the sling has been used in different hitch configurations
>   b. replacement links have been used to repair the chain
>   c. the sling has been used for more than one year
>   d. strands in the supporting core have weakened
> 5. A piece of rigging hardware used to couple the end of a wire rope to eye fittings, hooks, or other connections is a(n) _____.
>   a. eyebolt
>   b. hitch
>   c. shackle
>   d. U-bolt
> 6. A lifting clamp is most likely to be used to move loads such as _____.
>   a. steel plates
>   b. piping bundles
>   c. concrete blocks
>   d. plastic tubing
> 7. Chain hoists are able to lift heavy loads by utilizing a _____.
>   a. rope and pulley system
>   b. rigger's strength
>   c. stationary counterweight
>   d. gear system
> 8. Before attempting to lift a load with a chain hoist, make sure that the _____.
>   a. hoist is secured to a come-along
>   b. load is properly balanced
>   c. tag lines are properly anchored
>   d. tackle is connected to its power source
> 9. A hitch configuration that allows slings to be connected to the same load without using a spreader beam is a _____.
>   a. double-wrap hitch
>   b. choker hitch
>   c. bridle hitch
>   d. basket hitch
> 10. To make the emergency stop signal that is used by riggers, extend both arms _____.
>   a. horizontally with palms down and quickly move both arms back and forth
>   b. directly in front and then move both arms up and down repeatedly
>   c. vertically above the head and wave both arms back and forth
>   d. horizontally with clenched fists and move both arms up and down
>
> 00106-15 Introduction to Basic Rigging Module Six 31

NCCER Standardized Curricula

NCCER's training programs comprise more than 80 construction, maintenance, pipeline, and utility areas and include skills assessments, safety training, and management education.

Boilermaking
Cabinetmaking
Carpentry
Concrete Finishing
Construction Craft Laborer
Construction Technology
Core Curriculum: Introductory Craft Skills
Drywall
Electrical
Electronic Systems Technician
Heating, Ventilating, and Air Conditioning
Heavy Equipment Operations
Highway/Heavy Construction
Hydroblasting
Industrial Coating and Lining Application Specialist
Industrial Maintenance Electrical and Instrumentation Technician
Industrial Maintenance Mechanic
Instrumentation
Insulating
Ironworking
Masonry
Millwright
Mobile Crane Operations
Painting
Painting, Industrial
Pipefitting
Pipelayer
Plumbing
Reinforcing Ironwork
Rigging
Scaffolding
Sheet Metal
Signal Person
Site Layout
Sprinkler Fitting
Tower Crane Operator
Welding

Maritime

Maritime Industry Fundamentals
Maritime Pipefitting
Maritime Structural Fitter

Green/Sustainable Construction

Building Auditor
Fundamentals of Weatherization
Introduction to Weatherization
Sustainable Construction Supervisor
Weatherization Crew Chief
Weatherization Technician
Your Role in the Green Environment

Energy

Alternative Energy
Introduction to the Power Industry
Introduction to Solar Photovoltaics
Introduction to Wind Energy
Power Industry Fundamentals
Power Generation Maintenance Electrician
Power Generation I&C Maintenance Technician
Power Generation Maintenance Mechanic
Power Line Worker
Power Line Worker: Distribution
Power Line Worker: Substation
Power Line Worker: Transmission
Solar Photovoltaic Systems Installer
Wind Turbine Maintenance Technician

Pipeline

Pipeline Fundamentals
Pipeline Corrosion Control
Pipeline Electrical and Instrumentation
Pipeline Maintenance
Pipeline Mechanical
Pipeline Operations (Control Center, Gas, Liquid)

Safety

Field Safety
Safety Orientation
Safety Technology

Supplemental Titles

Applied Construction Math
Tools for Success

Management

Construction Workforce Development Professional
Fundamentals of Crew Leadership
Mentoring for Craft Professionals
Project Management
Project Supervision

Spanish Titles

Acabado de concreto: nivel uno (*Concrete Finishing Level One*)
Aislamiento: nivel uno (*Insulating Level One*)
Albañilería: nivel uno (*Masonry Level One*)
Andamios (*Scaffolding*)
Carpintería: Formas para carpintería, nivel tres (*Carpentry: Carpentry Forms, Level Three*)
Currículo básico: habilidades introductorias del oficio (*Core Curriculum: Introductory Craft Skills*)
Electricidad: nivel uno (*Electrical Level One*)
Herrería: nivel uno (*Ironworking Level One*)
Herrería de refuerzo: nivel uno (*Reinforcing Ironwork Level One*)
Instalación de rociadores: nivel uno (*Sprinkler Fitting Level One*)
Instalación de tuberías: nivel uno (*Pipefitting Level One*)
Instrumentación: nivel uno, nivel dos, nivel tres, nivel cuatro (*Instrumentation Levels One through Four*)
Orientación de seguridad (*Safety Orientation*)
Paneles de yeso: nivel uno (*Drywall Level One*)
Seguridad de campo (*Field Safety*)

Acknowledgments

This curriculum was revised as a result of the farsightedness and leadership of the following sponsors:

American Petroleum Institute
Central NDT
Construction Craft Academy
CP Masters
EMS USA, Inc.
Enbridge Pipelines, Inc.
Enhance Co.
Enterprise
Flint Energy Services, Inc.
Jomax Construction
Kinder Morgan, Inc.
L.E. Bell Construction Co. Inc.
Marathon Pipe Line
Midwestern Contractors
OneOK
Plains All-American Pipeline
Praxair Services
Protech EIS
Safety Advantage LLC
Southland Safety
Sunoco Logistics, L.P.
Sunoco Pipeline
Thorco Holdings, LLC
URS/Flint Energy

This curriculum would not exist were it not for the dedication and unselfish energy of those volunteers who served on the Authoring Team. A sincere thanks is extended to the following:

Gary Affolter
Carl Allebach
Travis Ashby
Steve Barker
Art Birx
James Carroll
Leslie Carroll
Gordon Carson
Bill Emery
Paul Fontenot II
Paula Gates-Rippe
Trobi Hall
Sid Harrell
David Hoffman
Kimberly-Joy Harris
Les Harrison
Lisa Hartnup
Benjamin Johnson
Dan Lake
Tom Magness
Wallace McCracken
Tim Mills
Steve Montoy
Michael Nordstrom
Tom Pickthall
Donnie Pittsinger
William Rado
David Reeder
Tom Reeder
Randall Rogers
Dan Rohret
Dan Ruane
DeWayne Schapansky
Mark Seitz
Blaine Shellhammer
Lynn Simpson
Cameron Strother
Charles Whiten
Rick Wilenzick
Michael Williams
Erin Worthington
Donavon Wright
Will Wright

NCCER Partners

American Council for Construction Education
American Fire Sprinkler Association
Associated Builders and Contractors, Inc.
Associated General Contractors of America
Association for Career and Technical Education
Association for Skilled and Technical Sciences
Construction Industry Institute
Construction Users Roundtable
Design Build Institute of America
GSSC – Gulf States Shipbuilders Consortium
ISN
Manufacturing Institute
Mason Contractors Association of America
Merit Contractors Association of Canada
NACE International
National Association of Women in Construction
National Insulation Association
National Technical Honor Society
National Utility Contractors Association
NAWIC Education Foundation
North American Crane Bureau
North American Technician Excellence
Pearson
Prov
SkillsUSA®
Steel Erectors Association of America
U.S. Army Corps of Engineers
University of Florida, M. E. Rinker Sr., School of Construction Management
Women Construction Owners & Executives, USA

NCCER Business Partners

Contents

CT9_1-17

Install Bonds

Explains how stray current interference affects cathodic protection, and it describes the procedures used to install bonds to eliminate stray current interference. (Module ID CT9_1-17; 5 Hours)

CT9_2-17

Install Galvanic Anodes

Describes how stray current interference affects cathodic protection. It presents the steps for the general procedures used to install galvanic anodes for stray current mitigation. (Module ID CT9_2-17; 5 Hours)

CT9_3-17

Install Rectifiers

Describes how stray current interference affects cathodic protection. It presents the steps for the general procedures used to install rectifiers for stray current mitigation. (Module ID CT9_3-17; 5 Hours)

CT9_4-17

Install Impressed Current Groundbeds

Explains how stray current interference affects cathodic protection. It describes the general procedures for installing impressed current groundbeds to mitigate stray current interference. (Module ID CT9_4-17; 5 Hours)

CT9_5-17

Repair Shorted Casings

Explains how shorted casings can compromise the cathodic protection of a pipeline. It describes how to test for shorted casings, and presents the general procedures for clearing, filling, repairing, monitoring, and replacing a shorted casing. (Module ID CT9_5-17; 5 Hours)

CT9_6-17

Install Electrical Insulating Device

Describes how electrical insulation/isolation is used in cathodic protection (CP) systems. It explains how to test for electrical isolation and describes procedures for installing various electrical insulating devices. (Module ID CT9_6-17; 5 Hours)

CT10-1-17

Insert and Remove Coupons

Covers the applications and use of corrosion coupons and describes the procedure for installing and removing coupons. (Module ID CT10_1-17; 5 Hours)

CT10_2-17

Monitor Probes (Online)

Describes various probes used to monitor conditions in a pipeline system, and provides a typical procedure for probe insertion and retrieval. (Module ID CT10-2-17; 5 Hours)

CT11_0-17

Monitoring and Controlling the Injection Rate of the Corrosion Inhibitor

Describes chemical injection systems used to inject corrosion inhibitors into a pipeline system, and it explains the general procedures for monitoring and adjusting the inhibitor injection rates. (Module ID CT11_0-17; 5 Hours)

AOCFG-17

Abnormal Operating Conditions – Field and Gas

Overview

Pipeline operators and contractors have programs, procedures, training, safety equipment, and warning devices for practically every facet of pipeline operations. Situations that are outside of normal operating conditions may arise at any time, and field personnel must know what to do to help keep pipeline operations safe for everyone. In some cases, an abnormal operating condition (AOC) is a routine matter that can be quickly and safely addressed by personnel. In other cases, an AOC can result in an emergency situation. In this module, personnel will learn how to recognize, properly react to, and properly report AOCs that may occur during pipeline operations.

Trainees with successful module completions may be eligible for credentialing through the NCCER Registry. To learn more, go to **www.nccer.org** or contact us at **1.888.622.3720**. Our website has information on the latest product releases and training, as well as online versions of our *Cornerstone* magazine and Pearson's product catalog.

Your feedback is welcome. You may email your comments to **curriculum@nccer.org**, send general comments and inquiries to **info@nccer.org**, or fill in the User Update form at the back of this module.

This information is general in nature and intended for training purposes only. Actual performance of activities described in this manual requires compliance with all applicable operating, service, maintenance, and safety procedures under the direction of qualified personnel. References in this manual to patented or proprietary devices do not constitute a recommendation of their use.

AOCFG-17
Abnormal Operating Conditions – Field and Gas

Objectives

When you have completed this module, you will be able to do the following:

1. Explain how to recognize abnormal operating conditions for both gas and liquid pipelines.
   a. Explain how to recognize abnormal pipeline facility conditions.
   b. Explain how to recognize the activation of a safety device.
   c. Explain how to recognize a communications failure and a control system failure.
   d. Explain how to recognize power interruptions.
   e. Explain how to recognize a fire, explosion, and natural disaster occurring with the pipeline.
   f. Explain how to recognize the unexpected release of hazardous liquid or gas.
   g. Explain how to recognize unexplained pressure/flow rate changes.
2. Describe the proper reaction to abnormal operating conditions for both gas and liquid pipelines.
   a. Describe the proper reaction to abnormal pipeline facility conditions.
   b. Describe the proper reaction to the activation of a safety device.
   c. Describe the proper reaction to a communications failure and a control system failure.
   d. Describe the proper reaction to power interruptions.
   e. Describe the proper reaction to a fire, explosion, and natural disaster occurring with the pipeline.
   f. Describe the proper reaction to the unexpected release of hazardous liquid or gas.
   g. Describe the proper reaction to unexplained pressure/flow rate changes.

Performance Tasks

This is a knowledge-based module; there are no performance tasks.

Trade Terms

Aboveground span
Abnormal operating condition (AOC)
American National Standards Institute (ANSI)
Breakout tank
Disbondment
Encroachment
Flame ionization detection unit
Highly volatile liquid (HVL)
Incipient fire
Intrinsically safe/explosionproof
Lower explosive limit (LEL)
Maximum allowable operating pressure (MAOP)
Maximum operating pressure (MOP)
Naturally occurring radioactive materials (NORMs)
One-Call notification system
Pipeline and Hazardous Materials Safety Administration (PHMSA)
Pipeline facility
Pipeline operator
Redundant system
Right-of-way (ROW)
Self-contained breathing apparatus (SCBA)
Supervisory control and data acquisition (SCADA)
Surge pressure
Uninterruptible power system (UPS)

Industry-Recognized Credentials

If you're training through an NCCER-accredited sponsor, you may be eligible for credentials from NCCER's Registry. The ID number for this module is AOCFG-17. Note that this module may have been used in other NCCER curricula and may apply to other level completions. Contact NCCER's Registry at 888.622.3720 or go to **www.nccer.org** for more information.

Contents

Topics to be presented in this module include:

Figures and Tables

Section One

1.0.0 Recognizing Abnormal Operating Conditions

Objective

Explain how to recognize abnormal operating conditions for both gas and liquid pipelines.

a. Explain how to recognize abnormal pipeline facility conditions.
b. Explain how to recognize the activation of a safety device.
c. Explain how to recognize a communications failure and a control system failure.
d. Explain how to recognize power interruptions.
e. Explain how to recognize a fire, explosion, and natural disaster occuring with the pipeline.
f. Explain how to recognize the unexpected release of hazardous liquid or gas.
g. Explain how to recognize unexplained pressure/flow rate changes.

Trade Terms

Aboveground span: A length of pipeline that is entirely aboveground and supported by pipe supports.

Abnormal operating condition (AOC): A condition identified by the operator that may indicate (a) a malfunction of a component or (b) a deviation from normal operations that may indicate a condition exceeding design limits or result in a hazard to persons, property, or the environment.

American National Standards Institute (ANSI): An organization that adopts standards for a variety of products, services, and processes in the United States to ensure consistency.

Breakout tank: A tank used to (a) relieve surges in a hazardous liquid pipeline system or (b) receive and store hazardous liquid transported by a pipeline for reinjection and continued transportation by pipeline.

Disbondment: Separation of the protective coating from the pipeline.

Encroachment: Activity carried out by, and/or equipment belonging to outside parties on the pipeline right-of-way (ROW). A particular encroachment might be authorized or not, so each situation requires specific handling.

Flame ionization detection unit: A device that detects some hazardous gases in the atmosphere.

Highly volatile liquid (HVL): A hazardous liquid which will form a vapor cloud when released to the atmosphere and which has a vapor pressure exceeding 276 kPa (40 psia) at 37.8°C (100°F).

Intrinsically safe/explosion-proof: Describes electrical equipment that is designed specifically for use in explosive or flammable locations so that a spark cannot ignite the atmosphere.

Lower explosive limit (LEL): The lowest concentration of a gas or a vapor in the atmosphere capable of producing fire in the presence of an ignition source. Different products have different LELs.

Maximum allowable operating pressure (MAOP): The maximum pressure at which a pipeline or segment of a pipeline may be operated (gas).

Maximum operating pressure (MOP): The maximum pressure at which a pipeline or segment of a pipeline may be normally operated (liquids).

One-Call notification system: A centralized communication system established by utilities, government agencies, and operators of underground facilities. This system provides one telephone number for contractors and the general public to call to notify of their intent to use equipment on or near a pipeline. It is also in place for any individual to call if pipeline damage is discovered. The One-Call system is generally set up on a state-by-state basis.

Pipeline and Hazardous Materials Safety Administration (PHMSA): An agency of the US Department of Transportation responsible for developing and enforcing regulations for pipeline operations in the United States.

Pipeline facility: New and existing pipe, rights-of-way, and any equipment, facility, or building used in the transportation of hazardous liquids or carbon dioxide, or new and existing pipelines, rights-of-way, and any equipment, facility, or building used in the transportation of gas or in the treatment of gas during the course of transportation.

Pipeline operator: The company that operates the pipeline, as opposed to an individual person who works for the company that operates the pipeline.

Redundant system: A system that duplicates the operations of a system so that, in the event of a failure, a backup system is available.

Right-of-way (ROW): The area along the pipeline which has been legally permitted to be used by the pipeline operator to construct, operate, and maintain the pipeline. The ROW extends the entire length of the pipeline.

Supervisory control and data acquisition (SCADA): A computer-based system or systems used by a controller in a control room that collects and displays information about a pipeline facility and may have the ability to send commands back to the pipeline facility.

Surge pressure: Pressure produced by a change in velocity of the moving stream that results from shutting down a pump station or pumping unit, closure of a valve, or any other blockage of the moving stream.

Uninterruptible power system (UPS): A system that provides continuity of power when the normal power supply is interrupted. A UPS provides AC power to the system it serves.

An abnormal operating condition (AOC) is defined under federal regulations as a condition identified by the pipeline operator that may indicate (1) a malfunction of a component or (2) a deviation from normal operations that may indicate a condition exceeding design limits or result in a hazard to persons, property, or the environment. AOCs are defined as such by federal pipeline safety law 49 *United States Code (U.S.C.)* 60102 and pipeline safety regulations 49 *Code of Federal Regulations (CFR)* 192 (for transportation of gas) and 49 *CFR* 195 (for transportation of liquids). The federal agency that regulates pipeline maintenance is the Pipeline and Hazardous Materials Safety Administration (PHMSA) in Washington, DC.

Personnel in the field are responsible for being aware of pipeline surroundings at all times so they can recognize AOCs when they occur. Types of AOCs, along with examples and proper reactions, are listed in *Table 1*. Note that the reactions are general.

Some pipeline equipment is fitted with warning devices to alert personnel to problems. Sometimes field workers will be notified of AOCs by control center personnel. In other cases, field workers must use their skills and training to observe the physical signs of AOCs. Field personnel play a critical role as the "eyes, ears, and nose" of the pipeline operator. They must be able to recognize the signs of AOCs and report them appropriately. They must stay in close communication with control center personnel who may need to take the proper action remotely. In the most serious cases, the control center may need to shut down the pipeline.

One of the most important responsibilities of field personnel is to be able to recognize the unexpected release of hazardous gases or liquids. In some cases, early recognition can prevent releases from turning into fires and explosions. Dead vegetation, dead wildlife (*Figure 1*), and an oily sheen on water are just a few of the signs that a release has occurred. *Table 2* illustrates the ways that field personnel must use their senses to be aware of the signs of releases.

1.1.0 Recognizing Abnormal Pipeline Facility Conditions

Any problem with the physical structure of the pipeline is defined as an abnormal pipeline facility condition. A pipeline facility is defined as new and existing pipe, right-of-way (ROW), and any equipment, facility, or building used in the transportation of hazardous liquids or carbon dioxide, or new and existing pipelines, rights-of-way, and any equipment, facility, or building used in the transportation of gas or in the treatment of gas during the course of transportation. Examples of components of a pipeline facility include compressors (for gas pipelines) and pump stations (for liquid pipelines).

The following are examples of abnormal pipeline facility conditions:

- Exposed pipeline that should be covered
- Pipeline system damage
- Missing or damaged pipeline supports
- Corrosion
- Coating separating from the pipeline (disbondment)
- Unauthorized activity and/or equipment on the pipeline right-of-way (unauthorized encroachment)
- Missing signs or markers
- Sunken floating tank roof
- Incorrect data or records caused by equipment malfunction
- Incorrect American National Standards Institute (ANSI) pressure rating of a component or material type
- Damage caused by fire or explosion
- Malfunctions due to human error

Work crews, line patrols, or other field personnel are often the first to discover abnormal pipeline facility conditions. However, a report may come in as a result of a call made to the One-Call notification system or from a person calling the pipeline operator directly. When a report comes in, personnel must obtain complete information so that they can quickly find the problem and address it. Many pipeline operators use a form to capture the following information:

- Caller's name and telephone number
- Date and time of the call

Table 1A Abnormal Operating Conditions: Examples and General Reactions (1 of 2)

Abnormal Operating Condition	Examples	Appropriate Reaction
Abnormal pipeline facility condition	• Exposed pipe • Evidence of corrosion • Incorrect ANSI pressure rating of a component or material type • Disbondment (coatings separating from the pipe) • Missing signs or markers • Unauthorized activity and/or equipment on the right-of-way (unauthorized encroachment) • Incorrect data or records caused by equipment malfunction • Mechanical/hydraulic system failures • Fire and explosion • Pipeline system damage, such as: – Missing or damaged pipeline supports – Damaged or inoperable pipeline components (pipe, valves, pumps, compressors, meters, tank gauges) due to external force – Gouges, dents, pitting, wrinkles, buckles, or cracks – Cracked or corroded flange or bolts – Sunken floating tank roof – Damaged or malfunctioning pontoon – Defective tank roof seal – Damaged tank shell – Damaged pipe due to natural disaster	Follow standard operating procedures and make notifications per company policy. Isolate and/or shut down the pipeline segment or facility if necessary. Complete repairs as necessary. Complete the required reporting.
Activation of a safety device	• Emergency or abnormal shutdown alarm • High-pressure shutdown alarm • Activation of a pressure relief valve • Activation of a smoke or fire detector • Activation of a hazardous atmosphere alarm • Activation of an overfill protection device • High sump-level shutdown alarm • High tank-level shutdown alarm • High pump-case pressure alarm • Seal failure alarm	Follow standard operating procedures and make notifications per company policy. Isolate and/or shut down the pipeline segment or facility if necessary. Complete the required reporting.
Communications failures	Personnel communication tools: • Lost use of radio communications • Lost use of landline telephone • Lost use of cell phone System communications: • Facility outages • SCADA communication outage • System programming errors/incorrect command logic • Loss of satellite communications • Loss of microwave signals • Stale data (communications going from the control center to the field, but not being received by the control center from the field)	Switch to alternate equipment or backup systems. Follow standard operating procedures and make notifications per company policy. Complete the required reporting.

Table 1B Abnormal Operating Conditions: Examples and General Reactions (2 of 2)

Power interruptions	• Transformer failure • Downed power lines • Broken leads or wires • Incorrect wire termination • Station or terminal power problems (due to motor overload or a tripped breaker, for example) • Weather conditions, such as an ice storm	Follow standard operating procedures. Switch to an alternate control system or power source if necessary. Notify the power company of the outage. Make additional notifications per company policy. Complete the required reporting.
Fire, explosion, or natural disaster	• Electrical fire • Ignition of a hazardous vapor • Unintentional detonation • Fire, explosion, or pipeline movement caused by natural disaster, such as: – Lightning strike – Flooding – Landslide – Earthquake –Tornado	Follow company emergency action plans. Call 911 if appropriate. Isolate and/or shut down the pipeline segment or facility if necessary. Limit access to the area as much as possible. Make notifications per company policy. Complete the required reporting.
Unexpected release of a hazardous gas or liquid	Signs include: • Vapor clouds or frost balls • Smell of hydrocarbon release • Dead vegetation/contaminated soil • Dead wildlife • Rainbow sheen on water Hazardous products include: • Hydrogen sulfide • Carbon dioxide • Propane • Natural gas • Refined fuels/products Note that the controlled transfer, sampling, venting, draining, and handling of hazardous products are not considered AOCs.	Follow company emergency action plans. Call 911 if appropriate. Isolate and/or shut down the pipeline segment or facility if necessary. Make notifications per company policy. Complete the required reporting.
Unexplained flow rate/ pressure change	Any change from the expected flow or pressure, such as: • Increased pressure • Decreased pressure • No pressure • High-pressure limit • Low-pressure limit • High flow • Low flow • No flow	Follow standard operating procedures and make notifications per company policy. Isolate and/or shut down the pipeline segment or facility if necessary. Complete the required reporting.

- Location of the problem (obtain the latitude/ longitude of the location from a global positioning system [GPS] device or a phone, if possible)
- Nearby landmarks such as highways, streams, mile markers, or any other structure that can help repair crews pinpoint the location
- The nature of the damage and any conditions observed at the site

1.1.1 Pipeline System Damage

Accidents that can damage the pipeline and its components occasionally happen. Personnel working on the right-of-way may strike the pipeline without realizing it. An excavator operator may accidentally hit the line. Although the locations of pipelines are clearly marked with signs or markers, construction work near the pipeline may hide markings. Ships' anchors may hit offshore

67107-13_F01.EPS

Figure 1 Dead wildlife may be a sign of an unexpected release.

pipelines. In some parts of the country, hunters have accidentally shot pipelines. Intentional damage due to vandalism, terrorist activity, or other criminal activity may also occur.

Pipeline damage may be discovered by field personnel, control center personnel, or a third party. Field personnel may notice physical signs of damage while out in the field. The control center may receive the indication of pipeline damage on the control center console as an unexplained change in operating conditions.

Physical signs of damage to the pipeline and its components may include gouges, cracks, pitting, dents, buckles, and inoperable valves, pumps, or tank gauges. Flanges or bolts may be cracked or show significant corrosion. Tank shells or pipeline supports may show damage. A sunken roof or a damaged pontoon can also indicate damage. Less obvious signs of damage could include a rainbow sheen on the water near submerged pipelines, which may indicate a release. The release of a highly volatile liquid (HVL) may be discovered by observing a frost ball or vapor cloud along the ROW of a liquid pipeline.

Table 2 Recognizing Warning Signs of Hazardous Gases or Liquids

Sense	Examples of Warning Signs
Sight	• Dead, discolored vegetation, often near healthy vegetation • Dead wildlife • Dry or frozen spots in soil during warm weather • Oily sheen on water/"rainbow" sheen • Vapor fog • Blowing dirt around the pipeline • Bubbles in standing water
Sound	Hissing, blowing, or roaring sound
Smell	Gaseous, skunk-like, or "rotten egg" odor is sometimes recognizable. Never rely on sense of smell only; a release can be completely odorless.

1.1.2 Damaged Pipeline Supports

Damage to pipeline supports can be a serious problem and can lead to a shutdown. In other cases, pipeline operations may continue while repairs are made. Personnel may recognize pipeline support damage while performing maintenance or inspection tasks in the field. Examples of damage include evidence of corrosion, visible cracking, sagging, or buckling of a support, or bolts or other components of the support that have broken or become loose. Damage to supports may be caused by corrosion, settling of soil, erosion due to wind or water, landslides, washouts, or earthquakes.

Any aboveground span must adhere to specific regulations. These spans refer to lengths of pipeline that are entirely aboveground and supported by pipe supports (see *Figure 2*). 49 *CFR* Part 195.110 requires that pipeline systems be designed to handle anticipated external loads from such conditions as earthquakes, vibration, thermal expansion, and contraction. In addition, the pipe and other components must be supported in such a way that the support does not cause excess stress to the pipe wall. 49 *CFR* Part 195.254 states each component of the pipeline that is aboveground must be protected from the forces exerted by the anticipated loads.

1.2.0 Recognizing the Activation of a Safety Device

Safety devices are a part of the pipeline system and are designed to sense and alert personnel to potential danger and/or hazardous conditions.

67107-13_F02.EPS

Figure 2 Aboveground span of pipeline on pipe supports.

Hazardous conditions can include changes in pressure that may exceed pipeline design limits as well as the release of harmful gases. The purpose of a safety device is to prevent injury or accidents. Safety devices include but are not limited to pressure relief systems, pressure limiting devices, high pump-case-pressure alarms, seal failure alarms, overfill protection devices, tank-level alarms, main-line pressure shutdown alarms, and hazardous atmosphere alarms.

When a safety device activates, it may sound an alarm. The device may ring loudly, beep continuously, or flash a light. A sensor on the device may also send a signal to the control center, where it appears as an alert on the control center console. Never ignore a safety alarm and never turn off a safety alarm until the problem has been checked and cleared by authorized personnel. Notify the control center when an alarm sounds, to maintain communication about appropriate steps to take.

1.2.1 Activation of a Pressure Relief System

The activation of relief devices along the pipeline system is a time-sensitive occurrence. Relief valves reduce pressure by discharging product. Depending on the product, it may be discharged into the atmosphere, tanks, flares, or sumps. Relief valves (see *Figure 3*) can be installed on pipeline or breakout tanks. The purpose of a pressure relief device is to protect the pipeline or pipeline facility from overpressure. Pressure relief valves are usually self-actuating at preset limits determined by the operator. However, according to DOT regulations, the pressure in a pipeline cannot exceed 110 percent of the operating pressure, so all preset limits must be within acceptable parameters.

1.2.2 Unintended Valve Closure or Shutdown

The unintended closure of a main-line block valve, remote block valve, manual block valve, pump-station suction or discharge valve, and certain terminal valves (manifold/tank) may cause a pressure change which could activate a pressure-relief system. Breakout tanks can be used to temporarily absorb surge pressure from the main-line and manifold pressure-relief valves to lower pressure on the line. According to 49 *CFR* 195, relief venting must be provided for each breakout tank. Pressure relief devices must be provided for each low-pressure and high-pressure breakout tank. There are different specifications for low-pressure and high-pressure tanks.

The unintended closure of the valves could also result in a shutdown of the pipeline. In these cases, personnel should inspect the equipment to determine why closure occurred, and reset, open, or close the valve when it is safe to do so. Such occurrences may not cause an AOC, but if the event causes the pipeline to exceed design limits in any way, its classification changes to abnormal.

1.3.0 Recognizing a Communications Failure

Communications refers to both the tools that personnel use to communicate with each other and with communication systems. Communications failures include the failure of communications among personnel, or the failure of communications between the control center and equipment on the pipeline.

1.3.1 Failure of Personnel Communication Tools

Pipeline personnel use landline telephones, cell phones, and radios to communicate with one another on the job. If using a cell phone near the pipeline facility, personnel must make sure it is intrinsically safe/explosion-proof. Personnel must keep their equipment in good working order and test their equipment regularly to ensure that it is working properly.

A breakdown in personnel communications equipment is easy to recognize. Personnel cannot hear messages or make phone calls. Some equipment may beep or flash a liquid crystal display to alert users to a problem. These tools are usually reliable, but breakdowns can occur. If radio communication goes down, personnel must address the problem immediately. A communications breakdown can also occur due to weather-related conditions and location.

67107-13_F03.EPS

Figure 3 Relief valve.

1.3.2 System Communications Failures

System communications refers to the computer equipment located in the control center and on some pipeline equipment. Personnel in the control center use the **supervisory control and data acquisition (SCADA)** system to manage pipeline operations. They receive signals from and send signals to pipeline equipment that may be many miles away. The SCADA system allows them to monitor pressure, volume, and temperature and keep products flowing safely and on schedule. If the system fails, control center personnel lose the ability to communicate with equipment on the pipeline. Some systems are set up to switch to a **redundant system**—a backup that comes on line if the main system fails.

Some failures of communications systems show up as warnings or messages on the control center console. A system failure may include any of the following events:

- Random facility outages
- Partial or complete facility outages
- Loss of the SCADA system
- Loss of satellite communications
- Loss of microwave signals
- Stale data
- Loss of signal point within the system

Mechanical and hydraulic failures are related to system failures. For example, the SCADA system sends a signal to a piece of equipment, but the equipment does not respond to the signal. In some cases, the problem is a result of a programming or command logic error. In other cases, equipment may not work because of a mechanical, electrical, or hydraulic problem. Examples of monitoring device failures include, but are not limited to, frozen valves, electrical component burnouts, and compressor breakdowns. A valve may have a mechanical failure that prevents it from being operated remotely by the control system. Sometimes electrical components or connections burn out, preventing the equipment from responding to a command by the system to start up or shut down.

1.4.0 Recognizing Power Interruptions

Electric power is essential to pipeline operations. Nearly every part of the pipeline depends on electricity in some way. A loss of power impacts the abilities of the control center or pipeline facility operations to safely control and monitor the pipeline facility. Power interruptions that result from bad weather or a problem at the local utility company, such as transformer failure or downed power lines, are not uncommon.

Some pipeline operators have an **uninterruptible power system (UPS)** that turns on if electrical power is off for more than a few seconds. UPS systems supply emergency alternating current (AC) power to pipeline components to keep the pipeline operating until power is restored. Instead of electricity, these systems run on engines fueled with gasoline, diesel, or natural gas. When power fails, an AC power failure alarm may be sent to the control center. To ensure that the backup system has turned on, control center personnel check to see whether they are receiving updated information from the pipeline.

If power is lost, electronic indicators will turn off. Some compressors and remote control valves may not operate. The SCADA system may switch off and pumps may shut down. In addition, lights, heat, and air conditioning may fail. An area-wide power outage may cause a loss of telephone communications. Local power outages may shut down only a portion of the pipeline system. Power outages can also cause pumps at the receiving terminal to stop operating, causing a gas or liquid shipment to overflow in tanks or be unable to be shipped further without pumps.

Some equipment is protected from power failure by batteries that supply temporary low voltage power. These batteries run on direct current (DC).

The following types of equipment can operate on DC power:

- Fire detection equipment and alarms
- Fire suppression systems
- Emergency lighting systems
- SCADA
- Alarm panels
- Control modules
- Process control panels
- Specialized compressor units

> **NOTE**
> Equipment plugged into electric sockets uses AC power. AC power reverses its direction at regular intervals. Batteries supply DC power. DC power flows in one direction.

1.5.0 Recognizing Fire, Explosion, or Natural Disaster

Fire, explosion, or natural disaster can occur at any time. Some events may be anticipated, but many cannot. Field personnel must be trained to recognize these events.

1.5.1 Fire

Personnel must take steps to minimize the likelihood of fire and explosions in a pipeline facility. Two examples include using intrinsically safe/explosion-proof electronic equipment and checking atmospheric conditions before starting any hot work. Always follow company procedures related to the prevention of fire and explosion. Detection devices installed along the pipeline and in pipeline facility buildings alert personnel in case of a fire. These devices send an alarm when their sensors detect heat or flame. Such equipment includes thermal detectors, flame detectors, and smoke detectors.

> **NOTE**
> Before performing any work that may produce an ignition source, verify that a hazard analysis has been performed. Examples of such work include welding or cutting, opening electrical boxes, or operating non-explosionproof tools in hazardous classified areas.

A thermal detector is a heat-sensitive unit. It is fitted with electrical contacts that stay open as long as conditions are normal. When the unit senses rapidly rising temperatures, the contacts start to close. When the contacts connect, the unit may send an alarm, which may be audible, to the control center and the fire control panel in the workstation. If a thermal detector activates, personnel should evacuate the area and, investigate the cause of the alarm (if appropriately trained according to company procedures).

A flame detector (*Figure 4*) is a light-sensitive unit. It is fitted with an ultraviolet detector that turns on an alarm when the unit senses light from a flame. The unit is sensitive enough to detect a flame even in the earliest stages of a fire. Some flame detectors are connected to a firefighting system that may dump fire suppression material into the area. Personnel can also dump this material over a fire manually by using pull switches.

A smoke detector can be a light-sensitive unit or an ionization detector, or it can be a combination of both types. Ionization detectors are more sensitive to fires in the early smoldering stage. Light detectors are more sensitive to the flaming stage of fires.

An ionization smoke detector detects ionization in air caused by smoke. If any smoke particles pass into the chamber of the detector, the ions will attach to the particles, and so become less able to carry the current. An electronic circuit detects the current drop and sounds the alarm. The light detector operates by detecting light beams. In the absence of smoke, the light passes in front of the detector in a straight line. When smoke enters the chamber across the path of the light beam, light is scattered by smoke particles, directing it at the sensor that triggers the alarm.

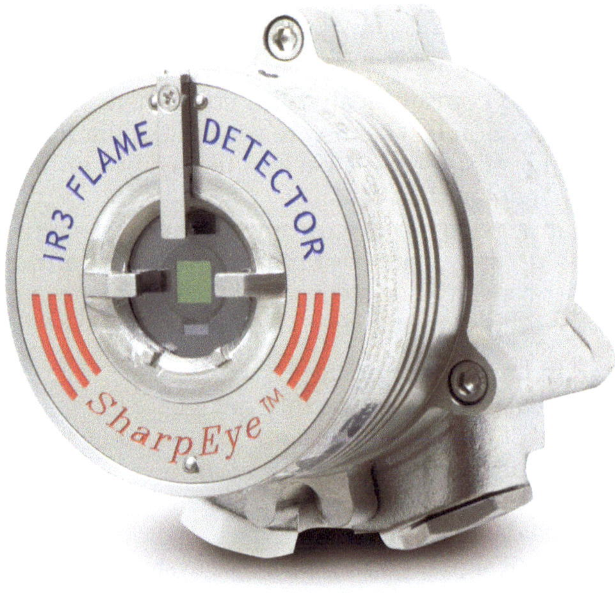

67107-13_F04.EPS

Figure 4 Flame detector.

> **CAUTION**
> Before using any fire suppression material, personnel must know how to handle it. Be sure to read the material safety data sheet (MSDS)/safety data sheet (SDS) for information about how to use the material safely.

Monitoring devices that detect a lower explosive limit (LEL) produce alarms to indicate that there is a risk of explosion. The LEL is the lowest concentration of a gas or a vapor in the atmosphere capable of producing fire in the presence of an ignition source. Different products have different LELs. The monitoring device must be set to detect the particular LEL in an enclosed area. If an LEL alarm activates while welding is taking place, the first action is to stop the welding immediately. Always check atmospheric conditions before reentering an area after an LEL alarm has activated.

1.5.2 Explosion

Pipeline operators install a number of warning devices to warn personnel about conditions that could cause explosions. However, operators cannot control every circumstance, such as intentional damage, that could lead to an explosion on the pipeline. An explosion on a pipeline registers in the control center as an unexplained change in pressure. An explosion may occur in any enclosed space where flammable vapors can accumulate to dangerous levels. Examples of such spaces include deep pipeline trenches and pipeline facilities.

A natural gas explosion may occur when natural gas leaks into an enclosed space. The odorant added to the gas sometimes alerts people to the problem. However, gas can sometimes accumulate undetected. For example, the building may be vacant or the odors may be too far away from people inside the building. In addition, products may not have an odorant added during transport. Be aware that a spark from electrical equipment can cause the gas to explode.

> **WARNING!**
> Personnel should never rely on their sense of smell to detect a natural gas leak. Products do not always have a detectable odor.

For an explosion to take place in an enclosed space, there must be a source of ignition and fuel. The vapors creating the fuel for an explosion can be of various types. Each type of flammable vapor or gas has a different lower and upper explosive limit (LEL/UEL). For example, when natural gas escapes into the air, its molecules generally dissipate or scatter because natural gas is lighter than air. However, in enclosed areas natural gas can accumulate, and if there is an ignition source an explosion may occur.

1.5.3 Natural Disasters

Natural disasters include earthquakes, hurricanes, tornadoes, landslides, wildfires, and flooding. These events may cause soil movement, which may result in forces sufficient to cause pipeline failure or product release. Even heavy rains resulting in saturated soil conditions may cause soil movement in some areas.

Pipeline operators generally have designated personnel to monitor weather conditions that can develop into a natural disaster. Control center personnel monitor for earthquakes with seismic indicators. In addition, they may obtain real-time information on the internet. Up-to-date earthquake information is available from the National Earthquake Information Center, which is located in Denver, Colorado. Personnel may track weather alerts and warnings through the National Weather Service website. In case of a power failure, personnel can continue to receive information through the National Weather Service's National Weather Radio or other electronic devices. A list of weather and natural disaster information sources is included in *Appendix A*.

Operators should also have field personnel designated to patrol the ROW to monitor for soil movement and to ensure the accessibility of pipeline components, such as valves, that are needed in case of emergency response actions such as shutdown, isolation, and containment in case of a pipeline failure.

1.6.0 Recognizing the Unexpected Release of Hazardous Gas or Liquid

Pipeline releases are often noticed when someone sees the physical signs. Depending on the product, the release may result in a pool of liquid on the ground, a frost ball, vapor cloud, mist in the air, dead or discolored vegetation, or a rainbow sheen on water. Other signs include dead fish in waterways or other dead wildlife. Gas leaks may also cause a whistling or hissing sound. Natural gas leaks may be identified by a loud, roaring noise. Note that the controlled transfer, sampling, venting, draining, and handling of hazardous products are not AOCs.

Personnel may use a combustible gas indicator or a flame ionization detection unit to detect hazardous gases such as hydrogen sulfide (H_2S), propane, natural gas, and in some cases, carbon dioxide (CO_2). An unexpected release may lead to a hazardous atmosphere, an explosion, or other serious condition. Additional PPE may be required with an unexpected release. Always follow company procedures when dealing with any release.

Personnel must not rely on their sense of smell to detect a hazardous gas for the following reasons:

- Some hazardous gases are odorless. Products may not have an odorant added to alert workers to the presence of gas.
- A low concentration of some hazardous gases can deaden the sense of smell.
- A person's sense of smell may be deadened by a medical problem.

1.7.0 Recognizing Unexplained Flow Rate / Pressure Changes

Flow rate and pressure are important indicators of both normal and abnormal operating conditions. Minor changes in pressure may occur normally in the pipeline. However, significant deviations from the expected flow or pressure are considered AOCs. Unexplained pressure changes include absences of, increases in, and decreases in pressure. The maximum internal pressure permitted during the operation of a gas pipeline is called the **maximum allowable operating pressure (MAOP)** and is defined in 49 *CFR* 192. The maximum internal pressure permitted during the operation of a liquid pipeline is called the **maximum operating pressure (MOP)** and is defined in 49 *CFR* 195.

Pressure-indicating devices located near mainline valves, compressor stations, and custody transfer stations monitor pressure and send readings to the control center. When an unexpected change in pressure occurs, an alarm may alert control center personnel, who then notify field personnel. Sudden pressure increases or decreases might indicate a problem such as the failure of a valve or a transmitter. Personnel must be familiar enough with the pipeline system, equipment, product characteristics, and environmental factors to quickly identify any important changes in pressure and flow.

Internal Combustion Engines

Pipeline facilities/systems that process or store flammable liquids or gases are particularly vulnerable to the risk of fire and explosion. But did you know that the internal combustion engines that are typically used in pipeline facilities can ignite flammable liquids and gases? Internal combustion engines include vehicles of any kind (trucks, cranes, excavators, fire engines, and ambulances) as well as stationary or portable engines such as compressors, generators, pumps, and welders.

Be sure that you know and understand your company's specific safety policies about internal combustion engines to ensure your own safety, the safety of other personnel, and the safety of the entire pipeline facility. To prevent internal combustion engines from becoming an ignition source, pipeline operators are responsible for implementing safety measures such as:

- Ensuring that materials and equipment are stored and used in accord with applicable Occupational Safety and Health Administration (OSHA) standards
- Ensuring that permanently mounted internal combustion engines are not stored where flammable vapors or gases could exist
- Establishing and marking safe boundaries and traffic routes for vehicles and portable equipment through hazardous areas
- Installing flammable gas and vapor detectors in areas where hazards could exist
- Providing training to personnel about the signs and hazards of flammable vapors and gas, the safe handling of internal combustion engines, and all relevant emergency procedures.

Additional Resources

It is assumed that you will learn to perform these procedures under the direct supervision of your crew leader, foreman, or instructor. For independent study, refer to the recommended Additional Resources at the back of this module.

1.0.0 Section Review

1. Which of the following is an abnormal pipeline facility condition?
   a. Aboveground pipe on pipe supports
   b. Missing signs or markers
   c. Valves operating properly
   d. Authorized encroachment

2. Safety devices include _____.
   a. hazardous atmosphere alarms
   b. tank shells
   c. SCADA
   d. compressors

3. SCADA serves as a _____.
   a. regulatory agency
   b. notification system for all utilities
   c. device worn to provide breathable air where the atmosphere is hazardous
   d. command system that assists control center personnel

4. An uninterruptible power system (UPS) supplies _____.
   a. temperature readings
   b. emergency backup power
   c. nuclear power
   d. pressure readings

5. A thermal detector, unlike a flame detector, senses _____.
   a. rapidly rising temperatures
   b. ultraviolet light
   c. flames
   d. pipeline pressure

6. Which of the following is a reason that personnel should *not* rely on their sense of smell to detect a hazardous gas release?
   a. Hazardous gases always have an odor.
   b. Low concentrations of hazardous gases heighten the sense of smell.
   c. A person's sense of smell may be deadened by a medical problem.
   d. Personnel can hear a gas release.

7. The maximum operating pressure of a gas pipeline is called _____.
   a. MOP
   b. MAOP
   c. surge pressure
   d. compression

Section Two

2.0.0 Reacting to Abnormal Operating Conditions

Objective

Describe the proper reaction to abnormal operating conditions for both gas and liquid pipelines.

a. Describe the proper reaction to abnormal pipeline facility conditions.
b. Describe the proper reaction to the activation of a safety device.
c. Describe the proper reaction to a communications failure and a control system failure.
d. Describe the proper reaction to power interruptions.
e. Describe the proper reaction to a fire, explosion, and natural disaster occuring with the pipeline.
f. Describe the proper reaction to the unexpected release of hazardous liquid or gas.
g. Describe the proper reaction to unexplained pressure / flow rate changes.

Trade Terms

Incipient fire: A fire in its beginning stage.

Naturally occurring radioactive materials (NORMs): Materials with radioactive elements that can be extracted during oil and gas extraction from the earth and can get into product streams and become transported in the pipeline.

Self-contained breathing apparatus (SCBA): A device worn to provide breathable air for an individual in an area where the atmosphere is hazardous.

Time is of the essence when AOCs occur and field personnel are expected to take decisive action. Personnel must always follow established company-specific policies and procedures. As regulation 49 *CFR* 195.402 states, each pipeline operator must ensure its personnel maintain a thorough knowledge of its operations, maintenance, and emergency manuals. Personnel must make use of their training, skills, and good judgment to maintain safe operation until solutions are found.

2.1.0 Reacting to Abnormal Facility Conditions

Some abnormal pipeline facility conditions may require immediate action. Others may require scheduling repairs for a later date. If field personnel recognize an abnormal pipeline facility condition, they must notify the appropriate personnel so that repairs can be made safely and in a timely manner. If notified of a condition by a third-party call, it is important that personnel record information received from the third party. They must also keep in mind that a caller's description may not be entirely accurate. The problem may be different than described. In rare cases, a call might not be valid. After a report is confirmed, field personnel should generally follow these steps:

Step 1 Investigate and evaluate the report to determine the extent of the damage, if any.

Step 2 Carry out company safety procedures to protect people, property, and the environment if necessary.

Step 3 Notify the appropriate personnel so that proper notifications or repairs can be made.

2.1.1 Pipeline System Damage

When field personnel notice or accidently cause pipeline damage, they investigate the extent of the damage, note the location on the pipeline, and isolate the damage, as applicable. They must also contact the control center, according to company procedures. If an immediate hazard exists, contact 911 emergency personnel.

> **NOTE**
> Before excavation or other digging near a pipeline begins, responsible personnel must contact the One-Call notification system.

Some incidents may be serious enough to require an emergency shutdown of the pipeline. One example of a serious incident is someone losing control of a vehicle on the ROW, extensively gouging or denting a large section of pipeline or damaging a valve. Whenever pipeline damage results in a release and/or explosion, field personnel must notify the control center as soon as possible. In some cases, control center personnel may be able to drain a portion of the pipeline into a delivery system to reduce the amount of product released from the pipeline.

Once the pipeline has resumed normal operations, the pipeline operator conducts an internal investigation. This investigation focuses on evaluating control systems and mechanical equipment as well as personnel performance.

Each company has specific procedures for reacting to pipeline damage. These procedures explain how and when to contact the control center, emergency personnel, and local, state, and federal law enforcement agencies. Always follow company policy, make appropriate notifications per company policy, and complete the required reporting.

2.1.2 Damaged Pipeline Supports

The proper reaction to the discovery of damaged pipeline supports varies depending on the severity of the damage. If the damage to the support is critical or indicative of a possible release, the individual making the discovery should contact the control center or other appropriate personnel immediately. Notification is especially important if the pipe is in danger of excessive movement. The appropriate personnel must then follow company policy precisely to investigate the condition and dispatch qualified personnel to the scene. Until the arrival of qualified personnel, all other personnel should stay a safe distance away from the site, evacuate the area if necessary, and make the appropriate notifications as required by company policy.

In some cases, the pipeline operator may need to isolate the pipeline segment. Whether or not the segment needs to be isolated, a damage report must be submitted so that the repair work can be scheduled. The decision to continue or halt operations is a judgment call made by management and qualified inspection personnel. If the pipeline segment must be isolated, affected personnel must be notified.

Field personnel must know and follow their company's procedures. Those procedures are especially important when a condition requires a decision about whether to isolate a segment of the pipeline facility. Make notifications and complete the reporting as required by company policy.

2.2.0 Reacting to the Activation of a Safety Device

Personnel must respond to safety device alarms quickly and calmly. To react to an alarm, personnel should follow these steps:

Step 1 Notify affected personnel.

Step 2 Carry out company warning procedures to protect people, property, and the environment.

Step 3 Troubleshoot the problem and make necessary repairs if qualified to do so.

Step 4 Reset and test the warning device alarm.

Step 5 Make the appropriate notifications and complete the required reporting.

It is important to ensure that the alarm is not a false alarm. In case of a false alarm, the control center should notify affected personnel and have the alarm equipment checked. Generally, no further action is required when an alarm is false.

When a hazardous atmosphere alarm sounds, the proper reaction is somewhat different. In this case, personnel wearing the appropriate personal protective equipment (PPE) must follow these steps:

Step 1 Evacuate the area and don the appropriate PPE if available.

Step 2 Make the appropriate notifications per company policy.

During an emergency, personnel must focus on three key areas: ensuring personnel safety, reducing the damage to equipment, and protecting the environment. The general rule is to put people first. The goal is also to prevent small problems from turning into larger incidents. All reactions must be made safely. Personnel investigating the alarm must wear the appropriate PPE. If told by authorized personnel to evacuate an area, they must leave immediately.

If reacting to a seal failure alarm, personnel should investigate the alarm, check the station sump level if it is equipped with one, and notify the appropriate personnel.

If a pressure-relief system activates, an audible alarm may activate. Immediate action must be taken to reduce pressure. A pressure reduction is usually achieved by shutting down the upstream pumping station, closing a block valve, or shutting down the pipeline system. Field personnel may need to inspect the relief system, test it, and recalibrate it as necessary.

2.3.0 Reacting to a Communications Failure

The proper reaction to a communications failure depends on whether the failure involves a communications tool that personnel use or the failure of system communications.

2.3.1 Failure of Personnel Communication Tools

When communication tools fail, personnel should first check to see if the problem is a simple one that can be quickly and easily fixed. Immediately correct any communications failure, if possible. For example, some telephones have a volume control that may accidentally be turned off. The telephone may seem to be dead, but a simple adjustment to the volume corrects the problem. Other simple solutions include replacing a battery or recharging a unit.

Some communications problems are more serious. Weather or problems at the local utility company can knock out communications completely. Regardless of the outage, the best preparation is to ensure personnel are trained to follow these steps:

Step 1 Switch to alternate equipment if necessary.

Step 2 Notify the appropriate personnel per company policy.

Step 3 Complete any required reporting or documentation.

If communications are interrupted for long periods, it may not be possible to ask for instructions from a remote location. Training before an emergency situation occurs is the best way to ensure that field personnel know how to react properly. However, field personnel must do everything that they can to re-establish communications as soon as possible.

2.3.2 System Communications Failures

The control center will generally notify field personnel about system communications failures so that they can isolate, investigate, and react according to company policy. The proper reactions for field personnel may include some or all of the following activities:

- Switch to alternate or backup systems
- Isolate a pipeline segment if necessary
- Determine the scope of the problem
- Take corrective actions to restore the system
- Estimate how long the main system will be down
- Notify affected personnel about how long repairs will take
- Complete the required reporting or documentation

If the problem is with the SCADA system, or equipment cannot otherwise be operated from the control center, field personnel may have to operate equipment manually. The pipeline may be able to continue to operate while repairs are made. Operations are dependent on company procedures.

2.4.0 Reacting to Power Interruptions

The proper reaction to a power failure depends on which part(s) and how much of the pipeline is affected. Regardless of how the outage affects the pipeline facility, the appropriate personnel should always notify the power company of the outage as soon as possible. If only a single component or station is involved, control center and/or field personnel may be able to adjust line pressure and flow to keep the pipeline in operation. For example, if a power interruption occurs while a valve is in operation, field personnel may have to operate the valve manually. However, if a station is needed to keep pressure within normal operating limits, for example, control center personnel may have to isolate a pipeline segment. If a shutdown is not required, the gas or liquid shipment will continue to arrive, and the schedule must be followed and the tasks completed manually.

Each company has specific procedures for responding to a power interruption. Always follow company policy, make appropriate notifications per company policy, and complete the required documentation.

2.5.0 Reacting to a Fire, Explosion, or Natural Disaster

The pipeline operator must have a training program in place so that personnel know what to do in case of a fire, explosion, or natural disaster that affects a pipeline. Many companies conduct regular drills so that personnel can practice their responses to such events. Drills help ensure that personnel are prepared to handle emergencies safely and calmly. Each operator has its own procedures for emergencies. In general, trained personnel should know the following information:

- The location of fire suppression equipment and how to operate it; certain fire suppression equipment and PPE may require additional training
- The process for shutting down equipment manually if communication with the control center is lost
- The company's evacuation plan
- The procedures for contacting local emergency response agencies, if necessary

Depending on the situation, pipeline personnel may be able to safely handle an emergency without calling for outside help. Follow company procedures for contacting local emergency response agencies.

2.5.1 Fire

When a fire occurs, the safety of personnel is always top priority, so personnel should evacuate the area. Protection of property is always second.

Personnel must ensure that any injured personnel receive medical attention as soon as possible. Consider any alarm as an actual alarm and react appropriately and decisively.

Personnel must also eliminate the source of the fire, but only as authorized. For example, only qualified personnel are authorized to turn off valves. Company policy dictates procedures for emergency actions and must be followed. Procedures are site-specific and may allow for some flexibility in responding. However, personnel should not need to consult a manual in case of an alarm. They must be adequately trained and familiar with emergency procedures in advance. It may even be necessary to shut down the pipeline if required to control the emergency. Always notify the appropriate personnel and emergency agencies and complete any required documentation or reporting.

According to 49 *CFR* 192 and 195, operators shall establish and conduct a continuing training program to instruct emergency personnel in the proper use of firefighting procedures and equipment, in the identification of hazardous conditions likely to cause emergencies, in the steps necessary to control accidental releases, and in the characteristics of hazardous liquids or carbon dioxide being transported.

The proper reaction differs according to the type of fire in progress. If the fire is small or just beginning (**incipient fire**), personnel should use the available fire extinguishing equipment to control the situation. According to *OSHA 1910.157*, companies must provide designated personnel with training on fire extinguishing equipment as well as the hazards involved with incipient firefighting.

Use the proper type of fire extinguisher for each class of fire. It is important to know the extinguisher's fire class rating before using it. The fire class rating is printed at the bottom of the label on a fire extinguisher. There are four fire class ratings (see *Table 3*). Most extinguishers can be used for multiple types of fires.

If the fire is larger than the immediate workforce can contain, personnel should try to mitigate the situation, notify appropriate personnel and emergency agencies, and complete the required documentation and/or reporting. For example, during an electrical storm, if lightning strikes a full tank at a facility and the roof seal catches on fire, personnel should follow company policy and contact the fire department to contain the fire.

Once the fire is out, the air is cleared, and any injured personnel have received medical attention, the following actions are taken:

Table 3 Fire Class Ratings

Fire Class	Materials and Proper Fire Extinguisher
A	Involves ordinary combustibles, such as wood or paper. This type of fire is suppressed by cooling the fire. Class A extinguishers contain water. A Class A fire extinguisher can never be used on a Class B, C, or D fire.
B	Involves grease, flammable liquids, and gases. Class B extinguishers contain carbon dioxide or another material that smothers the fire by removing oxygen.
C	Involves electrical equipment. Class C extinguishers are designed to protect the firefighter from electric shock. This type of extinguisher smothers the fire.
D	Involves metals. Class D extinguishers contain a powder that either forms a crust around the burning metal or gives off gases that prevent oxygen from reaching the fire. Some metals may keep burning even after they have been coated with powder from a Class D extinguisher. The best way to fight these fires is to keep using the extinguisher so that the fire does not spread.

- Alarms are reset and tested.
- Equipment damage is assessed and reported to the appropriate personnel.
- Work orders are issued and required permits are obtained for repairs.
- Personnel performance is evaluated and corrective action is determined if necessary.
- Injuries or deaths are reported to the appropriate regulatory agencies if required.

The evaluation of personnel performance during a fire may reveal problems in personnel response. Management generally recommends retraining so that personnel can improve their responses to a fire.

A false alarm or an expected alarm due to maintenance activities requires no further action unless the control center determines that the system needs to be investigated. If an alarm has been verified as a false alarm, the appropriate personnel follow these steps:

Step 1 Notify affected personnel.

Step 2 Test alarms to ensure that they are operating properly.

Step 3 Reset alarms or ensure that they have automatically reset.

2.5.2 Explosion

Each operator has specific procedures for reacting to an explosion. In addition, the correct order of the steps to take may vary from operator to operator. A response to an explosion must be appropriate, decisive, and calm. If the main-line valves do not close automatically, designated personnel must initiate actions to ensure that the valves are closed. In general, the proper reactions to an explosion are:

- Evacuate and notify appropriate personnel
- Evaluate and assess the situation.
- Carry out company safety procedures to protect people, property, and the environment.
- Call local emergency response agencies if necessary.
- Confirm the location of all pipeline personnel near the explosion location.
- Confirm the location of any members of the public near the explosion location.
- Call local, state, or federal law enforcement if the cause indicates criminal or terrorist activity.
- Make the appropriate notifications and complete the required documentation and/or reporting.
- Conduct an incident investigation.

2.5.3 Natural Disaster

Much of the pipeline's ability to withstand extreme weather that can result in natural disaster is already present in its design and installation.

Each company has specific procedures for dealing with natural disasters. These procedures reflect local conditions. For example, hurricanes occur more frequently in Florida or the Gulf states. Therefore, companies located in these areas include hurricane response in their procedures.

The following are examples of proper reactions to weather-related problems:

- Monitor weather conditions such as tropical storms that may become hurricanes
- Carry out company safety procedures for protecting people, property, and the environment
- Adjust work schedules as needed to respond to emergency conditions
- Follow company procedures for responding to fire, explosion, or damage resulting from natural disasters
- Set up a command post if conditions warrant such a step
- Work with local emergency response agencies if required
- Conduct investigations when appropriate

2.5.4 Emergency Shutdowns

Events occurring on the pipeline that present immediate hazard to persons or property require an emergency shutdown. An event might be caused by human error, mechanical failure, electrical failure, or an accident on the pipeline. It might also be brought about by threats from weather or civil disobedience. Regardless of the reason, the line must be shut down quickly and safely. Each company has a standard operating procedure for emergency shutdowns.

In the case of a fire, explosion, or release, time for discussion may be very limited. Once the need is confirmed, decisive action must be taken. The pipeline must be isolated by control center personnel by safely shutting down the originating station and continuing downstream in a safe manner until all pumps are offline. Control center personnel will then close block valves to complete the shutdown and notify other personnel as needed. In case of a pump shutdown from low pressure, there may be some time to discuss the decision with appropriate management if the line has enough flexibility to withstand the stress.

Keep in mind that threats from natural disaster or civil disobedience may come without warning. If field personnel witness these kinds of threats to the pipeline, they should contact their supervisor or control center immediately. Affected facilities may need to be isolated. Once facilities are isolated, pipeline shutdown may follow.

In emergency conditions such as a possible release, the pipeline operator has full responsibility and authority to make shutdown decisions and take appropriate action. These actions are governed by the company release policy above all other considerations.

2.5.5 Evacuations

The safety of pipeline personnel and the public is the first priority when any AOC occurs. If there is any threat of injury or death to any persons, an evacuation of pipeline personnel must occur. Events that could result in the need for an evacuation are fire, explosion, release of hydrocarbon or other hazardous vapors, seismic activity, or other natural disasters.

Companies have standard procedures in place to guide the evacuation of personnel. By law, emergency response plans must be in place and evaluated annually. Facility layouts with evacuation maps are required to be posted in every area or department. Safe areas are designated for all personnel. When the command to evacuate is given, trained personnel may act as safety coordinators. Their job is to conduct a roll call to confirm that all personnel are safe. To assist in this process, personnel must report to their designated safe areas and identify themselves to the designated safety coordinator. If any personnel cannot be accounted for, authorized personnel must locate and rescue them if necessary.

Some emergencies may require an evacuation of the public located near the pipeline. The pipeline operator may work with the local authorities to provide details about the incident, determine whether a general evacuation is required, provide contact information for people residing or working near the pipeline, and provide personnel to help staff a command center if requested.

NOTE

Local emergency planning committees (LEPCs) exist in every jurisdiction in the United States. Law enforcement agencies, firefighting companies, local government agencies, hospitals, companies, and citizen's groups make up the memberships of LEPCs. Committee members meet on a regular basis to plan for emergencies. Planning for a general evacuation is just one example of the projects handled by an LEPC.

2.6.0 Reacting to the Unexpected Release of Hazardous Gas or Liquid

When a hazardous gas or liquid is released from a pipeline unexpectedly, personal safety should be the first consideration. Beyond that, the proper reaction to the unexpected release of a product depends on the characteristics of the product and the location of the release. Some gases and liquids can explode under certain conditions, and some are toxic to people and the environment.

NOTE

To identify the hazards associated with the chemical being released, consult the material safety data sheets (MSDS)/safety data sheets (SDS).

Regardless of location or the type of product, when a product is released personnel *must* take the following steps:

Step 1 Evacuate the area.

Step 2 Identify the source of the release and eliminate potential ignition sources.

Step 3 Vent the area if possible.

Step 4 Prevent anyone who is not wearing the appropriate PPE from entering the area (personnel must be trained to use PPE).

Step 5 Call for emergency rescue and medical help, if required.

Step 6 Contact the control center; appropriate personnel may need to isolate the pipeline.

Step 7 Have the power company turn off the power, if appropriate.

WARNING

When witnessing a release and there is a downed person in the area of the release, call 911 and evacuate the area. Do not attempt to rescue personnel.

NOTE

OSHA stipulates that employers must ensure that contract personnel are trained in the work practices necessary to perform their job safely.

Of equal importance are actions personnel must avoid. When reacting to a release, personnel *must not*:

- Open switchgear doors.
- Operate electrical switches or equipment in the facility, including lights, doors, louvers, or vents.
- Operate personal electrical equipment (including cell or landline telephones) unless the equipment is classified as intrinsically safe/explosion-proof.

Intrinsically safe/explosion-proof electrical equipment is designed so that a spark cannot ignite the atmosphere. It is made to meet the following specifications:

- Outside air cannot penetrate the casing.
- The casing must be strong enough to contain an explosion inside the equipment without igniting the surrounding atmosphere.
- The interior must be filled with nonflammable powder or inert gas to prevent ignition.

This equipment is determined to be safe by the National Electrical Manufacturers Association (NEMA), and is certified according to the standards of the *National Electric Code®*. It must also meet the requirements of Underwriters Laboratories (UL). Equipment that is intrinsically safe/explosion-proof should always be labeled as such.

> **WARNING**
>
> Electronic equipment must not be used in the presence of hazardous gases/vapors unless it is classified as intrinsically safe/explosion-proof. Failure to use intrinsically safe/explosion-proof equipment in a hazardous atmosphere may result in an explosion.

2.6.1 Releases in Enclosed Areas

If a product release occurs in an enclosed area, the source of the release must be determined quickly. It is also important that personnel move to a safe location as quickly as possible and then notify the appropriate personnel, who can operate equipment remotely. If trained, personnel may close the cutoff valve to mitigate the situation.

2.6.2 Releases in Outdoor Areas

Any release must be approached by personnel from the upwind side. Releases in outdoor areas include the additional concern for the safety of the general public, as well as the safety of the environment. Personnel should evacuate the area, go upwind, and take additional appropriate action per company policy. If an odor of hydrocarbon is detected even a few hundred feet from the source, contact control center personnel to shut down the pipeline. Report the incident and any action taken, and complete the required documentation.

2.6.3 Toxic Products

Personnel must take special precautions when dealing with the release of products that can become toxic. Examples include carbon dioxide (CO_2), hydrogen sulfide (H_2S), propane, anhydrous ammonia, and natural gas.

Carbon dioxide is a colorless, odorless, gaseous compound used as a propellant in many domestic applications, such as fire extinguishing equipment, and is a byproduct of combustion. Because it is heavier than air, it can collect in low-lying areas where it may displace oxygen; exposure can result in suffocation and death. A leak detector can detect a hazardous atmosphere, but without it, personnel can be overcome without realizing the problem. Some pipeline operations pump carbon dioxide commercially, and in such operations, the hazards are increased. To enter an area where a carbon dioxide release is suspected, personnel must use the proper gas detection meter, calibrated for the detection of carbon dioxide, and wear the appropriate PPE, such as **self-contained breathing apparatus (SCBA)** (see *Figure 5*).

67107-13_F05.EPS

Figure 5 Self-contained breathing apparatus.

> **WARNING**
>
> Carbon dioxide (CO_2) is a colorless, odorless gas that is heavier than air, so it can collect in low-lying areas where it may displace oxygen. Exposure can be fatal.

Hydrogen sulfide (H_2S) is a colorless gas that smells like rotten eggs. However, in fairly low concentrations, it can deaden the sense of smell. Exposure to high concentrations can be fatal.

> **WARNING**
>
> Exposure to high concentrations of hydrogen sulfide (H_2S) can be fatal. Even in low concentrations, it can cause headaches or eye irritation. In addition, it is flammable and may cause a flash fire. When a gas detection unit shows that H_2S is present, wear the appropriate PPE or immediately leave the area.

Anhydrous ammonia is a liquefied gas that has a sharp odor. Exposure in high concentration can cause death. Even exposure in low concentrations can cause severe reactions from mild irritation to severe burns. The reaction depends on the length of exposure and concentration amounts. If anhydrous ammonia is released, call emergency personnel.

> **WARNING**
>
> Exposure to high concentrations of anhydrous ammonia can cause death. Even in low concentrations, it can cause severe reactions. If detected, wear the appropriate PPE or immediately leave the area. If a release occurs, call emergency personnel.

Propane is a common natural gas that is colorless and heavier than air, so it can collect in low-lying areas where it may displace oxygen. Exposure can cause death. At high concentrations, it is flammable.

> **WARNING**
>
> Exposure to propane gas can cause death. At high concentrations, it is flammable. If detected, wear the appropriate PPE or immediately leave the area.

Naturally occurring radioactive materials (NORMs) may be present when oil and gas is extracted from the earth. These NORMs are then introduced into the product streams and transported in the pipeline. Natural gas liquids (NGLs), along with the processing associated with these products, may trap NORMs in the form of sludge, scale, and particulates. Filters, valves, scrapers, and even internal surfaces of pipelines may contain high levels of NORMs. Precautions should be taken to protect personnel by monitoring for NORMs in accordance with all local, state, federal, and company requirements. If high levels are detected, personnel should distance and/or shield themselves from the source and notify the appropriate personnel.

Only appropriately trained personnel may enter an area where a hazardous gas has been detected. These personnel must be trained and qualified to use SCBA. Other personnel must stay a safe distance away until the deadly atmosphere is cleared. Warning signs must be posted at access points so that personnel do not enter until the area has been inspected and tested.

2.7.0 Reacting to Unexplained Flow Rate/ Pressure Changes

If a change in pressure or flow rate occurs that is different from what is expected, alarms may be initiated to alert personnel to the problem. Personnel must act quickly to address the problem. Qualified personnel will investigate and evaluate the problem. Some problems may require field personnel to manually adjust components, such as valves and transmitters, or make repairs. More serious problems may result in the need to isolate a segment of the pipeline.

Additional Resources

It is assumed that you will learn to perform these procedures under the direct supervision of your crew leader, foreman, or instructor. For independent study, refer to the recommended Additional Resources at the back of this module.

2.0.0 Section Review

1. If a damaged pipeline support is discovered and the damage appears to be critical, the individual who made the discovery should _____.
   a. begin repairs immediately
   b. stay close to the site
   c. report it to the appropriate personnel
   d. assume it is safe

2. If a hazardous atmosphere alarm sounds, field personnel should first _____.
   a. call 911
   b. evacuate the area
   c. notify control center personnel
   d. verify that the alarm is not false

3. Field personnel may need to operate equipment in the field manually if there is a system failure.
   a. True
   b. False

4. When a power outage occurs, the appropriate pipeline personnel should contact _____.
   a. the One-Call notification system
   b. the power company
   c. OSHA
   d. local authorities

5. Fire extinguishing equipment can be used on an incipient fire.
   a. True
   b. False

6. Electronic equipment must not be used in the presence of hazardous gases unless it is classified as _____.
   a. OSHA approved
   b. intrinsically safe/explosion-proof
   c. NRC approved
   d. mechanically safe/disruption proof

7. When an unexplained flow rate or pressure change occurs, field personnel may need to _____.
   a. notify the One-Call system
   b. contact the power company
   c. make repairs
   d. remove pipe supports

SUMMARY

Pipeline operators have developed effective procedures and policies that enable personnel to respond quickly, safely, and effectively to AOCs. Usually an AOC is a routine matter that can be handled safely by pipeline personnel. Sometimes, however, an AOC can result in an emergency situation that may involve working with outside agencies.

In this module, you learned about the different types of AOCs that can occur on the pipeline and the proper reactions to each type of AOC. During your career, you may have to respond to some or all of the AOCs outlined in this module. When responding to these AOCs, you must follow your company's policies and procedures. Protecting people, property, and the environment should be your utmost concern.

Review Questions

1. By definition, an AOC is a condition that may exceed _____.
   a. company operating budget
   b. design limits
   c. safety laws
   d. size limits

2. The federal agency that regulates pipeline maintenance is _____.
   a. OSHA
   b. GAO
   c. EPA
   d. PHMSA

3. By definition, a pipeline facility includes the right-of-way.
   a. True
   b. False

4. Which of the following is *not* always an AOC?
   a. Exposed pipe
   b. Reduced pressure
   c. Lost radio communication
   d. Unexpected release of a hazardous gas

5. Pressure relief systems, tank-level alarms, and main-line pressure shutdown alarms are examples of _____.
   a. prevention devices
   b. safety devices
   c. hazardous atmosphere alarms
   d. fire alarms

6. The loss of SCADA is an example of which type of AOC?
   a. Hydraulic failure
   b. Mechanical failure
   c. Communications failure
   d. Power failure

7. In the event of a power failure, pipeline companies may be able to keep the pipeline running by using an _____.
   a. operations override power source
   b. uninterruptible power system
   c. interface power system
   d. indirect power system

8. In the control center, one indication of a pipeline explosion could be a(n) _____.
   a. change in product temperature
   b. valve malfunction
   c. power outage
   d. unexplained change in pressure

9. Which of the following is *not* true of hazardous gas?
   a. Odorants are often added.
   b. High concentrations can deaden a person's sense of smell.
   c. Some gases are odorless.
   d. Low concentrations can deaden a person's sense of smell.

10. Which of the following best defines an unexplained change in flow rate or pressure?
    a. Reduced pressure only
    b. No flow only
    c. High flow rate
    d. Any change from the intended flow or pressure

11. When pipeline support damage is discovered, notification is especially important if the pipe _____.
    a. is in danger of excessive movement
    b. has changed color
    c. is metal
    d. is corroded

12. When a safety device sounds an alarm, field personnel must _____.
    a. turn it off
    b. never ignore it
    c. reset it
    d. contact the local emergency agency

13. If personnel communication tools fail, personnel should first _____.
    a. check for problems that have simple solutions
    b. contact the control center
    c. call 911
    d. complete any required reporting

14. If SCADA fails, equipment may have to operated manually by _____.
    a. PHMSA
    b. field personnel
    c. control center personnel
    d. local authorities

15. When a power interruption occurs, personnel may be able to keep the pipeline in operation by adjusting the _____.
    a. MAOP
    b. MOP
    c. line pressure and flow
    d. product temperature

16. A false alarm may trigger an investigation of the system, which is determined by _____.
    a. the control center
    b. field personnel
    c. the job site supervisor
    d. SCADA

17. Before using a fire extinguisher, it is important to know its _____.
    a. fire class rating
    b. OSHA rating
    c. combustibility
    d. capacity

18. The pipeline operator may work with local authorities to facilitate an evacuation of the area surrounding a pipeline.
    a. True
    b. False

19. An area where a hazardous gas has been detected may only be entered by personnel trained in rescue operations and qualified to use a _____.
    a. combustible gas indicator
    b. flame ionization indicator
    c. self-contained breathing apparatus
    d. fire suppression kit

20. All of the following are examples of a pipeline facility *except* _____.
    a. compressors
    b. pump stations
    c. control center
    d. rights-of-way

Trade Terms Quiz

Fill in the blank with the correct term that you learned from your study of this module.

1. The computer-based system(s) used by a controller in a control room that collects information about a pipeline facility and may have the ability to send commands back to the pipeline facility is called_______.
2. The _______ is the area along the pipeline which has been legally permitted to be used by the pipeline to construct, operate, and maintain the pipeline.
3. A fire in its beginning stage is called a(n) _______.
4. _______ is the separation of the protective coating from the pipeline.
5. A(n) _______ is a system that provides continuity of power when the normal power supply is interrupted.
6. The maximum pressure at which a pipeline or segment of a pipeline may be operated under this part (gas) is called _______.
7. A(n) _______ is used to relieve surges in a hazardous liquid pipeline system or to receive and store hazardous liquid transported by a pipeline for injection and continued transportation by pipeline.
8. The lowest concentration of a gas or a vapor in the atmosphere capable of producing fire in the presence of an ignition source is the _______.
9. A _______ duplicates the operations of a system so that, in the event of a failure, a backup system is available.
10. A(n) _______ indicates a malfunction of a component or a deviation from normal operations that may indicate a condition exceeding design limits or result in a hazard to persons, property, or the environment.
11. A device that detects some hazardous gases in the atmosphere is a(n) _______.
12. The pressure produced by a change in the velocity of the moving stream that results from shutting down a pump station or pumping unit, closure of a valve, or any other blockage of the moving stream is called _______.
13. _______ is the maximum pressure at which a pipeline or segment of a pipeline may be normally operated under this part (liquids).
14. _______ is a centralized communication system established by utilities, government agencies, and operators of underground facilities.
15. A hazardous liquid which will form a vapor cloud when released to the atmosphere and which has a vapor pressure exceeding 276 kPa (40 psia) at 37.8°C (100°F) is classified as a(n) _______.
16. The government agency that is responsible for developing and enforcing regulations for pipeline operations in the United States is the _______.
17. Electrical equipment that is _______ is designed specifically for use in explosive or flammable locations so that a spark cannot ignite the atmosphere.
18. An organization that adopts standards for a variety of products, services, and processes in the United States to ensure consistency is the _______.
19. A _______ is a device worn by an individual to provide breathable air in an atmosphere that is hazardous.
20. The _______ is the company that operates the pipeline, as opposed to an individual person who works for the company that operates the pipeline.

21. New and existing pipe on the pipeline itself, the right-of-way (ROW), and any equipment, facility, or building used in the transportation of a liquid or gas are defined as part of a(n) _______.

22. Activity carried out by, and/or equipment belonging to outside parties on the pipeline ROW is referred to as _______.

23. A length of pipeline that is entirely aboveground and supported by pipe supports is called a(n) _______.

24. Materials with radioactive elements that can be extracted during oil and gas extraction from the earth and can get into product streams and become transported in the pipeline are called _______.

Trade Terms

Aboveground span
Abnormal operating condition (AOC)
American National Standards Institute (ANSI)
Breakout tank
Disbondment
Encroachment
Flame ionization detection unit
Highly volatile liquid (HVL)
Incipient fire
Intrinsically safe/explosion-proof
Lower explosive limit (LEL)
Maximum allowable operating pressure (MAOP)
Maximum operating pressure (MOP)
Naturally occurring radioactive materials (NORMs)
One-Call notification system
Pipeline and Hazardous Materials Safety Administration (PHMSA)
Pipeline facility
Pipeline operator
Redundant system
Right-of-way (ROW)
Self-contained breathing apparatus (SCBA)
Supervisory control and data acquisition (SCADA)
Surge pressure
Uninterruptible power system (UPS)

Trade Terms Introduced in This Module

Aboveground span: Length of pipeline that is entirely aboveground and supported by pipe supports.

Abnormal operating condition (AOC): A condition within a pipeline system, identified by the operator, that may indicate (1) a malfunction of a component or (2) a deviation from normal operations that may indicate a condition exceeding design limits or result in a hazard to persons, property, or the environment.

American National Standards Institute (ANSI): An organization that adopts standards for a variety of products, services, and processes in the United States to ensure consistency.

Breakout tank: a tank used to (a) relieve surges in a hazardous liquid pipeline system or (b) receive and store hazardous liquid transported by a pipeline for reinjection and continued transportation by pipeline.

Disbondment: Separation of the protective coating from the pipeline.

Encroachment: Activity carried out by, and/or equipment belonging to outside parties on the pipeline right-of-way (ROW). A particular encroachment might be authorized or not, so each situation requires specific handling.

Flame ionization detection unit: A device that detects some hazardous gases in the atmosphere.

Highly volatile liquid (HVL): A hazardous liquid which will form a vapor cloud when released to the atmosphere and which has a vapor pressure exceeding 276 kPa (40 psia) at 37.8°C (100°F).

Incipient fire: A fire in its beginning stage.

Intrinsically safe/explosion-proof: Describes electrical equipment that is designed specifically for use in explosive or flammable locations so that a spark cannot ignite the atmosphere.

Lower explosive limit (LEL): The lowest concentration of a gas or a vapor in the atmosphere capable of producing fire in the presence of an ignition source. Different products have different LELs.

Maximum allowable operating pressure (MAOP): The maximum pressure at which a pipeline or segment of a pipeline may be operated under this part (gas).

Maximum operating pressure (MOP): The maximum pressure at which a pipeline or segment of a pipeline may be normally operated under this part (liquids).

Naturally occurring radioactive materials (NORMs): Materials with radioactive elements that can be extracted during oil and gas extraction from the earth and can get into product streams and become transported in the pipeline.

One-Call notification system: A centralized communication system established by utilities, government agencies, and operators of underground facilities. This system provides one telephone number for contractors and the general public to call to notify of their intent to use equipment on or near a pipeline. It is also in place for any individual to call if pipeline damage is discovered. The One-Call system is generally set up on a state-by-state basis.

Pipeline and Hazardous Materials Safety Administration (PHMSA): An agency of the US Department of Transportation responsible for developing and enforcing regulations for pipeline operations in the United States.

Pipeline facility: New and existing pipe, rights-of-way and any equipment, facility, or building used in the transportation of hazardous liquids or carbon dioxide, or new and existing pipelines, rights-of-way, and any equipment, facility, or building used in the transportation of gas or in the treatment of gas during the course of transportation.

Pipeline operator: The company that operates the pipeline, as opposed to an individual person who works for the company that operates the pipeline.

Redundant system: A system that duplicates the operations of a system so that, in the event of a failure, a backup system is available.

Right-of-way (ROW): The area along the pipeline which has been legally permitted to be used by the pipeline operator to construct, operate, and maintain the pipeline. The ROW extends the entire length of the pipeline.

Self-contained breathing apparatus (SCBA): A device worn to provide breathable air for an individual in an area where the atmosphere is hazardous.

Supervisory control and data acquisition (SCADA): a computer-based system or systems used by a controller in a control room that collects and displays information about a pipeline facility and may have the ability to send commands back to the pipeline facility.

Surge pressure: Pressure produced by a change in velocity of the moving stream that results from shutting down a pump station or pumping unit, closure of a valve, or any other blockage of the moving stream.

Uninterruptible power system (UPS): A system that provides continuity of power when the normal power supply is interrupted. A UPS provides AC power to the system it serves.

Additional Resources

The following recommended resources can provide additional helpful information related to the requirements for this covered task.

American National Standard Institute (ANSI), 1899 L Street, NW, 11th Floor, Washington, DC 20036. **www.ansi.org**

American Petroleum Institute (API), 1220 L Street, NW, Washington, DC 20005. **www.api.org**

American Society for Mechanical Engineers (ASME), Two Park Avenue, New York, NY 10016-5990. **www.asme.org**

American Society for Testing and Materials (ASTM) Specification D396.

Association of Oil Pipe Lines (AOPL), 1808 Eye Street, NW, Suite 300, Washington, DC 20006. **www.aopl.org**

Code of Federal Regulations 49, Parts 192 and 195. **www.ecfr.gov**

International Society of Automation (ISA), 67 TW Alexander Drive, Research Triangle Park, NC 27709. **www.isa.org**

National Association of Corrosion Engineers (NACE), 1440 South Creek Drive, Houston, TX 77084. **www.nace.org**

National Electrical Code®. **www.nfpa.org**

Occupational Safety and Health Administration (OSHA), 200 Constitution Avenue, NW, Washington, DC 20210. **www.osha.gov**

Pipeline and Hazardous Material Safety Administration (PHMSA), East Building, 2nd Floor, 1200 New Jersey Avenue, SE, Washington, DC 20590. **www.phmsa.dot.gov**

US Bureau of Labor Statistics, Postal Square Building, 2 Massachusetts Avenue, NE, Washington, DC 20212. **www.bls.gov**

US Energy Information Administration, 100 Independence Avenue, SW, Washington DC 20585. **www.eia.gov**

Figure Credits

U.S. Geological Survey/photo by Don Becker, Module Opener

Energy Information Administration, Office of Oil & Gas, Natural Gas Division, Gas Transportation Information System, 2009, Figure 1

Photograph No. 5998083 (Photographer Eric Vance), September 2006; Records of the Environmental Protection Agency, 1944–2006, Record Group 412; National Archives and Records Administration–College Park, Figure 9

Cameron Strother, Southland Safety, Figure 15

Photo courtesy of STV, Figure 34

Courtesy of Liquid Controls, LLC, Figure 35

Photographs and illustrations provided courtesy of Daniel Measurement and Control, Inc. Daniel Senior and Simplex are registered trademarks of Daniel Measurement and Control, Figure 36

Courtesy of Sunoco Logistics, Figure 39

Appendix A

Weather and Natural Disaster Information Sources

Emergency Managers Weather Information. US Department of Commerce, National Oceanic and Atmospheric Administration, National Weather Service, 1325 East-West Highway, Silver Spring, MD 20910.**www.nws.noaa.gov/emwin/index.htm**

National Earthquake Information Center, US Geological Survey. Box 25046, DFC, MS 967 Denver, CO 80225. **www.earthquake.usgs.gov/regional/neic**

National Hurricane Center Tropical Prediction Center. 11691 SW 17th Street, Miami, FL 33165-2149. **www.nhc.noaa.gov**

National Weather Service and National Weather Service Weather Radio, National Oceanic and Atmospheric Administration (NOAA). 1325 East-West Highway, Silver Spring, MD 20910. **www.weather.gov**

Appendix B

Organizations of Interest to the Pipeline Industry

American Gas Association (AGA). 400 N. Capitol Street, NW, Washington, DC 20001. **www.aga.org**

American Petroleum Institute (API). 1220 L Street, NW, Washington, DC 20005. **www.aga.org**

Interstate Natural Gas Association of America (INGAA). 20 F Street, NW, Suite 450, Washington, DC 20001. **www.ingaa.org**

National Institute for Occupational Safety and Health (NIOSH). 395 E Street, SW, Suite 9200, Patriots Plaza Building, Washington, DC 20201. **www.cdc.gov/niosh**

National Response Center (NRC). c/o United States Coast Guard (CG-5335 - Stop 7581), 2100 2nd Street, SW, Washington, DC 20593-0001. **www.nrc.uscg.mil**

Occupational Safety and Health Administration (OSHA). 200 Constitution Avenue, NW, Washington, DC 20210. **www.osha.gov**

Pipeline Hazardous Material Safety Administration (PHMSA). East Building, 2nd Floor, 1200 New Jersey Ave., SE, Washington, DC 20590. **www.phmsa.dot.gov**

Section Review Answer Key

Answer	Section Reference	Objective
Section One		
1. b	1.1.0	1a
2. a	1.2.0	1b
3. d	1.3.2	1c
4. b	1.4.0	1d
5. a	1.5.1	1e
6. c	1.6.0	1f
7. b	1.7.0	1g
Section Two		
1. c	2.1.2	2a
2. b	2.2.0	2b
3. a	2.3.2	2c
4. b	2.4.0	2d
5. a	2.5.1	2e
6. b	2.6.0	2f
7. c	2.7.0	2g

NCCER CURRICULA — USER UPDATE

NCCER makes every effort to keep its textbooks up-to-date and free of technical errors. We appreciate your help in this process. If you find an error, a typographical mistake, or an inaccuracy in NCCER's curricula, please fill out this form (or a photocopy), or complete the online form at **www.nccer.org/olf**. Be sure to include the exact module ID number, page number, a detailed description, and your recommended correction. Your input will be brought to the attention of the Authoring Team. Thank you for your assistance.

Instructors – If you have an idea for improving this textbook, or have found that additional materials were necessary to teach this module effectively, please let us know so that we may present your suggestions to the Authoring Team.

NCCER Product Development and Revision
13614 Progress Blvd., Alachua, FL 32615

Email: curriculum@nccer.org
Online: www.nccer.org/olf

❑ Trainee Guide ❑ Lesson Plans ❑ Exam ❑ PowerPoints Other ____________

Craft / Level: Copyright Date:

Module ID Number / Title:

Section Number(s):

Description:

Recommended Correction:

Your Name:

Address:

Email: Phone:

CT1_1-17

Measure Structure-to-Soil Potentials

Overview

Each year, corrosion costs the pipeline industry millions of dollars. While the use of protective coatings helps minimize corrosion, most pipelines also use cathodic protection systems. Annual testing is conducted to measure the integrity of the cathodic protection system. One type of electrical testing involves measuring the structure-to-soil potential at specific points in the system. This module describes cathodic protection systems and explains how to measure structure-to-soil potentials.

Trainees with successful module completions may be eligible for credentialing through the NCCER Registry. To learn more, go to **www.nccer.org** or contact us at 1.888.622.3720. Our website has information on the latest product releases and training, as well as online versions of our *Cornerstone* magazine and Pearson's product catalog.

Your feedback is welcome. You may email your comments to **curriculum@nccer.org**, send general comments and inquiries to **info@nccer.org**, or fill in the User Update form at the back of this module.

This information is general in nature and intended for training purposes only. Actual performance of activities described in this manual requires compliance with all applicable operating, service, maintenance, and safety procedures under the direction of qualified personnel. References in this manual to patented or proprietary devices do not constitute a recommendation of their use.

CT1_1-17
MEASURE STRUCTURE-TO-SOIL POTENTIALS

Objectives

When you have completed this module, you will be able to do the following:

1. Describe cathodic protection systems and fundamental aspects of measuring structure-to-soil potentials.
   a. Describe cathodic protection system types, standards, and planning.
   b. Describe fundamental aspects of taking structure-to-soil readings.
2. Describe how to measure structure-to-soil potentials (CT1_1-17).
   a. Identify the prerequisites, competencies, and abnormal operating conditions associated with taking structure-to-soil readings.
   b. Describe the procedure for measuring structure-to-soil potentials.

Performance Task

Under the supervision of your instructor, you should be able to do the following:

1. Measure structure-to-soil potentials (CT1_1-17).
   - Identify potential abnormal operating conditions that may occur during performance of this CT, and know the appropriate actions to take in response to them.
   - Utilize the appropriate personal protective equipment according to relevant company procedures.
   - Select the instrumentation to be used.
   - Identify the correct test point locations where measurements will be taken.
   - Correctly connect the test leads to the voltage meter and reference cell.
   - Measure the structure-to-soil potential.
   - Field-analyze readings to ensure they are within the desired range of readings, including a check of the polarity.
   - Complete appropriate documentation as required by operator's procedures.

Trade Terms

Abnormal operating conditions (AOCs)
Anode
Casing vent
Casing-to-soil potential
Cathode
Confined space
Copper-copper sulfate electrode (CSE)
Electrical potential
Foreign structure
Galvanic
High input impedance voltage meter
Ions
IR drop
Polarity
Potential measurement
Rectifier
Reference electrode
Shielding
Structure-to-soil potential
Test leads
Test stations

Industry Recognized Credentials

If you are training through an NCCER-accredited sponsor, you may be eligible for credentials from NCCER's Registry. The ID number for this module is CT1_1-17. Note that this module may have been used in other NCCER curricula and may apply to other level completions. Contact NCCER's Registry at 888.622.3720 or go to **www.nccer.org** for more information.

Contents

Figures and Tables

SECTION ONE

1.0.0 CATHODIC PROTECTION SYSTEMS AND STRUCTURE-TO-SOIL READINGS

Objective

Describe cathodic protection systems and fundamental aspects of measuring structure-to-soil potentials.

a. Describe cathodic protection system types, standards, and planning.
b. Describe fundamental aspects of taking structure-to-soil readings.

Trade Terms

Anode: The electrode of an electrochemical cell at which oxidation occurs.

Casing vent: An aboveground metal pipe that is connected to a buried casing to provide access to and venting of the atmosphere within the casing.

Casing-to-soil potential: The voltage potential of a casing in soil or an electrolyte as it relates to a reference electrode.

Cathode: The electrode of an electrochemical cell at which reduction occurs.

Copper-copper sulfate electrode (CSE): A non-polarizing electrode used in corrosion control to measure the electrical potential of a metal structure to a surrounding electrolyte to determine the potential for corrosion damage or to monitor the effectiveness of existing control measures.

Electrical potential: The potential energy of a unit charge at any point in an electric circuit, measured with respect to a specified reference point in the circuit or to ground.

Foreign structure: A pipeline or other metal structure in the vicinity of a pipeline system under cathodic protection.

Galvanic: Pertaining to electricity that flows as a result of a chemical action.

High input impedance voltage meter: An electronic device that is used to measure structure-to-soil potential.

Ions: Electrically charged atoms (either positively charged or negatively charged).

IR drop: The potential difference between the top of the pipe and the top of the soil.

Polarity: The distinction between the positive and negative potential differences at the terminals of a meter or electrical device.

Potential measurement: The activity of measuring the structure-to-soil potential.

Rectifier: A device that converts AC current to DC current and, in an impressed cathodic protection system, supplies current to the structure in the soil.

Reference electrode: An electrode whose open-circuit potential is constant under similar conditions of measurement; it is used to measure the relative potentials of other electrodes.

Shielding: A cover designed to protect against mechanical damage, or the action of diverting cathodic protection current from its natural path.

Structure-to-soil potential: The potential difference between the surface of a buried or submerged metallic structure (such as a pipe) and the electrolyte that is measured with reference to an electrode in contact with the electrolyte.

Test stations: Specific areas or locations, sometimes called test points, where a connection to a structure is made so that a structure-to-soil potential can be measured.

Pipeline corrosion can cause catastrophic failures with potentially severe consequences for the environment and serious injury or even death to those in the vicinity of the failure. While **test stations** record data at specific points in the system, these stations do not record data from the entire pipeline. Testing is done annually to measure the integrity of the cathodic protection (CP) system. One type of electrical testing involves measuring the **structure-to-soil potential** at specific points in the system.

Four conditions must be present for corrosion to occur. There must be an **anode**, a **cathode**, an electrolyte, and a metallic path. These conditions can involve two separate structures or they may be present on one structure. If any one of the four conditions is eliminated, corrosion can be prevented.

In anodic areas, current flows from the steel in the structure into the surrounding electrolyte (soil and water), causing the structure to corrode. In cathodic areas, current flows from the electrolyte onto the structure surface, and corrosion is eliminated.

The amount of metal that is lost due to corrosion is directly proportional to the amount of current flow. For example, one ampere of direct current (DC) discharging into a soil electrolyte (the current flow path) can remove approximately 20 pounds (9 kilograms [kg]) of steel in one year.

Rates vary with different metals; normal corrosion currents are usually given in thousandths of amperes, or milliamperes (mA). This information can be used to calculate the amount of protection needed.

Protective coatings are one method of corrosion protection because they prevent contact between a structure and the electrolyte. However, coatings are not perfect. They often have holidays (voids) that allow current to flow from the structure to the electrolyte. Cathodic protection systems are also used to prevent corrosion. The goal of a cathodic protection system is to eliminate the potential difference between the anodic and cathodic areas of a structure.

1.1.0 Cathodic Protection System Types, Standards, and Planning

Cathodic protection is the use of direct current to oppose the discharge of corrosion current from a structure. In order to mitigate corrosion, a cathodic protection system induces current onto a structure from an electrolyte.

The rate of corrosion can be reduced if the exposed metal on the surface of a pipeline can be made to collect current, so direct current is forced onto the pipeline. This current shifts the potential of the pipeline in the active (negative) direction, resulting in a reduction in the corrosion rate of the metal. When the amount of applied current is adjusted properly, it overpowers the corrosion current from the anodic areas on a pipeline and generates a net current flow on the pipe surface at these points. The surface then becomes a cathode, and the corrosion rate is reduced.

When current is forced to flow onto the pipe at areas that were previously discharging current, the voltage provided by the cathodic protection system must be greater than the voltage from the corrosion cells.

The following are the two basic methods of cathodic protection:

- galvanic anodes
- impressed current

1.1.1 Galvanic Anode Systems

Galvanic anode systems are simple and reliable. In a galvanic system, a sacrificial anode is connected to a pipeline to provide cathodic current. The anode is sacrificed (corrodes) and in doing so, protects the pipeline. Galvanic anodes must be replaced when spent. Magnesium, zinc, and aluminum are common types of galvanic anodes.

Galvanic anode protection relies on the natural electrical potential between two metals to create current flow. Since the driving voltage is limited to the very small potential difference existing between the metals, the current output is relatively low.

Groups of galvanic anodes are buried in the earth and connected to a pipeline in what is known as a ground bed (*Figure 1*). In the process of discharging current, the anodes in the ground bed are consumed by corrosion. The rate of corrosion consumption is described in pounds per ampere-year. Galvanic anodes are normally used where relatively small amounts of current are needed and in soils with relatively low resistance.

There are two types of galvanic anode installations: single-anode systems and banks of anodes. A bank of anodes is a number of galvanic anodes on a single header cable. Single anodes are connected directly to the structure to be protected.

> **NOTE**
> Multiple single anodes can be directly connected as well, but this is not the preferred method. The ideal method for multiple anodes is to connect them in a bank to a test station.

Anode banks are used to protect sections of piping and produce voltage drops from point to point in the earth. These voltage drops spread outward from the anode bank. The difference in voltage is greatest close to the anode bank and becomes smaller with distance until it becomes insignificant; the point of insignificance is referred to as *remote earth*.

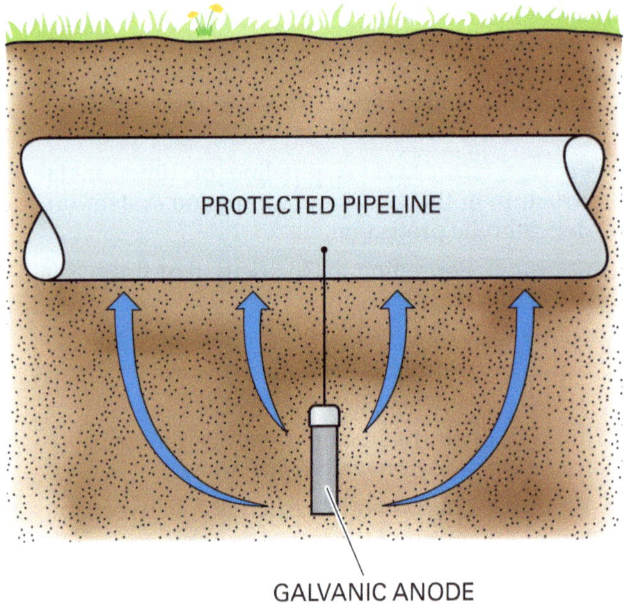

Figure 1 Diagram of galvanic anode circuit.

With a single anode, only the area close to the anodes is protected. An area of influence surrounds both the anode and the piping. As with an anode bank, a voltage drop is produced by the current flow from the anodes to the piping. Galvanic anodes can be used individually to protect small areas of exposed steel on coated structures. This practice is called *hot spotting* and is sometimes used in repair locations.

Another application of galvanic anodes is to control interference at the discharge point on a foreign pipeline. Galvanic anodes may also be added to an impressed current system if electrical **shielding** gets in the way of effective current distribution to remote points.

Among the advantages of using a galvanic anode system is that no external power source is needed. This type of system is easy to install and generates little or no stray current interference. Galvanic anode systems may be a good choice where conditions make it difficult to install impressed current systems.

One of the limitations of a galvanic anode system is its low current output. These systems may not be effective in a highly resistive environment. Also, if a structure is bare or poorly coated, a large number of anodes may be required. In addition, spent anodes may be difficult or expensive to replace. Due to its lower efficiency, the cost per unit ampere in a galvanic anode system is higher than that of an impressed current system. Galvanic system efficiency varies with the type of element and the environment in which it is used.

1.1.2 Impressed Current Systems

An impressed current system requires an external DC source such as a **rectifier** or battery. An anode material is placed in the electrolyte with the structure to be protected. The anode is then made more positive than the structure by connecting both the anode and the structure to a direct current supply. Any conductive material can be used as an impressed current anode, but since corrosion takes place at the anode, materials with very low consumption rates are most desirable. Impressed current anodes last longer than galvanic anodes.

Impressed current systems (*Figure 2*) can supply greater protection than galvanic anodes. The most common source of impressed current is utility power supplied through a rectifier, but solar power and thermoelectric generators may also be used.

The rectifiers used in impressed current systems convert higher-voltage alternating current (120/240 VAC) into lower-voltage direct current (typically 10 to 50 VDC). The output voltage of the unit is usually adjustable. Cathodic protection

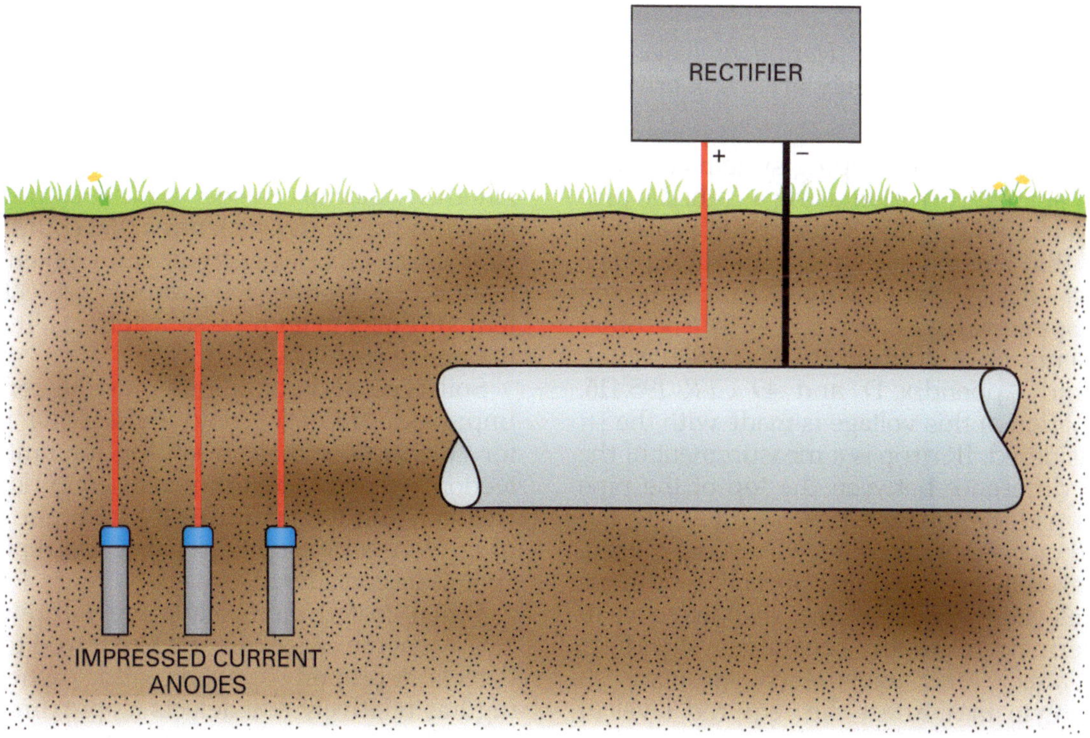

Figure 2 Diagram of impressed current circuit.

rectifiers are available in output capacities from one ampere to hundreds of amperes.

Impressed current protection is used where large amounts of current are required, such as for bare or poorly coated structures. It can also be used to overcome high electrolytic resistivity and stray current or cathodic interference problems. Impressed current systems can also be used to protect structures with dissipated galvanic anodes and for systems that include a number of different types of structures.

The disadvantages of impressed current systems include increased complexity and increased maintenance requirements. There is also a greater risk of generating stray current interference. In addition, overprotection may damage coatings or cause the pipeline to become brittle.

1.1.3 Cathodic Protection Standards

Pipeline technicians who are working with cathodic protection systems must always follow the latest revision of the following standards that have been developed to ensure that structures are fully protected from corrosion:

- NACE International Standard *SP-01-69, Standard Practice – Control of External Corrosion on Underground or Submerged Metallic Piping Systems*
- 49 *CFR*, Section 192, *Transportation of Natural and Other Gas by Pipeline*, Subpart I, Requirements for Corrosion Control
- 49 *CFR*, *Transportation of Hazardous Liquids by Pipeline*, Section 195, Subpart H, Corrosion Control

All underground steel piping systems must be protected against corrosion using a cathodic protection system. The system must create a cathodic potential that meets a negative (cathodic) voltage of at least −0.85V, with reference to a saturated copper-copper sulfate electrode (CSE) or other requirements as described in 49 *CFR*, Section 192.463 and Appendix D, and 49 *CFR* 195.416. Determination of this voltage is made with the IR drop considered. IR drop is a measurement of the potential difference between the top of the pipe and the top of the soil.

If another type of reference electrode is used, the values will be different. Special conditions such as elevated temperatures, disbonded coatings, thermal insulating coatings, shielding, bacterial attack, and unusual contaminants in the electrolyte may affect the required level of protection.

These criteria are based on the fact that when a pipeline is under cathodic protection, direct current flows from the conducting environment onto the pipeline, forcing the pipeline to assume a negative electrical polarity with respect to the environment. The value of −0.85V (with IR drop considered) is used for steel pipe as measured using a standard copper-copper sulfate reference electrode to contact the environment.

Potentials on steel pipe that are less negative than −0.85V (with IR drop considered) to a copper-copper sulfate electrode (CSE) indicate less than full cathodic protection. To maintain the minimum potential at distant locations, it is usually necessary to have higher negative potentials at drainage points.

Determining the correct level of protection is very important, as overprotection can damage coated structures. When cathodic protection potentials are more negative than necessary on coated pipelines, gaseous hydrogen is generated at coating defects in the steel surface. Hydrogen bubbles may cause the coating to lift around defects, increasing the current requirements for cathodic protection. This effect is most likely to occur in low-resistance environments.

1.1.4 Planning Cathodic Protection Systems

The two basic considerations when planning a cathodic protection system are determining the amount of protection needed overall and then selecting the type of system (galvanic or impressed current) that best fits the situation. In a galvanic anode system, the layout of the anodes must be carefully planned to ensure that the needed protection is provided. In an impressed current system, the external power source must be considered.

Another factor to consider is the proximity of any other underground structures that could produce stray currents. Stray currents can weaken the protection provided by the cathodic protection system and may even reverse its effects, increasing the rate of corrosion.

Soil type and conditions must be considered. Improved drainage could reduce the potential for corrosion and decrease the amount of protection needed. Environmental conditions, local weather, and temperature all play a part in planning. Access for testing, repairs, and replacement should be considered as well.

When coated piping fails prematurely due to corrosion, the cause can often be traced to coating defects. A properly applied coating on newer piping has relatively few defects (approximately one percent), but any defect is an opportunity for corrosion to take hold. Defects can be very difficult to locate, especially when the piping is buried.

To prevent costly failures and the repairs and downtime that inevitably follow, regular coating inspections must be performed throughout the life of the coating. *Figure 3* shows typical coating defects.

For buried piping, sophisticated holiday detectors can be used. In locations where coating defects exist, measurable voltage gradients can be detected at ground level. The larger the defect, the greater the current flow onto that area.

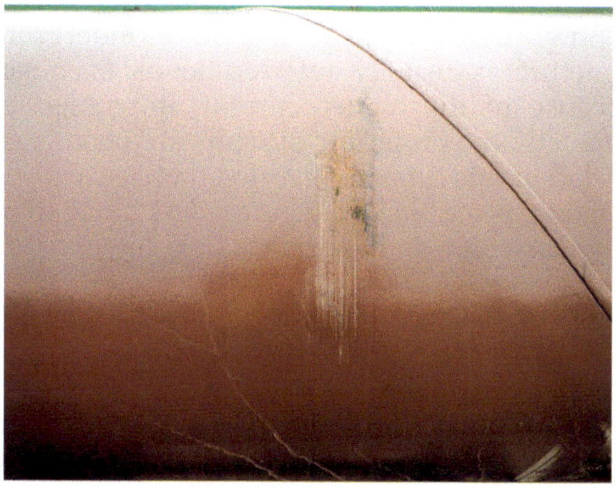

(A) SCRAPE

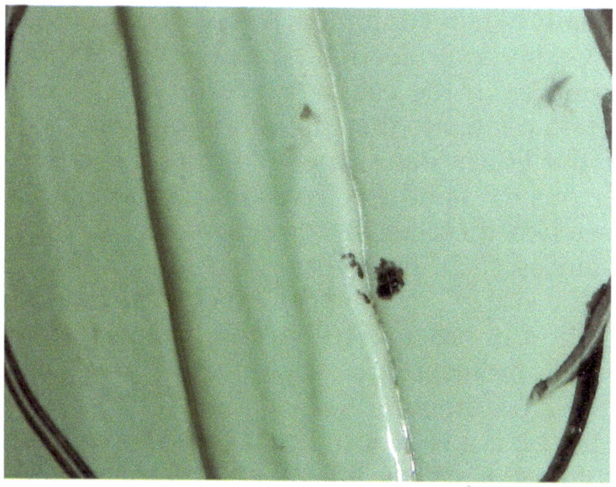

(B) HOLIDAY

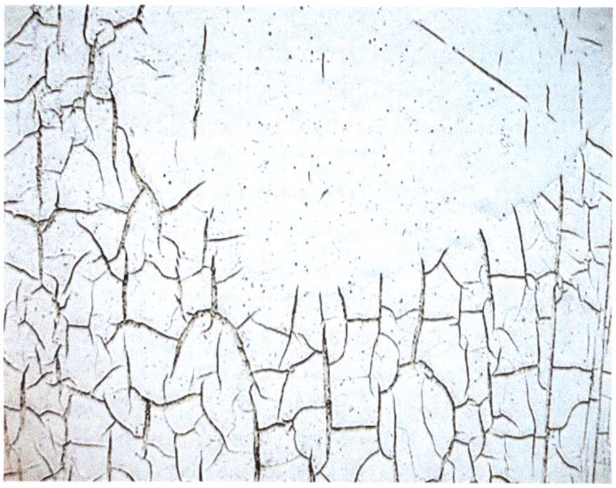

(C) CRACKING

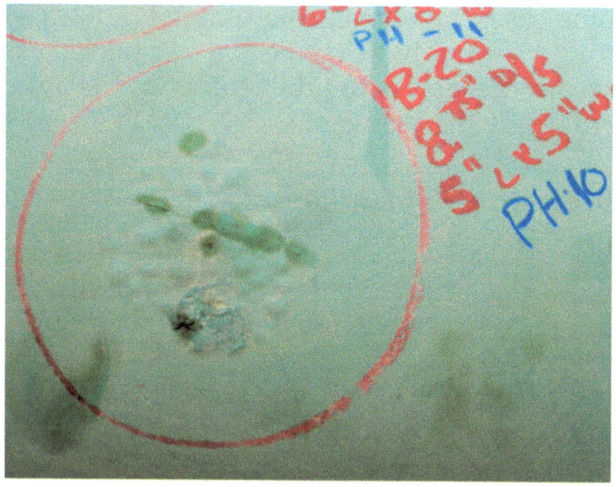

(D) BLISTERING

(E) PEELING

(F) DELAMINATION

Figure 3 Typical coating defects.

Although the type of coating may vary according to the environment, application, and piping material, the most widely used coating in the United States is fusion-bonded epoxy (FBE). It is also the most prevalent pipeline coating in Saudi Arabia, the United Kingdom, and Canada. FBE provides effective protection against corrosion, especially when combined with cathodic protection. Cathodic protection helps to overcome any deficiencies in the coating material. Deficiencies are often related to the application process.

In the FBE process, metallic iron is converted to ferrous ions and electrons. Problems with FBE arise when these ions are removed by chemical substances such as oxygen, chloride, and water, so FBE coatings must be combined with cathodic protection in order to fully protect the structure. Cathodic protection is very effective when used with FBE coatings.

If the piping is exposed or is subject to extreme conditions, a casing may be required. This is common where the piping runs through open trenches or under roads and railroad crossings. In some cases, an aboveground metal pipe called a casing vent might be connected to a buried casing to provide access to and venting of the atmosphere inside the casing.

1.1.5 Electrical Shielding

An electrical shield can be defined as any barrier that prevents or diverts the flow of cathodic protection current away from a protected structure. Electrical shielding can be of two types, either resulting from a nonmetallic insulating barrier that prevents current flow, or involving the diversion of current to other metallic (foreign) structures.

When underground metal structures are close together, an electrical shield may be produced that reduces the effect of the cathodic protection system. Pumping stations and tank farms are subject to this type of shielding. The severity of the shielding effect varies with conditions at each location. Uncoated metallic structures, especially those in direct contact, tend to increase the effects. Uncoated piping would be severely affected. Even grounding systems and rebar in buried concrete can affect the current flow. In these areas, a remote cathodic protection system is unlikely to provide adequate protection. A galvanic anode may be required to provide additional protection.

Measuring the potential between the earth and the structures will show whether the proximity of the structures has produced a significant shield. An example of electrical shielding in shown in *Figure 4*.

1.1.6 Stray Current from Other Structures

Stray currents occur when two or more cathodic protection systems interfere with one another. It should be company policy to notify all operators in an area when a cathodic protection system is planned or before making changes to an existing system.

Stray current damage to another underground structure is usually referred to as *interference*. Because of the higher current, impressed current cathodic protection systems are usually connected to this type of corrosion damage. Galvanic cathodic protection systems are less likely to produce corrosion damage due to their lower current.

1.2.0 Fundamentals of Structure-to-Soil Readings

Structure-to-soil readings are commonly required in pipeline corrosion testing. Soil potentials are commonly measured for the following types of structures:

- *Pipelines* – This includes all components of a pipeline facility through which a hazardous liquid is transported, including pipes, valves, pumps, and other assemblies. The pipeline facility encompasses piping and equipment used in production, extraction, recovery, lifting, stabilization, separation, or treating of petroleum. This includes piping between the production facilities and refineries.
- *Casings* – These are pipes that contain a carrier pipeline. Casings are typically used when a pipeline passes under railroad crossings or roads. They protect the pipeline from high load stresses. State and local regulations identify specific mandatory casing locations. Problems created by casings result from electrical contact between the coated pipeline and the bare casing. Contact with casings can interfere with the cathodic protection of pipelines due, in part, to the casing-to-soil potential. For this reason, transport pipes must be provided with sufficiently high, mechanically stable spacers to exclude electrical contacts. Designers and planners should use only the type of casing seals that resist ground water entering into the space between the casing and the pipeline. On cathodically protected pipelines, casings should be used only where required by law or where the pipeline cannot be laid at crossing points without risk of damage.
- *Tank bottoms* – An aboveground storage tank is any tank or other container that is aboveground, partially buried, bunkered, or in a subterranean vault, including floating fuel

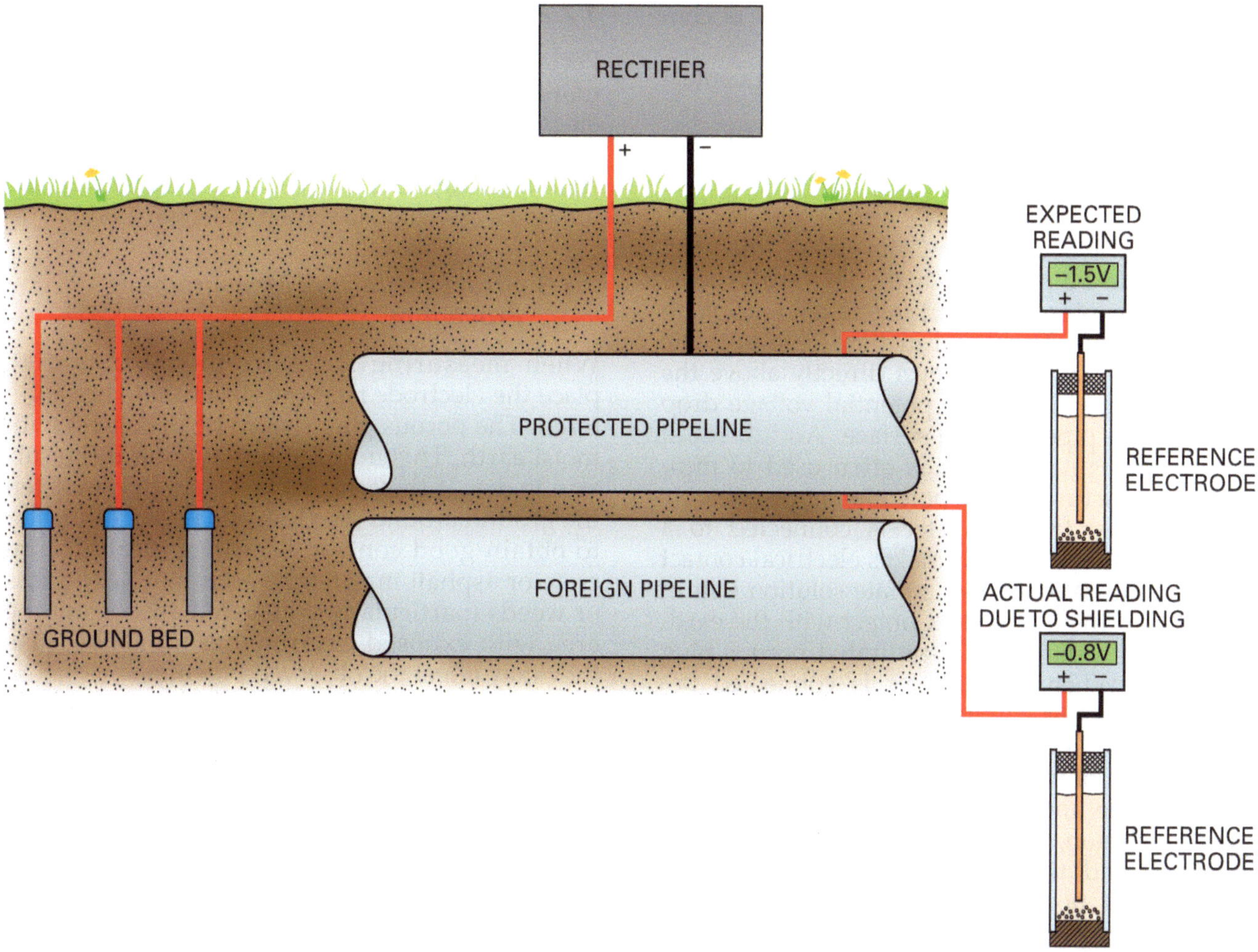

Figure 4 Electrical shielding.

systems. The tank bottom and any portion that is buried or submerged in electrolyte must be protected.

- *Foreign structures* – A **foreign structure** is any structure or pipeline that is not owned by the pipeline company. When steel structures contact the earth or foreign metallic structures, corrosion can occur and protection is required.
- *Well casings* – Wells drilled in oil and natural gas fields are fitted with casings for protection and stabilization. Concrete is poured into the space between the outer pipe, or casing, and the surrounding rock/soil. The concrete filling is not uniform in thickness and may contain voids. Pipeline well casings are subject to various types of corrosion, depending on the surrounding soil content and the degree of protection provided by the cement. Boreholes are often located close to one another, so they may be affected by the cathodic protection from other wells. In large oil fields and reservoirs, cathodic protection for all the wells is appropriate.
- *Underground storage tanks (USTs)* – These are stationary structures designed to contain petroleum. This classification includes all associated underground piping and connections. In reference to storage tanks, the term *underground* typically means that the volume, including the volume of underground pipes, is ten percent or more beneath the surface of the ground.

Two common types of reference electrodes are used to take measurements on pipeline structures. They are copper-copper sulfate electrodes (CSEs) and silver chloride electrodes.

1.2.1 Copper-Copper Sulfate Electrodes

The copper-copper sulfate electrode (CSE) (*Figure 5*) is the most common type of reference electrode used in pipeline corrosion work. It may be referred to as a copper sulfate electrode, copper sulfate reference, copper sulfate half-cell, or CuSO4 electrode. The essential components of a copper sulfate electrode include the following:

- Copper rod to which the instrument is connected
- Saturated solution of copper sulfate in contact with the rod
- Porous membrane in contact with the soil
- Surplus crystals of copper sulfate in the solution

To make a potential measurement, the electrode should be placed near the line at the same depth as the pipe; however, this is not feasible in most situations. The usual method is to take the measurement at surface level directly above the pipeline, allowing for the potential voltage drop between the pipe and the surface. A high input impedance voltage meter is often used to measure the structure-to-soil potential.

Electrically, the voltmeter is connected to a pure copper rod. The rod makes electrical contact with the saturated copper sulfate solution within the body of the electrode. Contact with the earth is made with the solution-saturated porous plug (*Figure 6*). When a voltmeter is connected between the copper sulfate reference electrode and the structure, the observed reading is actually a combination of two values: the potential between the electrode and the earth through the copper sulfate solution, and the potential between the structure and the earth. Each of these values represents a portion of the total cell potential between the piping and the copper rod of the electrode. The reference electrode-to-earth half-cell potential is a constant. The structure-to-earth half-cell potential is the variable that must be evaluated.

1.2.2 Silver Chloride (Salt Water) Electrodes

Portable salt-water reference electrodes are commonly made of zinc/silver-silver chloride. These electrodes are used for seawater installations, such as offshore facilities, harbor structures, and soils with high chloride content. A silver chloride reference electrode is shown in *Figure 7*.

1.2.3 Structure-to-Soil Measurement Procedure

When measuring structure-to-soil potential, place the electrode near the structure to be monitored. The porous plug should be in contact with moist earth. This may require digging if the surface is dry. If the area is extremely dry, moisten the ground around the electrode with fresh water to obtain good contact. Readings through concrete or asphalt may not be accurate. Keep grass or weeds (particularly when wet) away from the area with exposed electrode terminals because they may affect the observed potential.

Copper sulfate electrodes should be kept clean. When copper rods become encrusted, they may be cleaned using a nonmetallic abrasive pad. Rinse the rod thoroughly with distilled water before reassembling it. Ensure that the electrode is always full of saturated copper sulfate solution; use a few excess copper sulfate crystals or powder to ensure that the solution remains concentrated. Replace the solution with fresh solution periodically to avoid contamination. To avoid having the electrode dry out, keep the porous plug covered when the rod is not in use. If the electrode dries

Figure 5 Copper sulfate electrodes.

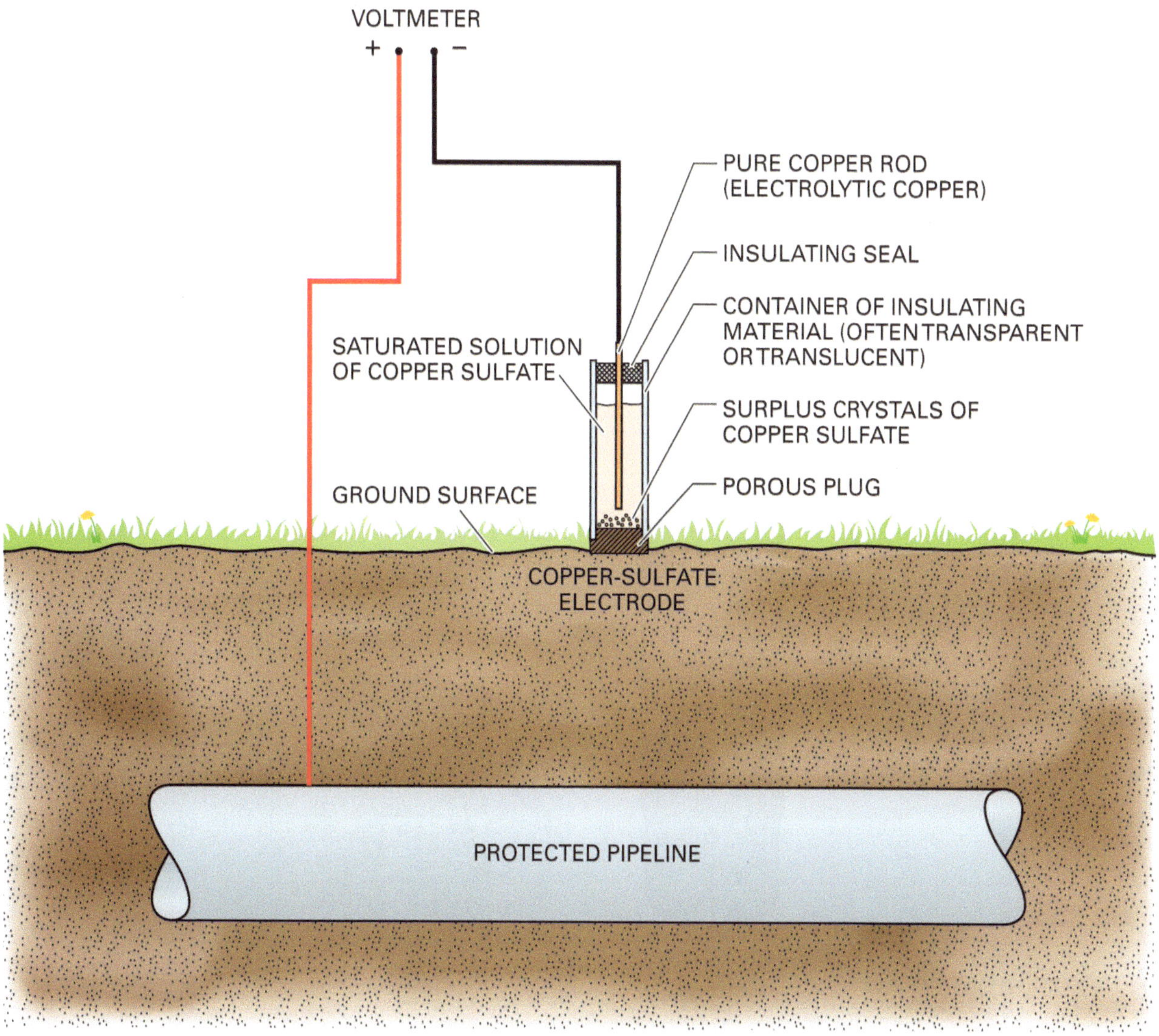

Figure 6 Voltmeter measurement of structure-to-soil potential.

out, re-wet the porous plug. Porous plugs seldom need to be replaced.

Each test kit should contain at least two copper sulfate electrodes that are matched in potential. The potentials should be checked periodically by measuring the potential difference between the two when placed side by side in soil or water. For best performance, the potential difference should not exceed the value indicated in the manufacturer's instructions (typically about 10 millivolts [mV]).

When using reference electrodes other than copper-copper sulfate, refer to the manufacturer's data for required adjustments.

Connecting the electrode to the negative terminal of a digital voltmeter and connecting the positive instrument terminal to the pipeline, should result in a negative reading. *Figure 8* shows voltmeter measurement of structure-to-soil potential.

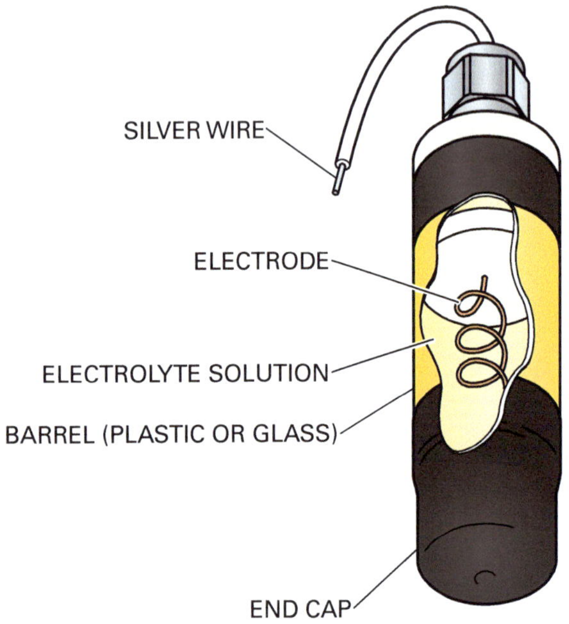

Figure 7 Silver chloride reference electrode.

Figure 8 Measuring structure-to-soil potential.

Cathodic Protection Digital Voltmeters

This DC cathodic protection voltmeter can be used to measure structure-to-soil potentials of buried or submerged metal structures, such as pipelines and storage tanks. It can also be used to measure structure-to-structure and structure-to-anode values. This meter has a connection for the half-cell to screw into the back of the voltmeter to measure structure-to-anode values.

STRUCTURE-TO-ANODE MEASUREMENT

Additional Resources

It is assumed that you will learn to perform these procedures under the direct supervision of your crew leader, foreman, or instructor. For independent study, refer to the recommended Additional Resources at the back of this module.

1.0.0 Section Review

1. When two or more cathodic protection systems interfere with one another, there is the likelihood for _____.
   a. short circuiting
   b. ground faults
   c. stray currents
   d. galvanic shielding

2. Copper-copper sulfate electrodes (CSEs) are used _____.
   a. to make potential measurements
   b. to monitor foreign structures
   c. to power rectifiers
   d. to provide galvanic protection

SECTION TWO

2.0.0 MEASURE STRUCTURE-TO-SOIL POTENTIALS

Objective

Describe how to measure structure-to-soil potentials (CT1_1-17).

a. Identify the prerequisites, competencies, and abnormal operating conditions associated with taking structure-to-soil readings.
b. Describe the procedure for measuring structure-to-soil potentials.

Performance Task

Measure structure-to-soil potentials (CT1_1-17).

- Identify potential abnormal operating conditions that may occur during performance of this CT, and know the appropriate actions to take in response to them.
- Utilize the appropriate personal protective equipment according to relevant company procedures.
- Select the instrumentation to be used.
- Identify the correct test point locations where measurements will be taken.
- Correctly connect the test leads to the voltage meter and reference cell.
- Measure the structure-to-soil potential.
- Field-analyze readings to ensure they are within the desired range of readings, including a check of the polarity.
- Complete appropriate documentation as required by operator's procedures.

Trade Terms

Abnormal operating conditions (AOCs): Conditions within a pipeline system, identified by the operator, that may indicate (1) a malfunction of a component or (2) a deviation from normal operations that may indicate a condition exceeding design limits or result in a hazard to persons, property, or the environment.

Confined space: A space that is large enough and so configured that an employee can bodily enter and perform assigned work; has limited or restricted means for entry or exit (such as tanks, vessels, vaults, etc.); and is not designed for continuous employee occupancy.

Test leads: Wires that are connected to a structure being tested that provide easy connection points for a meter.

Personnel taking structure-to-soil readings are required to have some basic competencies. They must have the ability to carefully follow the required procedures for measuring structure-to-soil potentials and recording the results. This section describes some of the basic skills needed by pipeline technicians and identifies abnormal operating conditions (AOCs) that personnel may need to respond to when taking structure-to-soil readings. It also describes the basic procedure used for measuring structure-to-soil potentials.

2.1.0 Prerequisites, Competencies, and Abnormal Operating Conditions

Personnel must possess a variety of skills in order to safely and accurately measure structure-to-soil potentials. Testing procedures require numerous steps which a pipeline technician must be competent to perform. Personnel must be aware of abnormal operating conditions (AOCs) that may occur during testing procedures and know how to properly respond to them.

2.1.1 Prerequisites and Competencies

Personnel assigned to measure structure-to-soil potentials must have a good understanding of hazards that can exist before and during this covered task, as well as what the task entails. Before measuring potentials, personnel must be familiar with hazards along the right-of-way (ROW), including those associated with ROW encroachments and unsupervised third-party construction. They must be familiar with the locations of test stations, and they must be able to properly choose, inspect, calibrate, and operate any tools or instruments to be used during the procedure. Pipeline work often involves trenched and excavated areas, as well as tank containment areas. For these reasons, personnel must be properly trained for confined space entry and excavation safety.

Personnel must be able to visibly inspect the pipe and determine if there are any unsafe integrity issues, such as evidence of a release or significant mechanical damage (gouges, scrapes, dents), and discontinue the work and make notifications if any are found. If old coating materials that may contain asbestos are encountered, personnel must make the proper notifications so that a qualified person can sample, test, and, if necessary, remove and dispose of any asbestos-containing material. For these and other safety reasons, personnel must be familiar with local emergency telephone numbers.

The task begins with equipment selections and ends with documenting the results. Once testing has begun, personnel must know how to position, manipulate, and read any equipment that might be used. If personnel encounter induced AC voltage, such as a condition that could occur if a pipeline is located near a high voltage AC electrical field, they must be knowledgeable about the effects and hazards associated with the voltage and know how to notify the appropriate company personnel. Personnel must understand that no job is complete until all tasks are documented and the proper notifications are made.

Personnel must have a basic understanding of additional tasks related to structure-to-soil potential measurements, such as conducting a close interval survey (CT1_2-17), testing to detect interference (CT1_3-17), inspecting and performing electrical testing of bonds (CT1_4-17), and inspecting and testing electrical isolation (CT1_5-17).

2.1.2 Abnormal Operating Conditions

Personnel measuring structure-to-soil potentials must be able to identify potential abnormal operating conditions (AOCs) and know the proper response to each, should they occur. The following are some potential AOCs, and the appropriate action personnel should take in response to them:

- *Damage to or unsupervised third party construction on a right-of-way is discovered* – Notify appropriate company personnel and, if the contractor continues to perform unsupervised work on site, discuss the dangers posed by the situation with the contractor.
- *Internal or external corrosion of the pipe is present and to the extent that structure-to-soil potential measurements would be impractical* – Document the condition according to company guidelines and notify personnel responsible for correcting the problem.
- *Voltage is less than minimum requirements* – Verify the cathodic protection level and implement mitigation if insufficient.
- *Erratic or floating readings* – Determine the cause of the erratic readings and repair the test leads or equipment, as needed.
- *Reverse polarity of readings* – Document and implement corrective actions.
- *Induced AC voltage is encountered* – Exercise caution around any equipment or structures and notify the appropriate company personnel.
- *Fire or explosion is caused by the ignition of a hydrocarbon liquid or gas* – Evacuate the area and notify the appropriate firefighting personnel. It may be necessary to call 911 in such an emergency.
- *Unexpected leaks of hazardous liquid or carbon dioxide because of a loss of pipe wall thickness* – Take steps to eliminate any possible ignition sources and notify the appropriate personnel, possibly including 911.

2.2.0 Procedure

When performing electrical testing, always follow company procedures and the manufacturer's instructions for the instrumentation in use. The general steps of a typical procedure for measuring structure-to-soil potential are as follows:

WARNING!

Always obtain and use the appropriate personal protective equipment (PPE) for each task. Stay alert to the possibility of unexpected leaks from the exposed pipeline such as hazardous liquid or carbon dioxide (CO_2). In the event of a leak, safely eliminate all ignition sources and immediately notify the appropriate personnel.

Step 1 Identify potential abnormal operating conditions that may occur, and know the appropriate actions to take in response to them.

Step 2 Utilize the appropriate personal protective equipment according to relevant company procedures.

Step 3 Select the instrumentation to be used (test leads, voltmeter, and reference electrode). Inspect the equipment for damage and proper operation before taking any measurements.

Step 4 Identify the test point locations where the measurements will be taken. Note that a structure may have several test points.

CAUTION

Always consult company procedures for the proper polarity connections. Company procedures should indicate which positive and which negative lead to connect to a particular source to obtain a correct reading.

Step 5 Connect the test leads to the voltmeter and the reference cell.

Step 6 Measure the structure-to-soil potential.

Step 7 Review the readings as they are taken to ensure that they fall within the desired range of readings, including a check of the polarity. Compare with previous values, if available.

Step 8 If the reading requires further evaluation, seek assistance from appropriate personnel for additional investigation and corrective actions before documentation.

Step 9 Complete appropriate documentation as required by operator's procedures.

Additional Resources

It is assumed that you will learn to perform these procedures under the direct supervision of your crew leader, foreman, or instructor. For independent study, refer to the recommended Additional Resources at the back of this module.

2.0.0 Section Review

1. Gouges, scrapes, and dents in pipeline surface material are common examples of _____.
   a. corrosive erosion
   b. mechanical damage
   c. substrate profiling
   d. adhesion enhancement

2. When measuring structure-to-soil potential, it is important to review the readings as they are taken to ensure that they are _____.
   a. all displaying an induced AC voltage as required
   b. automatically compiling an engineering coating analysis
   c. all within the desired range with the correct polarity
   d. displaying floating values according to NACE specifications

Summary

Pipeline corrosion can cause catastrophic failures with potentially serious consequences for the environment and personnel. While the use of protective coatings helps minimize corrosion by preventing contact between piping structures and the soil, most pipelines also use cathodic protection systems. Pipeline technicians conduct annual testing to ensure the integrity of cathodic protection systems. Personnel must be familiar with preparations, hazards, and procedures involved in measuring the structure-to-soil potential at specific points in a pipeline system. This module described cathodic protection systems and explained the general requirements and procedures for safely and effectively measuring structure-to-soil potentials.

Review Questions

1. A cathodic protection system turns the pipeline surface into a(n) _____.
   a. cathode
   b. heat sink
   c. anode
   d. electrolyte

2. Anode banks are used to provide cathodic protection to _____.
   a. small hot-spot areas
   b. aboveground structures
   c. sections of piping
   d. pipelines in saltwater

3. One characteristic of a typical rectifier unit is that it _____.
   a. decreases power to a galvanic anode
   b. has an AC ammeter and voltmeter
   c. converts DC current to AC current
   d. usually has an adjustable output voltage

4. When taking measurements between a steel pipe and a copper-copper sulfate electrode (CSE), adequate cathodic protection is indicated by a value of at least _____.
   a. –0.70V (with IR drop considered)
   b. –0.85V (with IR drop considered)
   c. –1.00V (with IR drop considered)
   d. –1.70V (with IR drop considered)

5. Stray current corrosion damage to an underground structure caused by a cathodic protection system on another structure is commonly referred to as _____.
   a. crossover damage
   b. interference
   c. proximity damage
   d. interception

6. The most common type of electrode in a pipeline cathodic protection system is the _____.
   a. copper-copper sulfate electrode (CSE)
   b. silver chloride electrode
   c. salt-water electrode
   d. zinc-chloride electrode

7. If extensive internal or external piping corrosion is found prior to measuring structure-to-soil potentials, a pipeline technician should _____.
   a. apply a cathodic patch over the damage and continue with the measurements
   b. make sure that the instruments are attached directly to the corroded area
   c. document the condition and notify personnel responsible for correcting the problem
   d. attach an approved anode bank to the structure and adjust the test instrument

8. The final step in any procedure for measuring structure-to-soil potential is to _____.
   a. inspect every piece of tested equipment for damage and proper operation
   b. accurately record all required information in accordance with company policy
   c. remove lockout/tagout barriers and restart all holiday detection equipment
   d. conduct a Barcol frequency test to certify the readings to NACE standards

Trade Terms Quiz

Fill in the blank with the correct term that you learned from your study of this module.

1. The distinction between the positive and negative potential differences at the terminals of a meter or electrical device is called ________.
2. Electricity that flows as a result of a chemical action is considered ________.
3. A non-polarizing electrode used in corrosion control to measure the electrical potential of a metal structure to a surrounding electrolyte to determine the potential for corrosion damage or to monitor the effectiveness of existing control measures is a(n) ________.
4. A cover designed to protect against mechanical damage, or the action of diverting cathodic protection current from its natural path is called ________.
5. The electrode of an electrochemical cell at which reduction occurs is a(n) ________.
6. Conditions within a pipeline system that may indicate (1) a malfunction of a component or (2) a condition exceeding design limits or resulting in a hazard to persons, property, or the environment are called ________.
7. The potential difference between the top of the pipe and the top of the soil is known as the ________.
8. Wires connected to a structure being tested that provide easy connection points for a meter are called ________.
9. The potential energy of a unit charge at any point in an electric circuit, measured with respect to a specified reference point in the circuit or to ground is called ________.
10. A device that converts AC current to DC current and, in an impressed cathodic protection system, supplies current to the structure in the soil is a(n) ________.
11. An aboveground metal pipe that is connected to a buried casing to provide access to and venting of the atmosphere within the casing is a(n) ________.
12. An electronic device that is used to measure structure-to-soil potential is a(n) ________.
13. The electrode of an electrochemical cell at which oxidation occurs is called a(n) ________.
14. Specific areas or locations where a connection to a structure is made so that a structure-to-soil potential can be measured are ________.
15. A pipeline or other metal structure in the vicinity of a pipeline system under cathodic protection is a(n) ________.
16. The potential difference between the surface of a buried or submerged metallic structure (such as a pipe) and the electrolyte that is measured with reference to an electrode in contact with the electrolyte is known as the ________.
17. An electrode whose open-circuit potential is constant under similar conditions of measurement, used to measure the relative potentials of other electrodes is a(n) ________.
18. A space that is large enough and so configured that an employee can bodily enter and perform assigned work; has limited or restricted means for entry or exit (such as tanks, vessels, vaults, etc.); and is not designed for continuous employee occupancy is called a(n) ________.

19. The activity of measuring the structure-to-soil potential is called ________.

20. Electrically charged atoms (either positively charged or negatively charged) are called ________.

21. The voltage potential of a casing in soil or an electrolyte as it relates to a reference electrode is called ________.

Trade Terms

Abnormal operating conditions (AOCs)
Anode
Casing vent
Casing-to-soil potential
Cathode
Confined space
Copper-copper sulfate electrode (CSE)
Electrical potential
Foreign structure
Galvanic
High input impedance voltage meter
Ions
IR drop
Polarity
Potential measurement
Rectifier
Reference electrode
Shielding
Structure-to-soil potential
Test leads
Test stations

Trade Terms Introduced in This Module

Abnormal operating conditions (AOCs): Conditions within a pipeline system, identified by the operator, that may indicate (1) a malfunction of a component or (2) a deviation from normal operations that may indicate a condition exceeding design limits or result in a hazard to persons, property, or the environment.

Anode: The electrode of an electrochemical cell at which oxidation occurs.

Casing vent: An aboveground metal pipe that is connected to a buried casing to provide access to and venting of the atmosphere within the casing.

Casing-to-soil potential: The voltage potential of a casing in soil or an electrolyte as it relates to a reference electrode.

Cathode: The electrode of an electrochemical cell at which reduction occurs.

Confined space: A space that is large enough and so configured that an employee can bodily enter and perform assigned work; has limited or restricted means for entry or exit (such as tanks, vessels, vaults, etc.); and is not designed for continuous employee occupancy.

Copper-copper sulfate electrode (CSE): A non-polarizing electrode used in corrosion control to measure the electrical potential of a metal structure to a surrounding electrolyte to determine the potential for corrosion damage or to monitor the effectiveness of existing control measures.

Electrical potential: The potential energy of a unit charge at any point in an electric circuit, measured with respect to a specified reference point in the circuit or to ground.

Foreign structure: A pipeline or other metal structure in the vicinity of a pipeline system under cathodic protection.

Galvanic: Pertaining to electricity that flows as a result of a chemical action.

High input impedance voltage meter: An electronic device that is used to measure structure-to-soil potential.

Ions: Electrically charged atoms (either positively charged or negatively charged).

IR drop: The potential difference between the top of the pipe and the top of the soil.

Polarity: The distinction between the positive and negative potential differences at the terminals of a meter or electrical device.

Potential measurement: The activity of measuring the structure-to-soil potential.

Rectifier: A device that converts AC current to DC current and, in an impressed cathodic protection system, supplies current to the structure in the soil.

Reference electrode: An electrode whose open-circuit potential is constant under similar conditions of measurement; it is used to measure the relative potentials of other electrodes.

Shielding: A cover designed to protect against mechanical damage, or the action of diverting cathodic protection current from its natural path.

Structure-to-soil potential: The potential difference between the surface of a buried or submerged metallic structure (such as a pipe) and the electrolyte that is measured with reference to an electrode in contact with the electrolyte.

Test leads: Wires that are connected to a structure being tested that provide easy connection points for a meter.

Test stations: Specific areas or locations, sometimes called test points, where a connection to a structure is made so that a structure-to-soil potential can be measured.

Additional Resources

The following recommended resources can provide additional helpful information related to the requirements for this covered task:

API Recommended Practice 1161, Third Edition, © 2014 American Petroleum Institute. API Publishing Services, 1220 L Street, NW, Washington, DC 20005. Distributed by Thompson Reuters (Scientific) LLC, **www.techstreet.com**.

49 *CFR* 192, Subpart I, **www.ecfr.gov**

49 *CFR* 195, Subpart H, **www.ecfr.gov**

NCCER Module AOCCC-17, *Abnormal Operating Conditions Control Center*

NCCER Module AOCFG-17, *Abnormal Operating Conditions Field and Gas*

NCCER Module 62105-14, *Introduction to Pipeline Documents*

NCCER Module 66101-14, *Introduction to the Pipeline Industry*

NCCER Module 75120-13, *Confined Spaces and Excavations*, from Field Safety

NCCER Module CT1_2-17, *Conduct Close Interval Survey*

NCCER Module CT1_3-17, *Test and Detect Interference*

NCCER Module CT1_4-17, *Inspect and Perform Electrical Test of Bonds*

NCCER Module CT1_5-17, *Inspect and Test Electrical Isolation*

NCCER Module CT14_1-17, *Locate Line*

API Recommended Practice 651, Cathodic Protection of Aboveground Petroleum Storage Tanks. Fourth Edition, 2014. Available from the American Petroleum Institute at **www.api.org**.

NACE International *SP0169-2013, Control of External Corrosion on Underground or Submerged Metallic Piping Systems*. 2013. Available from NACE International at **www.nace.org**.

Figure Credits

Tinker & Rasor, Module Opener, Figures 5, SA01, 8

Photo courtesy of Alan Kehr, Alan Kehr Anti-Corrosion, LLC, Figure 3A-B, 3F

Energy Maintenance Services, EMS USA, Figure 3C-E

Section Review Answer Key

Answer	Section Reference	Objective
Section One		
1. c	1.1.6	1a
2. a	1.2.1	1b
Section Two		
1. b	2.1.1	2a
2. c	2.2.0	2b

NCCER CURRICULA — USER UPDATE

NCCER makes every effort to keep its textbooks up-to-date and free of technical errors. We appreciate your help in this process. If you find an error, a typographical mistake, or an inaccuracy in NCCER's curricula, please fill out this form (or a photocopy), or complete the online form at **www.nccer.org/olf**. Be sure to include the exact module ID number, page number, a detailed description, and your recommended correction. Your input will be brought to the attention of the Authoring Team. Thank you for your assistance.

Instructors – If you have an idea for improving this textbook, or have found that additional materials were necessary to teach this module effectively, please let us know so that we may present your suggestions to the Authoring Team.

NCCER Product Development and Revision

13614 Progress Blvd., Alachua, FL 32615

Email: curriculum@nccer.org
Online: www.nccer.org/olf

❑ Trainee Guide ❑ Lesson Plans ❑ Exam ❑ PowerPoints Other ____________

Craft / Level: Copyright Date:

Module ID Number / Title:

Section Number(s):

Description:

Recommended Correction:

Your Name:

Address:

Email: Phone:

CT1_2-17

Conduct Close Interval Survey

Overview

Each year, corrosion costs the pipeline industry millions of dollars. Cathodic protection systems are used to help minimize corrosion. These systems must be tested on a regular basis to ensure their integrity, effectiveness, and regulatory compliance. One assessment method used to test cathodic protection systems on buried pipelines is a close interval survey. This module provides an overview of cathodic protection, describes types of close interval surveys, and explains the procedure for measuring the potential (voltage) difference between buried pipes and the surrounding soil.

Trainees with successful module completions may be eligible for credentialing through the NCCER Registry. To learn more, go to **www.nccer.org** or contact us at 1.888.622.3720. Our website has information on the latest product releases and training, as well as online versions of our *Cornerstone* magazine and Pearson's product catalog.

Your feedback is welcome. You may email your comments to **curriculum@nccer.org**, send general comments and inquiries to **info@nccer.org**, or fill in the User Update form at the back of this module.

This information is general in nature and intended for training purposes only. Actual performance of activities described in this manual requires compliance with all applicable operating, service, maintenance, and safety procedures under the direction of qualified personnel. References in this manual to patented or proprietary devices do not constitute a recommendation of their use.

CT1_2-17
Conduct Close Interval Survey

Objectives

When you have completed this module, you will be able to do the following:

1. Describe cathodic protection systems and identify different types of close interval surveys.
   a. Describe how cathodic protection systems minimize corrosion.
   b. Identify different types of close interval surveys.
2. Describe how to conduct a close interval survey (CT1_2-17).
   a. Identify the prerequisites, competencies, and abnormal operating conditions associated with conducting close interval surveys.
   b. Describe the procedure for conducting a close interval survey.

Performance Task

Under the supervision of your instructor, you should be able to do the following:

1. Conduct a close interval survey (CT1_2-17).
   - Identify potential abnormal operating conditions that may occur during performance of this CT, and know the appropriate actions to take in response to them.
   - Utilize the appropriate personal protective equipment according to relevant company procedures.
   - Identify the test point locations where connections will be made.
   - Correctly connect the test leads, the voltage meter or data logger, and reference cell.
   - Verify that current sources are operational (on for "On"/interrupted surveys and turned off/disconnected for depolarized survey).
   - Place the reference cell directly above the pipeline being surveyed.
   - Select the instrumentation to be used (survey wire, voltmeter, data logger, reference electrodes, etc.).
   - For interrupted surveys, install current interrupters at all identified current sources. They should be set at the operator-determined time cycle and synchronized.
   - Measure the structure-to-soil potential according to the desired intervals for this survey.
   - Verify data is recorded.
   - Complete appropriate documentation as required by operator's procedures.

Trade Terms

Anode
Casing-to-soil potential
Cathode
Close interval survey (CIS)
Confined space
Copper-copper sulfate electrode (CSE)
Current interrupter
Depolarized ("Off") survey
Electrical potential
Foreign structure
Galvanic
"Instant Off" potential
Interrupted ("On/Off") survey
Native survey
"On" survey
Polarity
Potential measurement
Rectifier
Reference electrode
Structure-to-soil potential
Test leads
Test stations

Industry Recognized Credentials

If you are training through an NCCER-accredited sponsor, you may be eligible for credentials from NCCER's Registry. The ID number for this module is CT1_2-17. Note that this module may have been used in other NCCER curricula and may apply to other level completions. Contact NCCER's Registry at 888.622.3720 or go to **www.nccer.org** for more information.

Contents

Figures

SECTION ONE

1.0.0 CATHODIC PROTECTION AND CLOSE INTERVAL SURVEYS

Objective

Describe cathodic protection systems and identify different types of close interval surveys.

a. Describe how cathodic protection systems minimize corrosion.
b. Identify different types of close interval surveys.

Trade Terms

Anode: The electrode of an electrochemical cell at which oxidation occurs.

Casing-to-soil potential: The voltage potential of a casing in soil or an electrolyte as it relates to a reference electrode.

Cathode: The electrode of an electrochemical cell at which reduction occurs.

Close interval survey (CIS): One of several assessment methods that use voltage measurements to determine the effectiveness of a cathodic protection system on buried pipelines. Also known as close interval potential surveys (CIPS).

Copper-copper sulfate electrode (CSE): A non-polarizing electrode used in corrosion control to measure the electrical potential of a metal structure to a surrounding electrolyte in order to determine the potential for corrosion damage or to monitor the effectiveness of existing control measures.

Depolarized ("Off") survey: A close interval survey that measures the potential difference between the structure and the ground surface after the cathodic protection current has been turned off and the structure-to-soil potential has stabilized.

Electrical potential: The potential energy of a unit charge at any point in an electric circuit, measured with respect to a specified reference point in the circuit or to ground.

Foreign structure: A pipeline or other metal structure in the vicinity of a pipeline system under cathodic protection.

Galvanic: Pertaining to electricity that flows as a result of a chemical action.

Interrupted ("On/Off") survey: A close interval survey that measures the potential difference between the structure and the ground surface as the cathodic protection current is switched on and off.

Native survey: A close interval survey that measures the difference in the electrical potential between a structure and the ground surface to establish a baseline for determining how much cathodic protection is needed for the structure.

"On" survey: A close interval survey that measures the potential difference between the structure and the ground surface as the cathodic protection current is applied.

Rectifier: A device that converts AC current to DC current and, in an impressed cathodic protection system, supplies current to the structure in the soil.

Reference electrode: An electrode whose open-circuit potential is constant under similar conditions of measurement; it is used to measure the relative potentials of other electrodes.

Structure-to-soil potential: The potential difference between the surface of a buried or submerged metallic structure (such as a pipe) and the electrolyte that is measured with reference to an electrode in contact with the electrolyte.

Test stations: Specific areas or locations, sometimes called test points, where a connection to a structure is made so that a structure-to-soil potential can be measured.

All underground steel piping systems must be protected against corrosion using a cathodic protection system. While test stations record data at specific points in a piping system, they do not record data from the entire pipeline. To measure the integrity and effectiveness of a cathodic protection system, annual or periodic testing must be performed at given intervals along the piping system. One such testing method is a close interval survey (CIS). This section provides a summary of how cathodic protection systems minimize corrosion, and identifies common types of close interval surveys.

1.1.0 Corrosion and Cathodic Protection

Pipeline corrosion can cause catastrophic failures with potentially severe consequences for the environment and serious injury or even death to those in the vicinity of the failure. Therefore, eliminating corrosion from piping systems is a critical aspect of the pipeline industry.

1.1.1 Corrosion

In order for pipeline corrosion to occur, the following four conditions must exist. There must be an **anode**, a **cathode**, an electrolyte, and a metallic path, defined as follows:

- *Anode* – A metal that will oxidize and corrode. In anodic areas, current flows from the steel in the pipeline structure into the surrounding soil and water (electrolyte), causing the structure to corrode.
- *Cathode* – A dissimilar metal, which is less likely to corrode than the anode, at which reduction occurs. In cathodic areas, current flows from the electrolyte onto the structure surface, and corrosion is eliminated.
- *Electrolyte* – The soil or water surrounding the pipeline structure.
- *Metallic path* – An electrical path connecting the anode and the cathode.

These conditions can involve two separate structures or they may be present on one structure. If any one of the four conditions is eliminated, corrosion can be prevented.

1.1.2 Cathodic Protection

Cathodic protection is the use of direct current to oppose the discharge of corrosion current from a structure. The goal of a cathodic protection system is to eliminate the potential difference between the anodic and cathodic areas of a pipeline. To mitigate corrosion, a cathodic protection system induces current onto the piping from an electrolyte (soil or water). By doing so, the system reduces the corrosion rate of the pipe surface by making it the cathode. When current is forced to flow onto the pipe at areas that were previously discharging current, the voltage provided by the cathodic protection system must be greater than the voltage from the corrosion cells.

The two basic methods of cathodic protection are **galvanic** anode systems and impressed current systems. In a galvanic system (*Figure 1*), a sacrificial anode is connected to a pipeline to provide cathodic current. The anode is sacrificed (corrodes) and in doing so, protects the pipeline. Galvanic anode protection relies on the natural **electrical potential** between two metals to create current flow. Since the driving voltage is limited to the very small potential difference existing between the metals, the current output is relatively low.

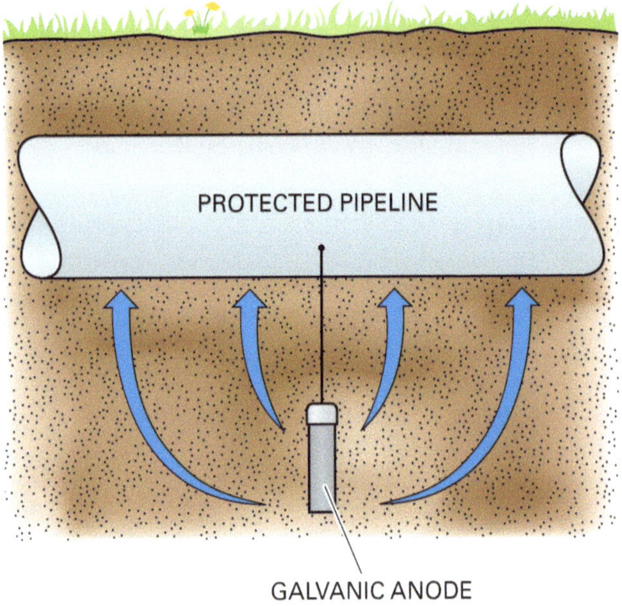

Figure 1 Diagram of galvanic anode circuit.

An impressed current system (*Figure 2*) requires an external DC source such as a **rectifier** or battery. An anode material is placed in the electrolyte with the structure to be protected. The anode is then made more positive than the structure by connecting both the anode and the structure to a direct current supply. Any conductive material can be used as an impressed current anode. However, corrosion takes place at the anode, so materials with very low consumption rates are most desirable. Impressed current anodes last longer than galvanic anodes.

Pipeline technicians who are working with cathodic protection systems must always follow the latest revision of the following standards that

Types of Galvanic Anode Systems

There are two basic types of galvanic anode installations: single anode systems and banks of anodes. Single anode systems protect only the area close to the anode. This practice is called *hot spotting* and is sometimes used in repair locations. A bank of anodes is a number of galvanic anodes on a single header cable. Anode banks are used to protect sections of piping and produce voltage drops from point to point in the earth.

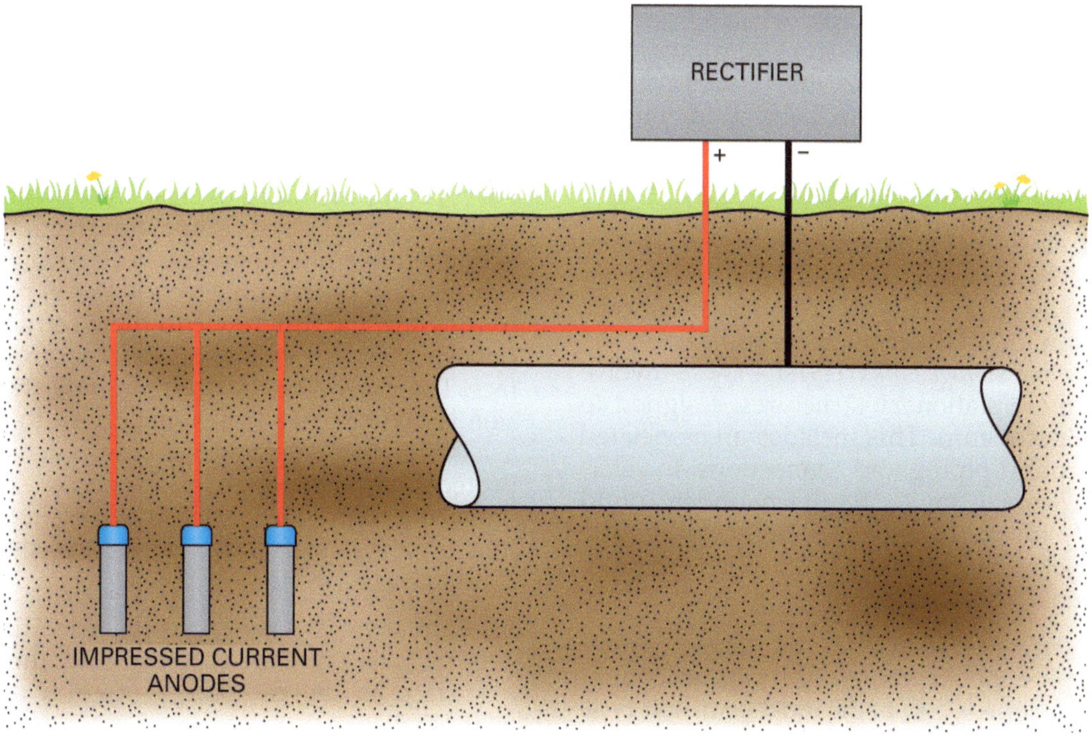

Figure 2 Diagram of impressed current circuit.

have been developed to ensure that structures are fully protected from corrosion:

- NACE International Standard *SP-01-69, Standard Practice – Control of External Corrosion on Underground or Submerged Metallic Piping Systems*
- 49 *CFR*, Section 192, *Transportation of Natural and Other Gas by Pipeline*, Subpart I, Requirements for Corrosion Control
- 49 *CFR*, *Transportation of Hazardous Liquids by Pipeline*, Section 195, Subpart H, Corrosion Control

Tests performed on cathodic protection systems to verify their integrity, effectiveness, and regulatory compliance are taken on various types of piping system structures. These structures include the following:

- *Pipelines* – These consist of all components of a pipeline facility through which a hazardous liquid is transported, including pipes, valves, pumps, and other assemblies. The pipeline facility encompasses piping and equipment used in production, extraction, recovery, lifting, stabilization, separation, or treating of petroleum. This includes piping between the production facilities and refineries.
- *Casings* – These are pipes that contain a carrier pipeline. Casings are typically used when a pipeline passes under railroad crossings or roads to protect the pipeline from high load stresses. State and local regulations identify specific mandatory casing locations. Problems created by casings result from electrical contact between the coated pipeline and the bare casing. Contact with casings can interfere with the cathodic protection of pipelines due, in part, to the **casing-to-soil potential**. For this reason, transport pipes must be provided with sufficiently high, mechanically stable spacers to exclude electrical contacts. Designers and planners should use only the types of casing seals that resist ground water from entering into the space between the casing and the pipeline. On cathodically-protected pipelines, casings should be used only where required by law or where the pipeline cannot be laid at crossing points without risk of damage.
- *Tank bottoms* – An aboveground storage tank is any tank or other container that is aboveground, partially buried, bunkered, or in a subterranean vault, including floating fuel systems. The tank bottom and any portion that is buried or submerged in electrolyte must be protected.
- *Foreign structures* – A **foreign structure** is any structure or pipeline that is not owned by the pipeline company. When steel structures contact the earth or foreign metallic structures, corrosion can occur and protection is required.
- *Well casings* – Wells drilled in oil and natural gas fields are fitted with casings for protection and stabilization. Cement is poured into the

space between the outer pipe, or casing, and the surrounding rock/soil. The cement filling is not uniform in thickness and may contain voids. Pipeline well casings are subject to various types of corrosion, depending on the surrounding soil content and the degree of protection provided by the cement. Boreholes are often located close to one another, so they may be affected by the cathodic protection from other wells. In large oil fields and reservoirs, cathodic protection for all the wells is appropriate.

- *Underground storage tanks (USTs)* – These involve one or more stationary structures designed to contain petroleum. This includes all associated underground piping and connections. In reference to storage tanks, the term underground typically means that the volume, including the volume of underground pipes, is ten percent or more beneath the surface of the ground.

A **reference electrode** is used to take measurements on pipeline structures. The **copper-copper sulfate electrode (CSE)** is the most common type of reference electrode used in pipeline corrosion work. It may be referred to as a copper-sulfate electrode. Some examples of copper-sulfate electrodes are shown in *Figure 3*.

Portable salt-water reference electrodes are commonly made of zinc/silver-silver chloride. These electrodes are used for seawater installations, such as offshore facilities, harbor structures, and soils with high chloride content. A silver chloride reference electrode is shown in *Figure 4*.

1.2.0 Close Interval Survey Methods

Close interval surveys, also known as *close interval potential surveys (CIPS)*, are used to measure the potential (voltage) difference between buried pipes and the surrounding soil. There are different types, or methods, of close interval surveys.

Before any type of cathodic protection current is applied to a piping system, a close interval survey method called a **native survey** is performed to measure the difference in the electrical potential between the structure and the ground surface. The native survey does not indicate if corrosion is taking place, it is simply used to establish a baseline for determining how much cathodic protection is needed for the structure.

Once a native survey has been completed, the following types of close interval surveys can be conducted, as needed:

- **"On" survey** – Measures the potential difference between the structure and the ground surface as the cathodic protection current is applied.
- **Interrupted ("On/Off") survey** – Measures the potential difference between the structure and the ground surface as the cathodic protection current is switched on and off.
- **Depolarized ("Off") survey** – Measures the potential difference between the structure and the ground surface after the cathodic protection current has been turned off and the **structure-to-soil potential** has stabilized.

Figure 3 Copper-copper sulfate electrodes.

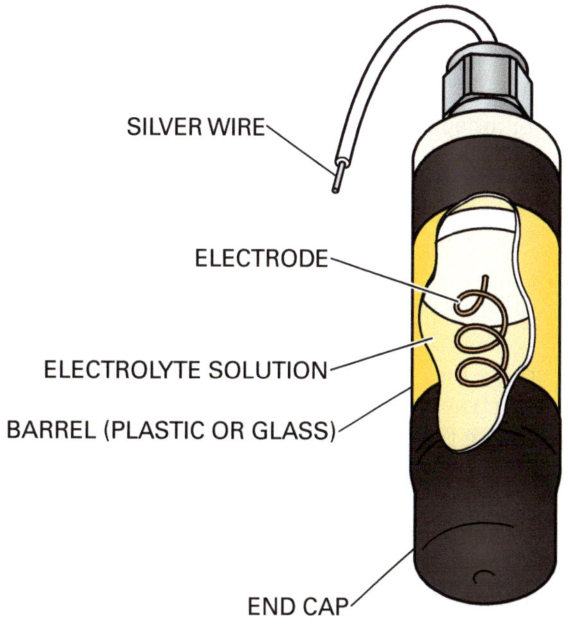

Figure 4 Silver chloride reference electrode.

Cathodic Protection Digital Voltmeters

This DC cathodic protection voltmeter can be used to measure structure-to-soil potentials of buried or submerged metal structures, such as pipelines and storage tanks. It can also measure structure-to-structure and structure-to-anode values. This meter has a connection for the half-cell to screw into the back of the voltmeter to measure structure-to-anode values.

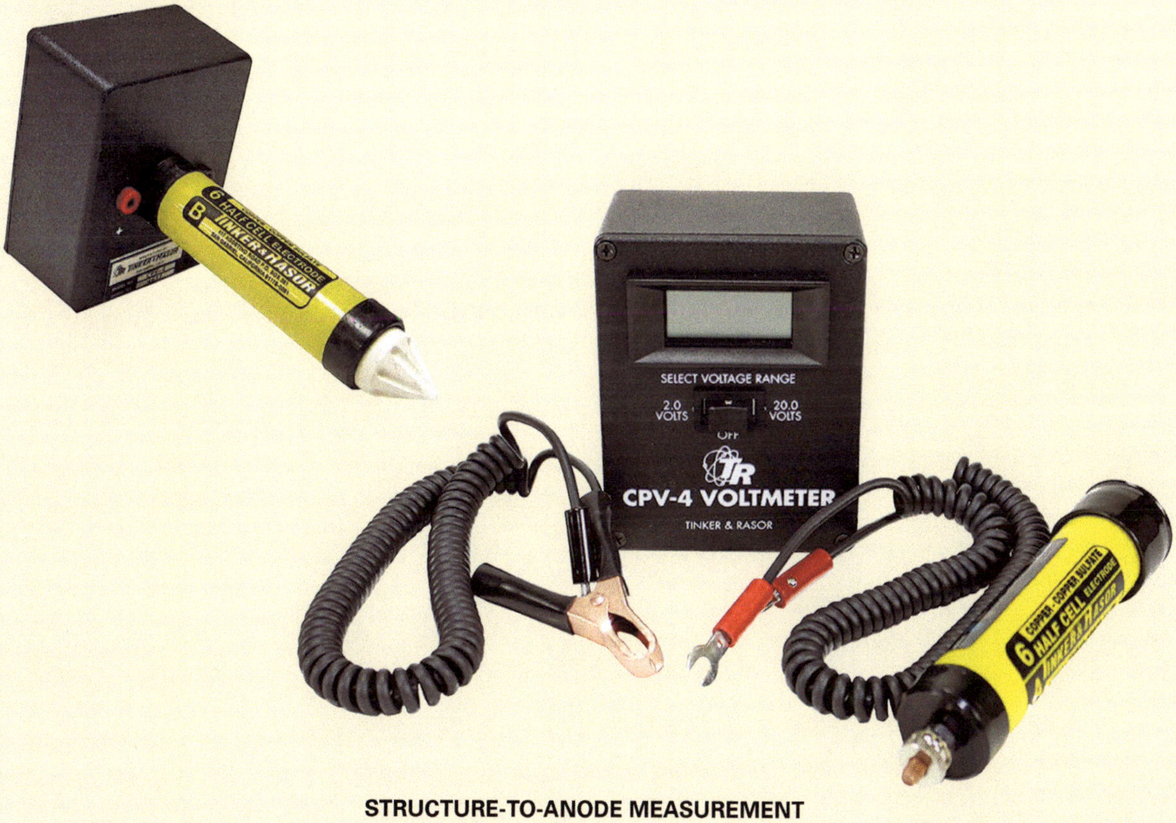

Additional Resources

It is assumed that you will learn to perform these procedures under the direct supervision of your crew leader, foreman, or instructor. For independent study, refer to the recommended Additional Resources at the back of this module.

1.0.0 Section Review

1. The basic goal of a cathodic protection system is to _____.
   a. attract DC voltage to increase the discharge of corrosion current from a structure
   b. induce current onto the piping from the anodic and cathodic areas of a foreign structure
   c. convert the surface area of a pipeline into a galvanic sacrificial anode to repel corrosion
   d. eliminate the potential difference between the anodic and cathodic areas of a pipeline

2. A type of close interval survey that measures the potential between a structure and the ground surface after the cathodic protection system current has been turned off and the structure-to-soil potential has stabilized is called a(n) _____.
   a. native survey
   b. depolarized ("Off") survey
   c. "On" survey
   d. interrupted survey

SECTION TWO

2.0.0 CONDUCT A CLOSE INTERVAL SURVEY

Objective

Describe how to conduct a close interval survey (CT1_2-17).

a. Identify the prerequisites, competencies, and abnormal operating conditions (AOCs) associated with conducting close interval surveys.
b. Describe the procedure for conducting a close interval survey.

Performance Task

Conduct a close interval survey (CT1_2-17).

- Identify potential abnormal operating conditions (AOCs) that may occur during performance of this CT, and know the appropriate actions to take in response to them.
- Utilize the appropriate personal protective equipment according to relevant company procedures.
- Identify the test point locations where connections will be made.
- Correctly connect the test leads, the voltage meter or data logger, and reference cell.
- Verify that current sources are operational (on for "On"/interrupted surveys and turned off/disconnected for depolarized survey).
- Place the reference cell directly above the pipeline being surveyed.
- Select the instrumentation to be used (survey wire, voltmeter, data logger, reference electrodes, etc.).
- For interrupted surveys, install current interrupters at all identified current sources. They should be set at the operator-determined time cycle and synchronized.
- Measure the structure-to-soil potential according to the desired intervals for this survey.
- Verify data is recorded.
- Complete appropriate documentation as required by operator's procedures.

Trade Terms

Abnormal operating conditions (AOCs): Conditions within a pipeline system, identified by the operator, that may indicate (1) a malfunction of a component or (2) a deviation from normal operations that may indicate a condition exceeding design limits or result in a hazard to persons, property, or the environment.

Confined space: A space that is large enough and so configured that an employee can bodily enter and perform assigned work; has limited or restricted means for entry or exit (such as tanks, vessels, vaults, etc.); and is not designed for continuous employee occupancy.

Current interrupter: A device used to interrupt the flow of DC current that cycles anodes, rectifiers, and other equipment.

"Instant Off" potential: The polarized half-cell potential of an electrode taken immediately after the current in a cathodic protection system is stopped.

Polarity: The distinction between the positive and negative potential differences at the terminals of a meter or electrical device.

Potential measurement: The activity of measuring the structure-to-soil potential.

Test leads: Wires that are connected to a structure being tested that provide easy connection points for a meter.

Personnel conducting a close interval survey are required to have some basic competencies. They must have the ability to carefully follow the required procedures for conducting a close interval survey and recording the results. This section describes some of the basic skills needed by pipeline technicians and identifies **abnormal operating conditions (AOCs)** that personnel may need to respond to when conducting close interval surveys. It also explains the basic procedures for conducting close interval surveys.

2.1.0 Prerequisites, Competencies, and Abnormal Operating Conditions

Personnel must possess a variety of skills in order to safely and successfully conduct close interval surveys. Testing procedures require numerous steps, which a pipeline technician must be competent to perform. Personnel must be aware of abnormal operating conditions (AOCs) that may occur during testing procedures and know how to properly respond to them.

2.1.1 Prerequisites and Competencies

Personnel assigned to conduct close interval surveys must have a good understanding of hazards that can exist before and during this covered task, as well as what the task entails. Before conducting surveys, personnel must be familiar with hazards along the right-of-way (ROW), including those associated with ROW encroachments and unsupervised third-party construction. They must be familiar with the locations of test stations, and they must be able to properly choose, inspect, calibrate, and operate any tools or instruments to be used during the procedure.

Pipeline work often involves trenched and excavated areas, as well as tank containment areas. For these reasons, personnel must be properly trained for confined space entry and excavation safety.

In the case of buried piping, this task begins when the pipe is first exposed and prepared for cleaning. For aboveground piping, the task begins when the component is accurately identified.

Personnel must be able to visibly inspect the pipe and determine if there are any unsafe integrity issues, such as evidence of a release or significant mechanical damage (gouges, scrapes, dents), and discontinue the work and make notifications if any are found. If old coating materials that may contain asbestos are encountered, personnel must make the proper notifications so that a qualified person can sample, test, and, if necessary, remove and dispose of any asbestos-containing material (ACM). For these and other safety reasons, personnel must be familiar with local emergency telephone numbers.

Once testing has begun, personnel must know how to position, manipulate, and read any equipment that might be used. If personnel encounter induced AC voltage, which could occur if a pipeline is located near a high voltage AC electrical field, they must be knowledgeable about the effects and hazards associated with the voltage and know how to notify the appropriate company personnel. Personnel must understand that no job is complete until all tasks are documented and the proper notifications are made.

Personnel must have a basic understanding of additional tasks related to structure-to-soil potential measurement, such as measuring structure-to-soil potentials (CT1_1-17), testing and detecting interference (CT1_3-17), inspecting and performing electrical testing of bonds (CT1_4-17), and inspecting and testing electrical isolation (CT1_5-17).

2.1.2 Abnormal Operating Conditions

Personnel conducting close interval surveys must be able to identify potential abnormal operating conditions (AOCs) and know the proper response to each, should they occur. The following are some potential AOCs, and the appropriate action personnel should take in response to them:

- *Damage to or unsupervised third party construction on a right-of-way is discovered* – Notify appropriate company personnel and, if the contractor continues to perform unsupervised work on site, discuss the dangers posed by the situation with the contractor.
- *Internal or external corrosion of the pipe is present and to the extent that structure-to-soil potential measurements would be impractical* – Document the condition according to company guidelines and notify personnel responsible for correcting the problem.
- *Erratic or floating readings occur* – Determine the cause of the erratic readings and repair the test leads or equipment, as needed.
- *Reverse polarity of readings occurs* – Document and notify the appropriate operator personnel immediately.
- *Induced AC voltage is encountered* – Exercise caution around any equipment or structures and notify the appropriate company personnel.
- *Fire or explosion caused by the ignition of a hydrocarbon liquid or gas* – Evacuate the area and notify the appropriate firefighting personnel. It may be necessary to call 911 in such an emergency.
- *Unexpected leaks of hazardous liquid or carbon dioxide because of a loss of pipe wall thickness* – Take steps to eliminate any possible ignition sources and notify the appropriate personnel, possibly including 911.

2.2.0 Procedure

When conducting a close interval survey, always follow company procedures and the manufacturer's instructions for the meter in use. The general steps for a typical procedure for conducting a close interval survey is as follows:

WARNING!

Always obtain and use the appropriate personal protective equipment (PPE) for each task. Stay alert to the possibility of unexpected leaks from the exposed pipeline such as hazardous liquid or carbon dioxide (CO_2). In the event of a leak, safely eliminate all ignition sources and immediately notify the appropriate personnel.

Step 1 Identify potential abnormal operating conditions (AOCs) that may occur, and know the appropriate actions to take in response to them.

Step 2 Utilize the appropriate personal protective equipment according to relevant company procedures.

Step 3 Identify the test point locations where connections will be made.

Step 4 Correctly connect the test leads, voltmeter or data logger, and reference cell.

Step 5 Verify that current sources are operational (power on for "On"/interrupted survey and turned off/disconnected for depolarized survey).

Step 6 Place the reference cell directly above the pipeline being surveyed. The reference electrode must be in contact with the electrolyte to obtain accurate results.

Step 7 Select the instrumentation to be used (survey wire, voltmeter, data logger, reference electrodes, etc.). Inspect the equipment for damage and proper operation before taking any measurements.

Step 8 For interrupted surveys, install a **current interrupter** at all identified current sources using synchronized timing as determined by operator procedures. Synchronized current interrupters are necessary to determine **"instant Off" potential**.

Step 9 Measure the structure-to-soil potential according to the desired intervals for the survey.

Step 10 Verify that data is being recorded. Lack of data may be a sign of equipment failure or a faulty electrode location.

Step 11 Document your findings on the appropriate company inspection form and make notifications according to company policy.

Additional Resources

It is assumed that you will learn to perform these procedures under the direct supervision of your crew leader, foreman, or instructor. For independent study, refer to the recommended Additional Resources at the back of this module.

2.0.0 Section Review

1. Gouges, scrapes, and dents in pipeline surface material are common examples of _____.
   a. corrosive erosion
   b. mechanical damage
   c. substrate profiling
   d. adhesion enhancement

2. When conducting a close interval survey, the reference cell should be _____.
   a. disconnected until valid readings begin to appear
   b. grounded at a safe distance from the test area
   c. placed directly above the pipeline being surveyed
   d. displaying the voltage induced from the AC rectifier

Summary

Pipeline corrosion can cause catastrophic failures with potentially severe consequences for personnel, as well as the environment. While the use of protective coatings helps minimize corrosion by preventing contact between piping structures and the soil, most pipelines also use cathodic protection systems. Pipeline technicians conduct annual or periodic testing to measure the integrity of the cathodic protection system. Part of a technician's job is to be familiar with preparations, hazards, and procedures involved in conducting close interval surveys at specific points in a pipeline system. This module reviewed cathodic protection, identified different types of close interval surveys, and explained the requirements and procedures for safely and effectively conducting close interval surveys on piping.

Review Questions

1. When current flows from the steel in a pipeline structure into the surrounding electrolyte (soil) causing the structure to corrode, the area is _____.
   a. pathodic
   b. anodic
   c. cathodic
   d. metabolic

2. A cathodic protection system turns the pipeline surface into a(n) _____.
   a. electrolyte
   b. heat sink
   c. anode
   d. cathode

3. The most common type of electrode in a pipeline cathodic protection system is the _____.
   a. copper-copper sulfate electrode
   b. silver chloride electrode
   c. salt-water electrode
   d. zinc-chloride electrode

4. A close interval survey that is used to establish a baseline for determining how much cathodic protection is needed for a structure is a(n) _____.
   a. "Off" survey
   b. depolarized survey
   c. native survey
   d. interrupted survey

5. If extensive internal or external piping corrosion is found prior to conducting a close interval survey, a pipeline technician should _____.
   a. apply a cathodic patch over the damage and continue with the measurements
   b. make sure that the instruments are attached directly to the corroded area
   c. document the condition and notify personnel about correcting the problem
   d. attach an approved anode bank to the structure and adjust the test instrument

6. The final step in any procedure for conducting a close interval survey is to _____.
   a. inspect every piece of tested equipment for damage and proper operation
   b. accurately record all required information in accordance with company policy
   c. remove lockout/tagout barriers and restart all holiday detection equipment
   d. conduct a Barcol frequency test to certify the readings to NACE standards

Trade Terms Quiz

Fill in the blank with the correct term that you learned from your study of this module.

1. A close interval survey that measures the potential difference between the structure and the ground surface after the cathodic protection current has been turned off and the structure-to-soil potential has stabilized is a(n) ________.
2. A close interval survey that measures the potential difference between the structure and the ground surface as the cathodic protection current is switched on and off is a(n) ________.
3. An electrode whose open-circuit potential is constant under similar conditions of measurement, and which is used to measure the relative potentials of other electrodes is a(n) ________.
4. Electricity that flows as a result of a chemical action is considered ________.
5. Specific areas or locations where a connection to a structure is made so that a structure-to-soil potential can be measured are ________.
6. A non-polarizing electrode used in corrosion control to measure the electrical potential of a metal structure to a surrounding electrolyte to determine the potential for corrosion damage or to monitor the effectiveness of existing control measures is a(n) ________.
7. The activity of measuring the structure-to-soil potential is called ________.
8. The potential energy of a unit charge at any point in an electric circuit, measured with respect to a specified reference point in the circuit or to ground is called ________.
9. Conditions within a pipeline system that may indicate (1) a malfunction of a component or (2) a condition exceeding design limits or resulting in a hazard to persons, property, or the environment are called ________.
10. A close interval survey that measures the difference in the electrical potential between a structure and the ground surface to establish a baseline for determining how much cathodic protection is needed for the structure is called a(n) ________.
11. The electrode of an electrochemical cell at which reduction occurs is a(n) ________.
12. The polarized half-cell potential of an electrode taken immediately after the current in a cathodic protection system is stopped is called ________.
13. Wires that are connected to a structure being tested that provide easy connection points for a meter are called ________.
14. A device used to interrupt the flow of DC current that cycles anodes, rectifiers, and other equipment is called a(n) ________.
15. The distinction between the positive and negative potential differences at the terminals of a meter or electrical device is called ________.
16. A pipeline or other metal structure in the vicinity of a pipeline system under cathodic protection is a(n) ________.
17. The potential difference between the surface of a buried or submerged metallic structure (such as a pipe) and the electrolyte that is measured with reference to an electrode in contact with the electrolyte is known as the ________.
18. An assessment method that uses voltage measurements to determine the effectiveness of a cathodic protection system on buried pipelines is called a(n) ________.
19. A device that converts AC current to DC current and, in an impressed cathodic protection system, supplies current to the structure in the soil is a(n) ________.

20. The voltage potential of a casing in soil or an electrolyte as it relates to a reference electrode is called the ________.

21. A close interval survey that measures the potential difference between the structure and the ground surface as the cathodic protection current is applied is called a(n) ________.

22. A space that is large enough and so configured that an employee can bodily enter and perform assigned work; has limited or restricted means for entry or exit (such as tanks, vessels, vaults, etc.); and is not designed for continuous employee occupancy is called a(n) ________.

23. The electrode of an electrochemical cell at which oxidation occurs is called a(n) ________.

Trade Terms

Abnormal operating conditions (AOCs)
Anode
Casing-to-soil potential
Cathode
Close interval survey (CIS)
Confined space
Copper-copper sulfate electrode (CSE)
Current interrupter
Depolarized ("Off") survey
Electrical potential
Foreign structure
Galvanic
"Instant Off" potential
Interrupted ("On/Off") survey
Native survey
"On" survey
Polarity
Potential measurement
Rectifier
Reference electrode
Structure-to-soil potential
Test leads
Test stations

Trade Terms Introduced in This Module

Abnormal operating conditions (AOCs): Conditions within a pipeline system, identified by the operator, that may indicate (1) a malfunction of a component or (2) a deviation from normal operations that may indicate a condition exceeding design limits or result in a hazard to persons, property, or the environment.

Anode: The electrode of an electrochemical cell at which oxidation occurs.

Casing-to-soil potential: The voltage potential of a casing in soil or an electrolyte as it relates to a reference electrode.

Cathode: The electrode of an electrochemical cell at which reduction occurs.

Close interval survey (CIS): One of several assessment methods that use voltage measurements to determine the effectiveness of a cathodic protection system on buried pipelines. Also known as close interval potential surveys (CIPS).

Confined space: A space that is large enough and so configured that an employee can bodily enter and perform assigned work; has limited or restricted means for entry or exit (such as tanks, vessels, vaults, etc.); and is not designed for continuous employee occupancy.

Copper-copper sulfate electrode (CSE): A non-polarizing electrode used in corrosion control to measure the electrical potential of a metal structure to a surrounding electrolyte in order to determine the potential for corrosion damage or to monitor the effectiveness of existing control measures.

Current interrupter: A device used to interrupt the flow of DC current that cycles anodes, rectifiers, and other equipment.

Depolarized ("Off") survey: A close interval survey that measures the potential difference between the structure and the ground surface after the cathodic protection current has been turned off and the structure-to-soil potential has stabilized.

Electrical potential: The potential energy of a unit charge at any point in an electric circuit, measured with respect to a specified reference point in the circuit or to ground.

Foreign structure: A pipeline or other metal structure in the vicinity of a pipeline system under cathodic protection.

Galvanic: Pertaining to electricity that flows as a result of a chemical action.

"Instant Off" potential: The polarized half-cell potential of an electrode taken immediately after the current in a cathodic protection system is stopped.

Interrupted ("On/Off") survey: A close interval survey that measures the potential difference between the structure and the ground surface as the cathodic protection current is switched on and off.

Native survey: A close interval survey that measures the difference in the electrical potential between a structure and the ground surface to establish a baseline for determining how much cathodic protection is needed for the structure.

"On" survey: A close interval survey that measures the potential difference between the structure and the ground surface as the cathodic protection current is applied.

Polarity: The distinction between the positive and negative potential differences at the terminals of a meter or electrical device.

Potential measurement: The activity of measuring the structure-to-soil potential.

Rectifier: A device that converts AC current to DC current and, in an impressed cathodic protection system, supplies current to the structure in the soil.

Reference electrode: An electrode whose open-circuit potential is constant under similar conditions of measurement; it is used to measure the relative potentials of other electrodes.

Structure-to-soil potential: The potential difference between the surface of a buried or submerged metallic structure (such as a pipe) and the electrolyte that is measured with reference to an electrode in contact with the electrolyte.

Test leads: Wires that are connected to a structure being tested that provide easy connection points for a meter.

Test stations: Specific areas or locations, sometimes called test points, where a connection to a structure is made so that a structure-to-soil potential can be measured.

Additional Resources

The following recommended resources can provide additional helpful information related to the requirements for this covered task:

API Recommended Practice 1161, Third Edition, © 2014 American Petroleum Institute. API Publishing Services, 1220 L Street, NW, Washington, DC 20005. Distributed by Thompson Reuters (Scientific) LLC, **www.techstreet.com**.

49 *CFR* 192, Subpart I, **www.ecfr.gov**

49 *CFR* 195, Subpart H, **www.ecfr.gov**

NCCER Module AOCCC-17, *Abnormal Operating Conditions Control Center*

NCCER Module AOCFG-17, *Abnormal Operating Conditions Field and Gas*

NCCER Module 62105-14, *Introduction to Pipeline Documents*

NCCER Module 66101-14, *Introduction to the Pipeline Industry*

NCCER Module CT1_1-17, *Measure Structure-to-Soil Potentials*

NCCER Module CT1_3-17, *Test and Detect Interference*

NCCER Module CT1_4-17, *Inspect and Perform Electrical Test of Bonds*

NCCER Module CT1_5-17, *Inspect and Test Electrical Isolation*

NCCER Module CT14_1-17, *Locate Line*

API Recommended Practice 651, Cathodic Protection of Aboveground Petroleum Storage Tanks. Fourth Edition, 2014. Available from the American Petroleum Institute at **www.api.org**.

NACE International *SP0169-2013, Control of External Corrosion on Underground or Submerged Metallic Piping Systems*. 2013. Available from NACE International at **www.nace.org**.

Figure Credits

Tinker & Rasor, Module Opener, Figure 3, SA01

Section Review Answer Key

Answer	Section Reference	Objective
Section One		
1. d	1.1.2	1a
2. b	1.2.0	1b
Section Two		
1. b	2.1.1	2a
2. c	2.2.0	2b

NCCER CURRICULA — USER UPDATE

NCCER makes every effort to keep its textbooks up-to-date and free of technical errors. We appreciate your help in this process. If you find an error, a typographical mistake, or an inaccuracy in NCCER's curricula, please fill out this form (or a photocopy), or complete the online form at **www.nccer.org/olf**. Be sure to include the exact module ID number, page number, a detailed description, and your recommended correction. Your input will be brought to the attention of the Authoring Team. Thank you for your assistance.

Instructors – If you have an idea for improving this textbook, or have found that additional materials were necessary to teach this module effectively, please let us know so that we may present your suggestions to the Authoring Team.

NCCER Product Development and Revision

13614 Progress Blvd., Alachua, FL 32615

Email: curriculum@nccer.org
Online: www.nccer.org/olf

❑ Trainee Guide ❑ Lesson Plans ❑ Exam ❑ PowerPoints Other ____________________

Craft / Level: Copyright Date:

Module ID Number / Title:

Section Number(s):

Description:

Recommended Correction:

Your Name:

Address:

Email: Phone:

CT1_3-17

Test and Detect Interference

Overview

Pipeline coatings and cathodic protection systems are used extensively by the pipeline industry to minimize corrosion. However, when pipelines are located near high voltage power lines or similar foreign structures, they can be affected by the electromagnetic field created by the alternating current. AC interference can be hazardous and cause severe corrosion, especially at pipeline coating defects. This module provides an overview of corrosion and cathodic protection, and identifies foreign structures that can cause interference. It also describes the basic procedure used for conducting an interference test.

Trainees with successful module completions may be eligible for credentialing through the NCCER Registry. To learn more, go to **www.nccer.org** or contact us at 1.888.622.3720. Our website has information on the latest product releases and training, as well as online versions of our *Cornerstone* magazine and Pearson's product catalog.

Your feedback is welcome. You may email your comments to **curriculum@nccer.org**, send general comments and inquiries to **info@nccer.org**, or fill in the User Update form at the back of this module.

This information is general in nature and intended for training purposes only. Actual performance of activities described in this manual requires compliance with all applicable operating, service, maintenance, and safety procedures under the direction of qualified personnel. References in this manual to patented or proprietary devices do not constitute a recommendation of their use.

CT1_3-17
Test and Detect Interference

Objectives

When you have completed this module, you will be able to do the following:

1. Describe cathodic protection systems and identify foreign structures that can cause interference.
   a. Describe how cathodic protection systems minimize corrosion.
   b. Identify foreign structures that can cause interference.
2. Describe testing to detect interference (CT1_3-17).
   a. Identify the prerequisites, competencies, and abnormal operating conditions associated with testing for interference.
   b. Describe the procedure for testing to detect interference.

Performance Task

Under the supervision of your instructor, you should be able to do the following:

1. Test to detect interference (CT1_3-17).
   - Identify potential abnormal operating conditions that may occur during performance of this CT, and know the appropriate actions to take in response to them.
   - Utilize the appropriate personal protective equipment according to relevant company procedures.
   - Notify appropriate personnel of interference problems.
   - Select the proper instrumentation, test leads, and reference electrode and be prepared to use them correctly.
   - Assess the area for other cathodic protection systems or sources of electrical interference.
   - Measure the structure-to-soil potential.
   - Field-analyze the readings to ensure they are within the desired range.
   - Interrupt rectifiers to determine if interference exists.
   - If interference is found, take corrective action. This may involve making required notifications.
   - Complete appropriate documentation as required by operator's procedures.

Trade Terms

Abnormal operating conditions (AOCs)
AC current mitigation
Anode
Casing-to-soil potential
Cathode
Confined space
Copper-copper sulfate electrode (CSE)
Electrical potential
Foreign structure
Galvanic
Ions
Interference
Polarity
Rectifier
Reference electrode
Structure-to-soil potential
Test leads
Test stations

Industry Recognized Credentials

If you are training through an NCCER-accredited sponsor, you may be eligible for credentials from NCCER's Registry. The ID number for this module is CT1_3-17. Note that this module may have been used in other NCCER curricula and may apply to other level completions. Contact NCCER's Registry at 888.622.3720 or go to **www.nccer.org** for more information.

Contents

Figures

Section One

1.0.0 Cathodic Protection and Foreign Structure Interference

Objective

Describe cathodic protection systems and identify foreign structures that can cause interference.

a. Describe how cathodic protection systems minimize corrosion.
b. Identify foreign structures that can cause interference.

Trade Terms

Anode: The electrode of an electrochemical cell at which oxidation occurs.

Casing-to-soil potential: The voltage potential of a casing in soil or an electrolyte as it relates to a reference electrode.

Cathode: The electrode of an electrochemical cell at which reduction occurs.

Copper-copper sulfate electrode (CSE): A non-polarizing electrode used in corrosion control to measure the electrical potential of a metal structure to a surrounding electrolyte to determine the potential for corrosion damage or to monitor the effectiveness of existing control measures.

Electrical potential: The potential energy of a unit charge at any point in an electric circuit, measured with respect to a specified reference point in the circuit or to ground.

Foreign structure: A pipeline or other metal structure in the vicinity of a pipeline system under cathodic protection.

Galvanic: Pertaining to electricity that flows as a result of a chemical action.

Ions: Electrically charged atoms (either positively charged or negatively charged).

Interference: The effect upon cathodic protection levels of a structure due to the interaction between its cathodic protection system and the cathodic protection system of another structure.

Rectifier: A device that converts AC current to DC current and, in an impressed cathodic protection system, supplies current to the structure in the soil.

Reference electrode: An electrode whose open-circuit potential is constant under similar conditions of measurement; it is used to measure the relative potentials of other electrodes.

Underground steel piping systems are protected against corrosion by pipeline coatings and cathodic protection systems. Pipelines located near structures such as high-voltage AC power lines can be affected by the electromagnetic field created by the alternating current. The induced AC current not only creates a potential safety hazard, it can cause excessive pipeline corrosion—especially where there are defects in the coating. This section provides a review of corrosion and cathodic protection, and describes the impact of interference and the foreign structure types that can cause it.

1.1.0 Corrosion and Cathodic Protection

Pipeline corrosion can cause catastrophic failures with potentially severe consequences for the environment and serious injury or even death to those in the vicinity of the failure. Therefore, eliminating corrosion from piping systems is a critical aspect of the pipeline industry.

1.1.1 Corrosion

In order for pipeline corrosion to occur, the following four conditions must exist. There must be an anode, a cathode, an electrolyte, and a metallic path, defined as follows:

- *Anode* – A metal that will oxidize and corrode. In anodic areas, current flows from the steel in the pipeline structure into the surrounding soil and water (electrolyte), causing the structure to corrode.
- *Cathode* – A dissimilar metal, which is less likely to corrode than the anode, at which reduction occurs. In cathodic areas, current flows from the electrolyte onto the structure surface, and corrosion is eliminated.
- *Electrolyte* – The soil or water surrounding the pipeline structure.
- *Metallic path* – An electrical path connecting the anode and the cathode.

These conditions can involve two separate structures or they may be present on one structure. If any one of the four conditions is eliminated, corrosion can be prevented.

1.1.2 Cathodic Protection

Cathodic protection is the use of direct current to oppose the discharge of corrosion current from a structure. The goal of a cathodic protection system is to eliminate the potential difference between the anodic and cathodic areas of a pipeline. To mitigate corrosion, a cathodic protection system induces current onto the piping from an electrolyte (soil or water). By doing so, the system reduces the corrosion rate of the pipe surface by making it the cathode. When current is forced to flow onto the pipe at areas that were previously discharging current, the voltage provided by the cathodic protection system must be greater than the voltage from the corrosion cells.

Two basic methods of cathodic protection are **galvanic** anode systems and impressed current systems. In a galvanic system (*Figure 1*), a sacrificial anode is connected to a pipeline to provide cathodic current. The anode is sacrificed (corrodes) and in doing so, protects the pipeline. Galvanic anode protection relies on the natural **electrical potential** between two metals to create current flow. Since the driving voltage is limited to the very small potential difference existing between the metals, the current output is relatively low.

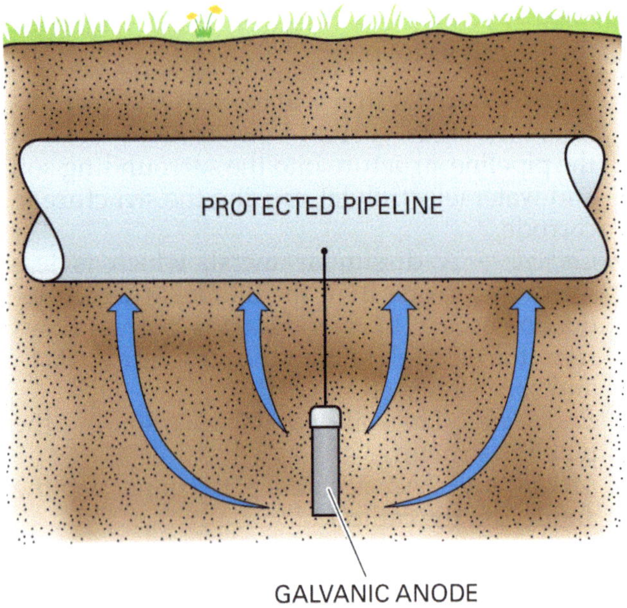

Figure 1 Diagram of galvanic anode circuit.

Galvanic Anode System Types

There are two basic types of galvanic anode installations: single anode systems and banks of anodes. Single anode systems protect only the area close to the anode. This practice is called *hot spotting* and is sometimes used in repair locations. A bank of anodes is a number of galvanic anodes on a single header cable. Anode banks are used to protect sections of piping and produce voltage drops from point to point in the earth.

An impressed current system (*Figure 2*) requires an external DC source such as a **rectifier** or battery. An anode material is placed in the electrolyte with the structure to be protected. The anode is then made more positive than the structure by connecting both the anode and the structure to a direct current supply. Any conductive material can be used as an impressed current anode. However, corrosion takes place at the anode, so materials with very low consumption rates are most desirable. Impressed current anodes last longer than galvanic anodes.

Pipeline technicians who are working with cathodic protection systems must always follow the latest revision of the following standards that have been developed to ensure that structures are fully protected from corrosion:

- NACE International Standard *SP-01-69, Standard Practice – Control of External Corrosion on Underground or Submerged Metallic Piping Systems*
- 49 *CFR*, Section 192, *Transportation of Natural and Other Gas by Pipeline*, Subpart I, Requirements for Corrosion Control
- 49 *CFR*, *Transportation of Hazardous Liquids by Pipeline*, Section 195, Subpart H, Corrosion Control

Tests performed on cathodic protection systems to verify their integrity, effectiveness, and regulatory compliance are taken on various types of piping system structures. These structures include the following:

- *Pipelines* – These consist of all components of a pipeline facility through which a hazardous liquid is transported, including pipes, valves, pumps, and other assemblies. The pipeline

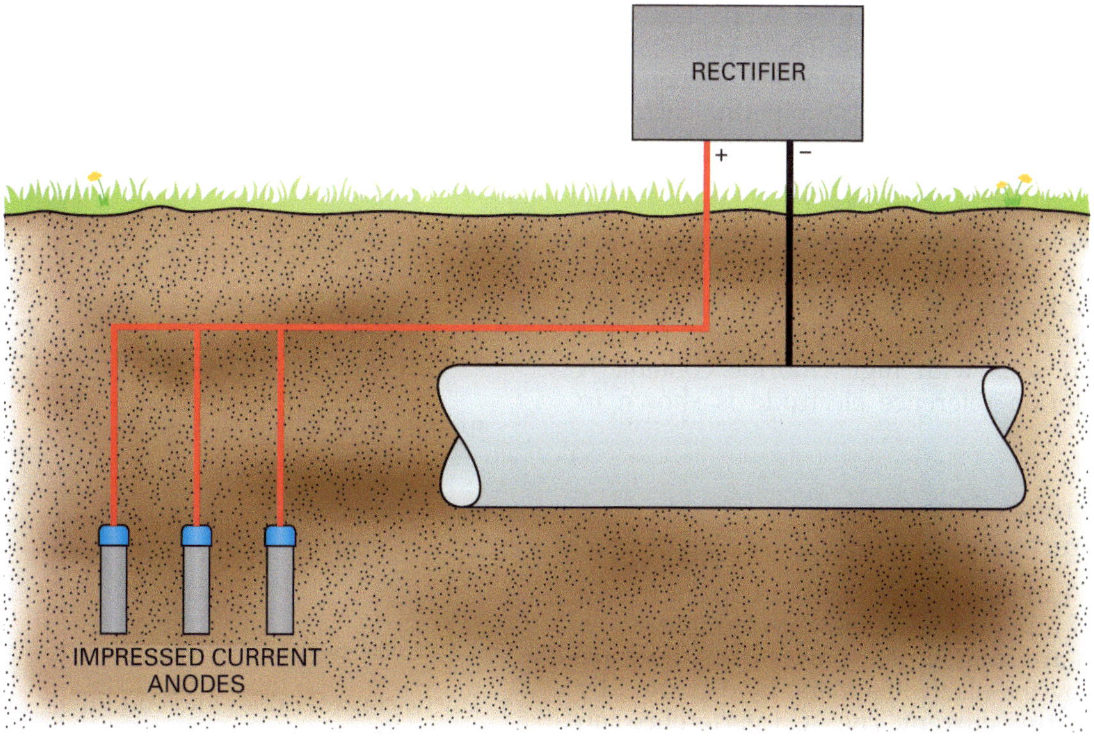

Figure 2 Diagram of impressed current circuit.

facility encompasses piping and equipment used in production, extraction, recovery, lifting, stabilization, separation, or treating of petroleum. This includes piping between the production facilities and refineries.

- *Casings* – These are pipes that contain a carrier pipeline. Casings are typically used when a pipeline passes under railroad crossings or roads to protect the pipeline from high load stresses. State and local regulations identify specific mandatory casing locations. Problems created by casings result from electrical contact between the coated pipeline and the bare casing. Contact with casings can interfere with the cathodic protection of pipelines due, in part, to the **casing-to-soil potential**. For this reason, transport pipes must be provided with sufficiently high, mechanically stable spacers to exclude electrical contacts. Designers and planners should use only the types of casing seals that resist ground water from entering into the space between the casing and the pipeline. On cathodically-protected pipelines, casings should be used only where required by law or where the pipeline cannot be laid at crossing points without risk of damage.
- *Tank bottoms* – An aboveground storage tank is any tank or other container that is aboveground, partially buried, bunkered, or in a subterranean vault, including floating fuel systems. The tank bottom and any portion that is buried or submerged in electrolyte must be protected.
- *Foreign structures* – A foreign structure is any structure or pipeline that is not owned by the pipeline company. When steel structures contact the earth or foreign metallic structures, corrosion can occur and protection is required.
- *Well casings* – Wells drilled in oil and natural gas fields are fitted with casings for protection and stabilization. Cement is poured into the space between the outer pipe, or casing, and the surrounding rock/soil. The cement filling is not uniform in thickness and may contain voids. Pipeline well casings are subject to various types of corrosion, depending on the surrounding soil content and the degree of protection provided by the cement. Boreholes are often located close to one another, so they may be affected by the cathodic protection from other wells. In large oil fields and reservoirs, cathodic protection for all the wells is appropriate.
- *Underground storage tanks (USTs)* – These involve one or more stationary structures designed to contain petroleum. This includes all associated underground piping and connections. In reference to storage tanks, the term underground typically means that the volume, including the volume of underground pipes, is 10 percent or more beneath the surface of the ground.

A reference electrode is used to take measurements on pipeline structures. The copper-copper sulfate electrode (CSE) is the most common type of reference electrode used in pipeline corrosion work. It may be referred to as a copper-sulfate electrode. Some examples of copper-sulfate electrodes are shown in *Figure 3*.

Portable salt-water reference electrodes are commonly made of zinc/silver-silver chloride. These electrodes are used for seawater installations, such as offshore facilities, harbor structures, and soils with high chloride content. A silver chloride reference electrode is shown in *Figure 4*.

1.2.0 Foreign Structures that Cause Interference

Pipelines can pick up interference current from a variety of foreign structures. For example, when pipelines are located near power lines, they are affected by the electromagnetic field created by the alternating current. The AC interference can be hazardous and can result in severe corrosion of pipelines. Voltages as low as 4 VAC (20 amps/m^2 – 100 amps/m^2) can result in corrosion. Any voltage higher than 15V is considered unsafe for pipeline work. If a voltage in excess of 15V is detected, work should be stopped and appropriate notifications should be made.

Foreign metal structures that can cause interference typically include the following:

- Water lines
- Telephone lines
- Electric power lines

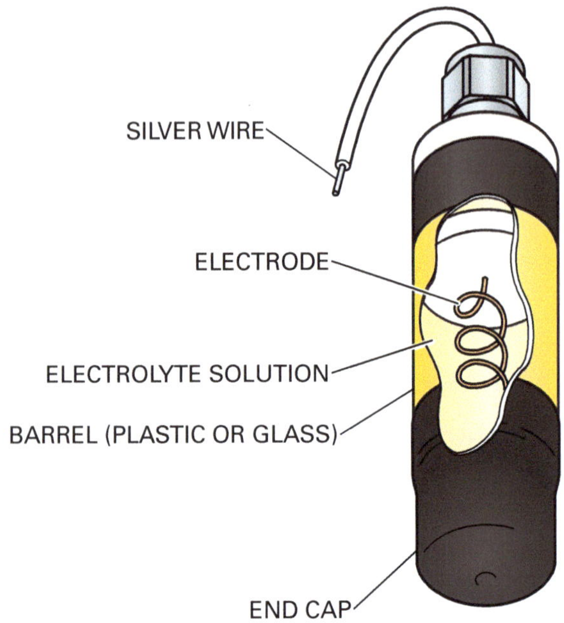

Figure 4 Silver chloride reference electrode.

- Traffic control cables and conduits
- Steam lines
- Pipe cables
- Street light cables
- Conduits
- Light standards
- Electric ground rods
- Sewage lines
- Manholes and drainpipes

Metallic objects such as reinforcing rod (rebar) or wire mesh in concrete can also cause interference. Also be aware of junk metal such as pipe and rails from abandoned facilities that may be present.

Figure 3 Copper-copper sulfate electrodes.

When installing cathodic protection systems, utility companies and the owners of nearby pipelines should be notified so that interference testing can be cooperatively conducted.

When a foreign structure is near a protected pipeline, it may pick up cathodic protection current. The foreign structure receives some cathodic protection by picking up this stray current, but when the current leaves, it releases ions and will eventually cause corrosion at the point of exit.

Pipeline mapping systems help gas utilities and pipeline companies locate contacts with foreign metal structures. The use of mapping systems significantly reduces the cost of troubleshooting cathodic protection on pipeline systems.

Cathodic Protection Digital Voltmeters

This DC cathodic protection voltmeter can be used to measure the structure-to-soil potential of buried or submerged metal structures, such as pipelines and storage tanks. It can also measure structure-to-structure and structure-to-anode values. This meter has a connection for the half-cell to screw into the back of the voltmeter to measure structure-to-anode values.

STRUCTURE-TO-ANODE MEASUREMENT

Additional Resources

It is assumed that you will learn to perform these procedures under the direct supervision of your crew leader, foreman, or instructor. For independent study, refer to the recommended Additional Resources at the back of this module.

1.0.0 Section Review

1. The basic goal of a cathodic protection system is to _____.
   a. attract DC voltage to increase the discharge of corrosion current from a structure
   b. induce current onto the piping from the anodic and cathodic areas of a foreign structure
   c. convert the surface area of a pipeline into a galvanic sacrificial anode to repel corrosion
   d. eliminate the potential difference between the anodic and cathodic areas of a pipeline

2. When a foreign structure is located near a pipeline that is protected by a cathodic protection system, the structure might _____.
   a. convert any DC stray current it receives to alternating current
   b. receive some cathodic protection by picking up stray current
   c. act as an amplifier and overwhelm the pipeline protection system
   d. store the current like a capacitor before releasing a voltage burst

SECTION TWO

2.0.0 TESTING FOR INTERFERENCE

Objective

Describe testing to detect interference (CT1_3-17).

a. Identify the prerequisites, competencies, and abnormal operating conditions (AOCs) associated with testing for interference.
b. Describe the procedure for testing to detect interference.

Performance Task

Test to detect interference (CT1_3-17).

- Identify potential abnormal operating conditions that may occur during performance of this CT, and know the appropriate actions to take in response to them.
- Utilize the appropriate personal protective equipment according to relevant company procedures.
- Notify appropriate personnel of interference problems.
- Select the proper instrumentation, test leads, and reference electrode and be prepared to use them correctly.
- Assess the area for other cathodic protection systems or sources of electrical interference.
- Measure the structure-to-soil potential.
- Field-analyze the readings to ensure they are within the desired range.
- Interrupt rectifiers to determine if interference exists.
- If interference is found, take corrective action. This may involve making required notifications.
- Complete appropriate documentation as required by operator's procedures.

Trade Terms

Abnormal operating conditions (AOCs): Conditions within a pipeline system, identified by the operator, that may indicate (1) a malfunction of a component or (2) a deviation from normal operations that may indicate a condition exceeding design limits or result in a hazard to persons, property, or the environment.

AC current mitigation: Engineered control methods used to limit extreme levels of induced current in pipelines.

Confined space: A space that is large enough and so configured that an employee can bodily enter and perform assigned work; has limited or restricted means for entry or exit (such as tanks, vessels, vaults, etc.); and is not designed for continuous employee occupancy.

Polarity: The distinction between the positive and negative potential differences at the terminals of a meter or electrical device.

Structure-to-soil potential: The potential difference between the surface of a buried or submerged metallic structure (such as a pipe) and the electrolyte that is measured with reference to an electrode in contact with the electrolyte.

Test leads: Wires that are connected to a structure being tested that provide easy connection points for a meter.

Test stations: Specific areas or locations, sometimes called test points, where a connection to a structure is made so that a structure-to-soil potential can be measured.

Personnel testing to detect interference are required to have some basic competencies. They must have the ability to carefully follow the required procedures for conducting interference testing and recording the results. This section describes some of the basic skills needed by pipeline technicians and identifies abnormal operating conditions (AOCs) that personnel may need to respond to when testing to detect interference. It also explains the basic procedures for testing to detect interference.

2.1.0 Prerequisites, Competencies, and Abnormal Operating Conditions

Personnel must possess a variety of skills in order to safely and successfully conduct interference testing. Testing procedures require numerous steps, which a pipeline technician must be competent to perform. Personnel must be aware of abnormal operating conditions (AOCs) that may occur during testing procedures and know how to properly respond to them.

2.1.1 Prerequisites and Competencies

Personnel assigned to conduct interference testing must have a good understanding of hazards that can exist before and during this covered task, as well as what the task entails. Before testing, personnel must be familiar with hazards along the right-of-way (ROW), including those associated with ROW encroachments and unsupervised third party construction. They must be familiar with the locations of test stations, and they must be able to properly choose, inspect, calibrate, and operate any tools or instruments to be used during the procedure.

Pipeline work often involves trenched and excavated areas, as well as tank containment areas. For these reasons, personnel must be properly trained for confined space entry and excavation safety.

Personnel must be able to visibly inspect the pipe and determine if there are any unsafe integrity issues, such as evidence of a release or significant mechanical damage (gouges, scrapes, dents), and discontinue the work and make notifications if any are found. If old coating materials that may contain asbestos are encountered, personnel must make the proper notifications so that a qualified person can sample, test, and, if necessary, remove and dispose of any asbestos-containing material (ACM). For these and other safety reasons, personnel must be familiar with local emergency telephone numbers.

This task begins with testing for direct current (DC) or alternating current (AC) interference and ends with documenting the results. Once testing has begun, personnel must know how to position, manipulate, and read any equipment that might be used. If personnel encounter induced AC voltage, which could occur if a pipeline is located near a high voltage AC electrical field, they must be knowledgeable about the effects and hazards associated with the voltage and know how to notify the appropriate company personnel. Personnel must understand that no job is complete until all tasks are documented and the proper notifications are made.

Personnel must have a basic understanding of additional tasks related to testing for interference, such as measuring the structure-to-soil potential (CT1_1-17), conducting a close interval survey (CT1_2-17), inspecting and performing electrical testing of bonds (CT1_4-17), and inspecting and testing electrical isolation (CT1_5-17).

2.1.2 Abnormal Operating Conditions

Personnel conducting interference testing must be able to identify potential abnormal operating conditions (AOCs) and know the proper response to each, should they occur. The following are some potential AOCs, and the appropriate action personnel should take in response to them:

Case History

Data Loggers

An intermittent high voltage was recorded on a pipeline with no readily apparent cause. The operator installed a data logger and recorded very high voltage (100–200V) on the pipeline that coincided with a train-charging station located above the line. The solution was to tie into an existing reverse current switch.

The Bottom Line: A data logger can be an invaluable tool when troubleshooting.

- *Damage to or unsupervised third party construction on a right-of-way is discovered* – Notify appropriate company personnel and, if the contractor continues to perform unsupervised work on site, discuss the dangers posed by the situation with the contractor.
- *Internal or external corrosion of the pipe is present and to the extent that structure-to-soil potential measurements would be impractical* – Document the condition according to company guidelines and notify personnel responsible for correcting the problem.
- *The reading is outside of expected parameters* – Check for possible causes of unexpected readings such as reverse polarity, broken bond, or change in cathodic system.
- *Induced AC voltage is encountered* – Exercise caution around any equipment or structures and notify the appropriate company personnel.
- *Fire or explosion is caused by the ignition of a hydrocarbon liquid or gas* – Evacuate the area and notify the appropriate firefighting personnel. It may be necessary to call 911 in such an emergency.
- *Unexpected leaks of hazardous liquid or carbon dioxide because of a loss of pipe wall thickness is discovered* – Take steps to eliminate any possible ignition sources and notify the appropriate personnel, possibly including 911.

2.2.0 Procedure

When testing for interference, always follow company procedures and the manufacturer's instructions for the meter in use. The general steps to perform when testing for interference are as follows:

WARNING! Always obtain and use the appropriate personal protective equipment (PPE) for each task. Stay alert to the possibility of unexpected leaks from the exposed pipeline such as hazardous liquid or carbon dioxide (CO_2). In the event of a leak, safely eliminate all ignition sources and immediately notify the appropriate personnel.

Step 1 Identify potential abnormal operating conditions (AOCs) that may occur, and know the appropriate actions to take in response to them.

Step 2 Utilize the appropriate personal protective equipment according to relevant company procedures.

Step 3 Select the instrumentation to be used (**test leads**, voltmeter, data logger, reference electrodes, etc.). Inspect the equipment for damage and proper operation before taking any measurements.

Step 4 Assess the area for other cathodic protection systems or sources of electrical interference.

CAUTION Always consult company procedures for the proper polarity connections. Company procedures should indicate which positive and which negative lead to connect to a particular source to obtain a correct reading.

Step 5 Measure the structure-to-soil potential.

Step 6 Review the readings as they are taken to ensure that they fall within the desired range. Compare the readings with previous values at the test location.

Step 7 Interrupt rectifiers to determine if interference exists. Interrupting a rectifier can help to detect its influence on other structures.

Step 8 If interference is found, **AC current mitigation** is required, which begins with AC modeling.

NOTE AC modeling takes into account soil resistivity, pipeline coating resistivity, utility peak load, standoff distance, and colocation. The mitigation system is designed using these parameters. Mitigation elements include zinc ribbon or gradient matting, copper cable, rods, copper solid-state decoupling devices, and isolation switches that only allow AC to flow, not DC. (This is normally accomplished by a company that specializes in such activities.)

Step 9 Document all findings on the appropriate company inspection form and make required notifications.

Additional Resources

It is assumed that you will learn to perform these procedures under the direct supervision of your crew leader, foreman, or instructor. For independent study, refer to the recommended Additional Resources at the back of this module.

2.0.0 Section Review

1. Gouges, scrapes, and dents in pipeline surface material are common examples of _____.
   a. corrosive erosion
   b. mechanical damage
   c. substrate profiling
   d. adhesion enhancement

2. When interference is found during interference testing, the AC current mitigation process should begin with _____.
   a. DC modeling
   b. interrupting all rectifiers
   c. AC modeling
   d. measuring the structure-to-soil potential

SUMMARY

Underground steel piping systems are protected against corrosion with pipeline coatings and cathodic protection systems. However, when pipelines are located near structures such as high-voltage AC power lines, they can be affected by the electromagnetic field created by the alternating current. The induced AC current is not only a potential safety hazard; it can also cause excessive pipeline corrosion, especially where there are coating defects. Technicians must understand corrosion and cathodic protection systems, and be able to identify foreign structures that can cause interference. This module provided an overview of corrosion and cathodic protection, and it described requirements and procedures for safely and effectively testing to detect interference.

Review Questions

1. When current flows from the steel in a pipeline structure into the surrounding electrolyte (soil) causing the structure to corrode, the area is _____.
   a. pathodic
   b. anodic
   c. cathodic
   d. metabolic

2. A cathodic protection system turns the pipeline surface into a(n) _____.
   a. electrolyte
   b. heat sink
   c. anode
   d. cathode

3. The most common type of electrode in a pipeline cathodic protection system is the _____.
   a. copper-copper sulfate electrode
   b. silver chloride electrode
   c. salt-water electrode
   d. zinc-chloride electrode

4. If an AC interference voltage in excess of 15V is detected during an interference test, _____.
   a. personnel should immediately begin an AC current initiation process
   b. all rectifiers should be replaced and structure-to-soil potentials re-tested
   c. work should be stopped and appropriate notifications should be made
   d. an interrupted close interval survey should be taken for verification

5. If extensive internal or external piping corrosion is found prior to testing for interference, a pipeline technician should _____.
   a. apply a cathodic patch over the damage and continue with the measurements
   b. make sure that the instruments are attached directly to the corroded area
   c. document the condition and notify personnel about correcting the problem
   d. attach an approved anode bank to the structure and adjust the test instrument

6. The final step in any procedure for testing to detect interference is to _____.
   a. inspect every piece of tested equipment for damage and proper operation
   b. accurately record all required information in accordance with company policy
   c. remove lockout/tagout barriers and restart all holiday detection equipment
   d. conduct a Barcol frequency test to certify the readings to NACE standards

Trade Terms Quiz

Fill in the blank with the correct term that you learned from your study of this module.

1. The potential difference between the surface of a buried or submerged metallic structure (such as a pipe) and the electrolyte that is measured with reference to an electrode in contact with the electrolyte is known as the ________.
2. The potential energy of a unit charge at any point in an electric circuit, measured with respect to a specified reference point in the circuit or to ground is called ________.
3. Engineered control methods used to limit extreme levels of induced current in pipelines is ________.
4. Specific areas or locations, sometimes called test points, where a connection to a structure is made so that a structure-to-soil potential can be measured are called ________.
5. A device that converts AC current to DC current and, in an impressed cathodic protection system, supplies current to the structure in the soil is a(n) ________.
6. A space that is large enough and so configured that an employee can bodily enter and perform assigned work; has limited or restricted means for entry or exit (such as tanks, vessels, vaults, etc.); and is not designed for continuous employee occupancy is called a(n) ________.
7. The distinction between the positive and negative potential differences at the terminals of a meter or electrical device is the ________.
8. Wires that are connected to a structure being tested that provide easy connection points for a meter are called ________.
9. The voltage potential of a casing in soil or an electrolyte as it relates to a reference electrode is called ________.
10. An electrode whose open-circuit potential is constant under similar conditions of measurement, and which is used to measure the relative potentials of other electrodes is a (n) ________.
11. A pipeline or other metal structure in the vicinity of a pipeline system under cathodic protection is a(n) ________.
12. A non-polarizing electrode used in corrosion control to measure the electrical potential of a metal structure to a surrounding electrolyte to determine the potential for corrosion damage or to monitor the effectiveness of existing control measures is a(n) ________.
13. Conditions within a pipeline system that may indicate (1) a malfunction of a component or (2) a condition exceeding design limits or resulting in a hazard to persons, property, or the environment are called ________.
14. The electrode of an electrochemical cell at which reduction occurs is a(n) ________.
15. The effect upon cathodic protection levels of a structure due to the interaction between its cathodic protection system and the cathodic protection system of another structure is called ________.
16. The electrode of an electrochemical cell at which oxidation occurs is called a(n) ________.
17. Electrically charged atoms (either positively charged or negatively charged) are ________.
18. Electricity that flows as a result of a chemical action is considered ________.

Trade Terms

- Abnormal operating conditions (AOCs)
- AC current mitigation
- Anode
- Casing-to-soil potential
- Cathode
- Confined space
- Copper-copper sulfate electrode (CSE)
- Electrical potential
- Foreign structure
- Galvanic
- Ions
- Interference
- Polarity
- Rectifier
- Reference electrode
- Structure-to-soil potential
- Test leads
- Test stations

Trade Terms Introduced in This Module

Abnormal operating conditions (AOCs): Conditions within a pipeline system, identified by the operator, that may indicate (1) a malfunction of a component or (2) a deviation from normal operations that may indicate a condition exceeding design limits or result in a hazard to persons, property, or the environment.

AC current mitigation: Engineered control methods used to limit extreme levels of induced current in pipelines.

Anode: The electrode of an electrochemical cell at which oxidation occurs.

Casing-to-soil potential: The voltage potential of a casing in soil or an electrolyte as it relates to a reference electrode.

Cathode: The electrode of an electrochemical cell at which reduction occurs.

Confined space: A space that is large enough and so configured that an employee can bodily enter and perform assigned work; has limited or restricted means for entry or exit (such as tanks, vessels, vaults, etc.); and is not designed for continuous employee occupancy.

Copper-copper sulfate electrode (CSE): A non-polarizing electrode used in corrosion control to measure the electrical potential of a metal structure to a surrounding electrolyte to determine the potential for corrosion damage or to monitor the effectiveness of existing control measures.

Electrical potential: The potential energy of a unit charge at any point in an electric circuit, measured with respect to a specified reference point in the circuit or to ground.

Foreign structure: A pipeline or other metal structure in the vicinity of a pipeline system under cathodic protection.

Galvanic: Pertaining to electricity that flows as a result of a chemical action.

Ions: Electrically charged atoms (either positively charged or negatively charged).

Interference: The effect upon cathodic protection levels of a structure due to the interaction between its cathodic protection system and the cathodic protection system of another structure.

Polarity: The distinction between the positive and negative potential differences at the terminals of a meter or electrical device.

Rectifier: A device that converts AC current to DC current and, in an impressed cathodic protection system, supplies current to the structure in the soil.

Reference electrode: An electrode whose open-circuit potential is constant under similar conditions of measurement; it is used to measure the relative potentials of other electrodes.

Structure-to-soil potential: The potential difference between the surface of a buried or submerged metallic structure (such as a pipe) and the electrolyte that is measured with reference to an electrode in contact with the electrolyte.

Test leads: Wires that are connected to a structure being tested that provide easy connection points for a meter.

Test stations: Specific areas or locations, sometimes called test points, where a connection to a structure is made so that a structure-to-soil potential can be measured.

Additional Resources

The following recommended resources can provide additional helpful information related to the requirements for this covered task:

API Recommended Practice 1161, Third Edition, © 2014 American Petroleum Institute. API Publishing Services, 1220 L Street, NW, Washington, DC 20005. Distributed by Thompson Reuters (Scientific) LLC, **www.techstreet.com**.

49 *CFR* 192, Subpart I, **www.ecfr.gov**

49 *CFR* 195, Subpart H, **www.ecfr.gov**

NCCER Module AOCCC-17, *Abnormal Operating Conditions Control Center*

NCCER Module AOCFG-17, *Abnormal Operating Conditions Field and Gas*

NCCER Module 62105-14, *Introduction to Pipeline Documents*

NCCER Module 66101-14, *Introduction to the Pipeline Industry*

NCCER Module CT1_1-17, *Measure Structure-to-Soil Potentials*

NCCER Module CT1_2-17, *Conduct Close Interval Survey*

NCCER Module CT1_4-17, *Inspect and Perform Electrical Test of Bonds*

NCCER Module CT1_5-17, *Inspect and Test Electrical Isolation*

NCCER Module CT14_1-17, *Locate Line*

API Recommended Practice 651, Cathodic Protection of Aboveground Petroleum Storage Tanks. Fourth Edition, 2014. Available from the American Petroleum Institute at **www.api.org**.

NACE International *SP0169-2013, Control of External Corrosion on Underground or Submerged Metallic Piping Systems*. 2013. Available from NACE International at **www.nace.org**.

Figure Credits

Tinker & Rasor, Module Opener, Figure 3, SA01

Section Review Answer Key

Answer	Section Reference	Objective
Section One		
1. d	1.1.2	1a
2. b	1.2.0	1b
Section Two		
1. b	2.1.1	2a
2. c	2.2.0	2b

NCCER CURRICULA — USER UPDATE

NCCER makes every effort to keep its textbooks up-to-date and free of technical errors. We appreciate your help in this process. If you find an error, a typographical mistake, or an inaccuracy in NCCER's curricula, please fill out this form (or a photocopy), or complete the online form at **www.nccer.org/olf**. Be sure to include the exact module ID number, page number, a detailed description, and your recommended correction. Your input will be brought to the attention of the Authoring Team. Thank you for your assistance.

Instructors – If you have an idea for improving this textbook, or have found that additional materials were necessary to teach this module effectively, please let us know so that we may present your suggestions to the Authoring Team.

NCCER Product Development and Revision
13614 Progress Blvd., Alachua, FL 32615
Email: curriculum@nccer.org
Online: www.nccer.org/olf

❑ Trainee Guide ❑ Lesson Plans ❑ Exam ❑ PowerPoints Other ____________

Craft / Level: Copyright Date:

Module ID Number / Title:

Section Number(s):

Description:

Recommended Correction:

Your Name:

Address:

Email: Phone:

CT1_4-17

Inspect and Perform Electrical Test of Bonds

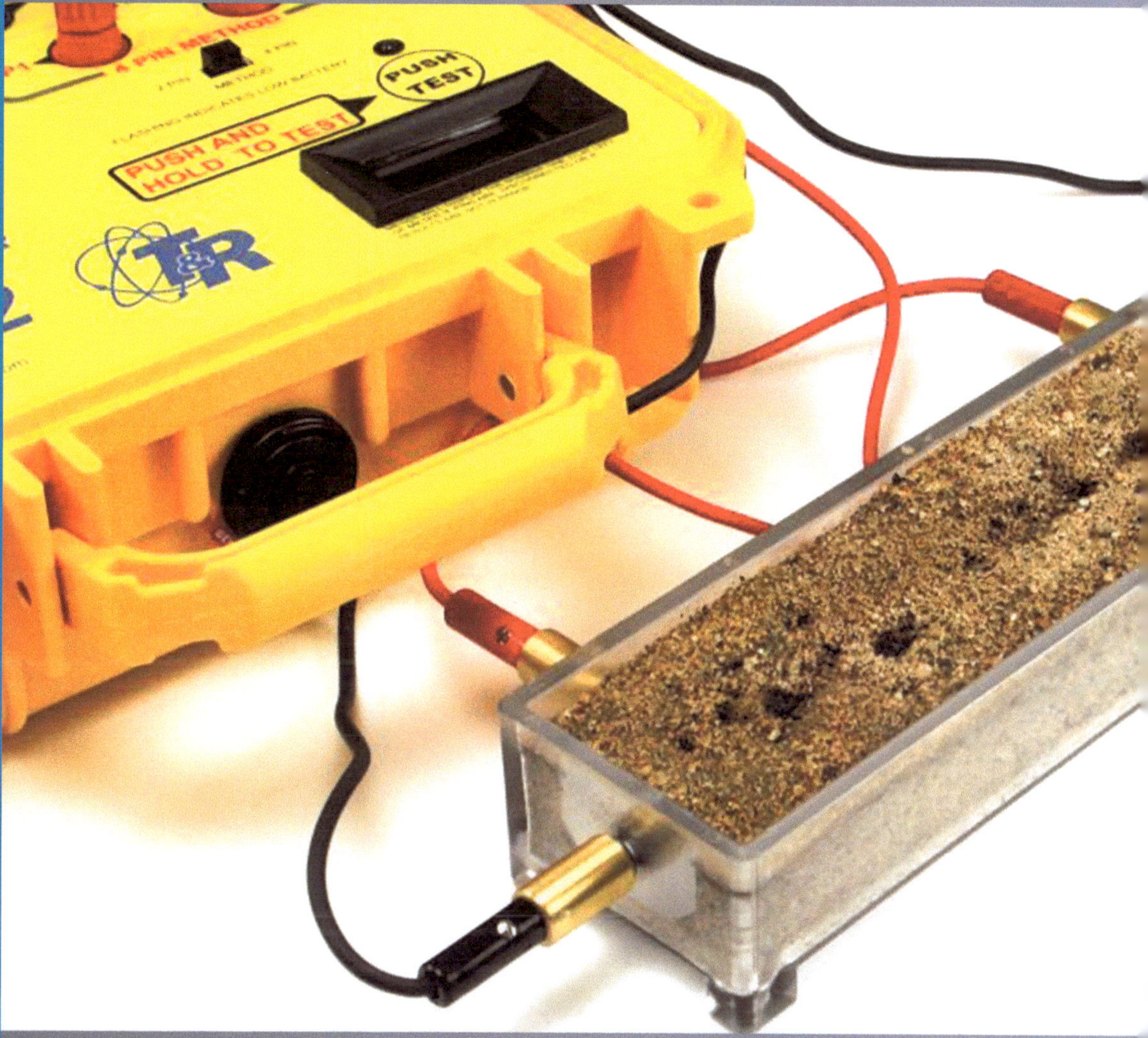

Overview

Pipelines near high-voltage power lines or similar foreign structures can be affected by the electromagnetic field created by the alternating current. The AC interference can be hazardous and cause severe corrosion, especially at pipeline coating defects. A metallic bond can be installed between a pipeline and a foreign structure In order to reduce the effects of interference. Bonds must be tested regularly to ensure that the connection is continuous. This module identifies foreign structures that can cause interference, and describes instruments used to measure electrical resistance. It also explains the basic procedure used for conducting a bond test.

Trainees with successful module completions may be eligible for credentialing through the NCCER Registry. To learn more, go to **www.nccer.org** or contact us at 1.888.622.3720. Our website has information on the latest product releases and training, as well as online versions of our *Cornerstone* magazine and Pearson's product catalog.

Your feedback is welcome. You may email your comments to **curriculum@nccer.org**, send general comments and inquiries to **info@nccer.org**, or fill in the User Update form at the back of this module.

This information is general in nature and intended for training purposes only. Actual performance of activities described in this manual requires compliance with all applicable operating, service, maintenance, and safety procedures under the direction of qualified personnel. References in this manual to patented or proprietary devices do not constitute a recommendation of their use.

CT1_4-17
Inspect and Perform Electrical Test of Bonds

Objectives

When you have completed this module, you will be able to do the following:

1. Identify foreign structures that can cause interference and describe test instruments used to measure electrical resistance.
   a. Identify foreign structures that can cause interference.
   b. Identify and describe resistance test instruments.
2. Describe how to inspect and perform electrical testing of bonds (CT1_4-17).
   a. Identify the prerequisites, competencies, and abnormal operating conditions associated with inspecting and performing electrical testing of bonds.
   b. Describe the procedure for conducting a bond test.

Performance Task

Under the supervision of your instructor, you should be able to do the following:

1. Inspect and perform electrical testing of bonds (CT1_4-17).
   - Identify potential abnormal operating conditions that may occur during performance of this CT, and know the appropriate actions to take in response to them.
   - Utilize the appropriate personal protective equipment according to relevant company procedures.
   - Identify the bond locations where measurements will be taken.
   - Conduct a visual inspection of the bond test station for physical damage to the bond station, a damaged shunt, loose connections, disconnected wires, arcing across terminals, etc.
   - Select the correct instrumentation (volt-ohmmeter, ammeter, test leads, reference cell) and be prepared to use it properly.
   - Make connections with the test equipment to take and record readings.
   - Measure the potentials for each of the structures at the bond location.
   - Identify the shunt type and size.
   - Measure the direction and magnitude of current flow between the structures.
   - Field-analyze the readings to ensure they are within a desired range, including a check of the polarity.
   - Complete appropriate documentation as required by operator's procedures.

Trade Terms

Abnormal operating conditions (AOCs)
Ammeter
Anode
Bond
Confined space
Electrical isolation
Foreign structure
Interference
Ions
Megohm
Multimeter
Ohms-cm (ohms per centimeter)
Polarity
Reference electrode
Shunt
Structure-to-soil potential
Test leads
Test stations
Wenner (four-pin) method

Industry Recognized Credentials

If you are training through an NCCER-accredited sponsor, you may be eligible for credentials from NCCER's Registry. The ID number for this module is CT1_4-17. Note that this module may have been used in other NCCER curricula and may apply to other level completions. Contact NCCER's Registry at 888.622.3720 or go to **www.nccer.org** for more information.

Contents

Figures

Section One

1.0.0 Electrical Resistance Measurements

Objective

Identify foreign structures that can cause interference and describe test instruments used to measure electrical resistance.

a. Identify foreign structures that can cause interference.
b. Identify and describe resistance test instruments.

Trade Terms

Ammeter: A meter that measures amperes. It usually comes as direct-connect type and clamp-on type.

Anode: The electrode of an electrochemical cell at which oxidation occurs.

Bond: The electrical connection of two structures, which usually have two different cathodic protection levels.

Foreign structure: A pipeline or other metal structure in the vicinity of a pipeline system under cathodic protection.

Interference: The effect upon cathodic protection levels of a structure due to the interaction between its cathodic protection system and the cathodic protection system of another structure.

Ions: Electrically charged atoms (either positively charged or negatively charged).

Megohm: A unit of resistance equal to 1,000,000 ohms.

Multimeter: A meter that allows measurement of different quantities and whose functions are made available through a selector switch. Most basic multimeters include an ammeter, voltmeter, and ohmmeter.

Ohms-cm (ohms per centimeter): Resistance of a cube of material (earth/soil) with the sides of the cube being measured in centimeters.

Test leads: Wires that are connected to a structure being tested that provide easy connection points for a meter.

Wenner (four-pin) method: Use of four metal probes (pins) placed equal distances apart in a straight line. A current is induced and current/resistance readings are taken at the appropriate points and recorded. Results are applied into the proper formula and charted to meet company specifications.

Underground steel piping systems located in the vicinity of a **foreign structure**, such as high-voltage AC power lines, can be affected by the electromagnetic field created by the alternating current. The induced AC current can create a potential safety hazard, and can also cause excessive pipeline corrosion, especially where there are coating defects.

Voltages as low as 4 VAC (20 amps/m^2 – 100 amps/m^2) can result in corrosion. Anything higher than 15V is considered unsafe for pipeline work. If a voltage in excess of 15V is detected, work should be stopped and appropriate notifications should be made. Pipeline technicians need to be familiar with foreign structures that can cause **interference** and instrumentation that can be used to test the metallic **bond** between piping systems and these structures.

1.1.0 Foreign Structures that Cause Interference

Pipelines can pick up interference current from a variety of foreign structures. Foreign metal structures that can cause interference typically include the following:

- Water lines
- Telephone lines
- Electric power lines
- Traffic control cables and conduits
- Steam lines
- Pipe cables
- Street light cables
- Conduits
- Light standards
- Electric ground rods
- Sewage lines
- Manholes and drainpipes

Metallic objects such as reinforcing rod (rebar) or wire mesh in concrete can also cause interference. Also be aware of junk metal such as pipe and rails from abandoned facilities that may be present.

When installing cathodic protection systems, utility companies and the owners of nearby pipelines should be notified so that interference testing can be cooperatively conducted.

When a foreign structure is near a protected pipeline, it may pick up cathodic protection current. The foreign structure receives some cathodic protection by picking up this stray current, but when the current leaves, it releases **ions** and will eventually cause corrosion at the point of exit.

Pipeline mapping systems help gas utilities and pipeline companies locate contacts with foreign metal structures. The use of mapping systems significantly reduces the cost of troubleshooting cathodic protection on pipeline systems.

1.2.0 Resistance Test Instruments

To minimize the effects of interference between a pipeline and a foreign structure, a metallic bond is installed between them. This provides electrical continuity between the two structures. Regular inspections and testing are necessary to ensure that the connection is continuous.

Testing of bonds is normally accomplished by measuring electrical resistance. Resistance can be measured using a megohmmeter, the ohmmeter or insulation resistance function of a **multimeter**, or a soil resistance meter. Each type of meter is suited to specific applications and resistance measurements.

1.2.1 Megohmmeters

A **megohm** is a unit of resistance equal to 1,000,000 ohms. A megohmmeter, or megger, (*Figure 1*) is an instrument used to measure very high resistance values (up to 2,000 megohms or more). A megger can be used to locate open circuits or short circuits. It consists of a DC current meter movement, a low-voltage DC power source (usually a battery), and current-limiting resistors connected in series.

Before measuring the resistance of an unknown resistor or electrical circuit, the **test leads** are shorted together. This causes current to flow through the meter movement and the pointer to deflect to the right. This indicates that there is zero resistance present across the input terminals of the ohmmeter. Most ohmmeters have a zero adjustment knob on the face of the meter. This knob is used to correct for the fact that, as the batteries of the meter weaken, their output voltage decreases and the meter will no longer deflect to full scale. Ohmmeters should always be zeroed before use.

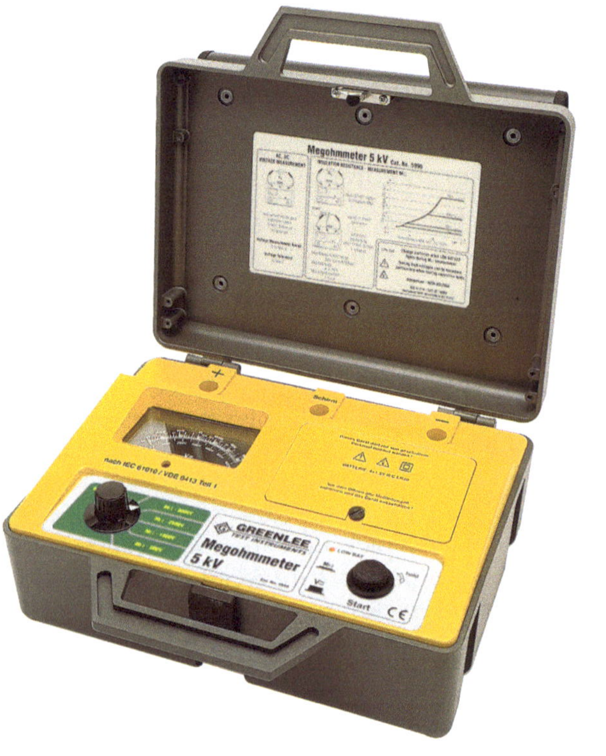

Figure 1 Megohmmeter.

WARNING!

Before using an ohmmeter, the circuit must be turned off and verified as de-energized using a voltmeter. Failure to de-energize the circuit could cause a damaging current to flow through the meter. This can cause damage to the meter and/or circuit, as well as personal injury.

Each component in the circuit should be tested individually by removing the component from the circuit and connecting the ohmmeter leads across it when making resistance measurements. If the component is not removed from the circuit, the component can be effectively isolated by disconnecting one of its leads from the circuit. However, ensure that other components are not in parallel with the component being tested, as this will lead to an incorrect reading.

1.2.2 Multimeters

A multimeter (*Figure 2*) is also referred to as a volt-ohm-milliammeter (VOM). It is a multipurpose instrument that can be used to measure voltage, current, resistance, and other circuit values. The instrument shown in *Figure 2* includes a voltmeter, **ammeter**, ohmmeter, and insulation tester (megohmmeter). There are many different models of the basic multimeter, and they can be analog or digital. For specific instructions on the use of a multimeter, refer to the applicable owner's manual.

1.2.3 Soil Resistance Meters

Among the first steps to be taken when planning an **anode** bank or an impressed current ground bed is performing a soil resistivity test. Soil resistivity is measured in **ohms-cm (ohms per centimeter)**. One of the most common means of testing the resistivity of soil is the **Wenner (four-pin) method**. In this method, four metal pins are driven into the earth. The pins must be equidistant from one another and must form a straight line. The pins are spaced at a distance equal to the soil depth to be tested. Current flows between the two outside pins, so the voltage drop between the two inner pins provides the resistance.

A resistance test meter can be used to measure the voltage drop at the same time that it supplies the necessary current for the outer pins. *Figure 3* shows a typical soil resistance meter. This meter can be used with either stainless steel soil pins (*Figure 4*) or soil boxes (*Figure 5*).

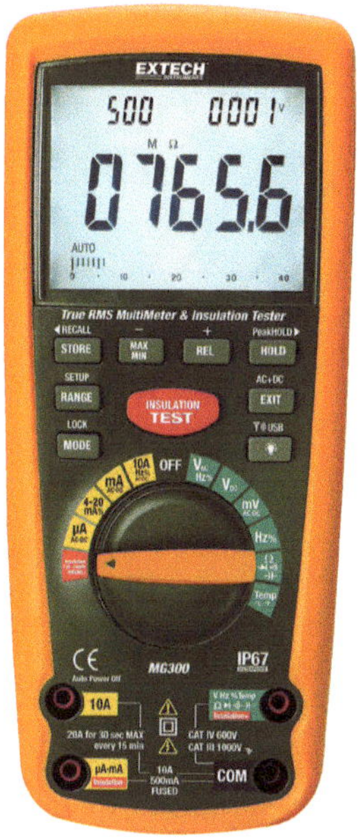

Figure 2 Multimeter.

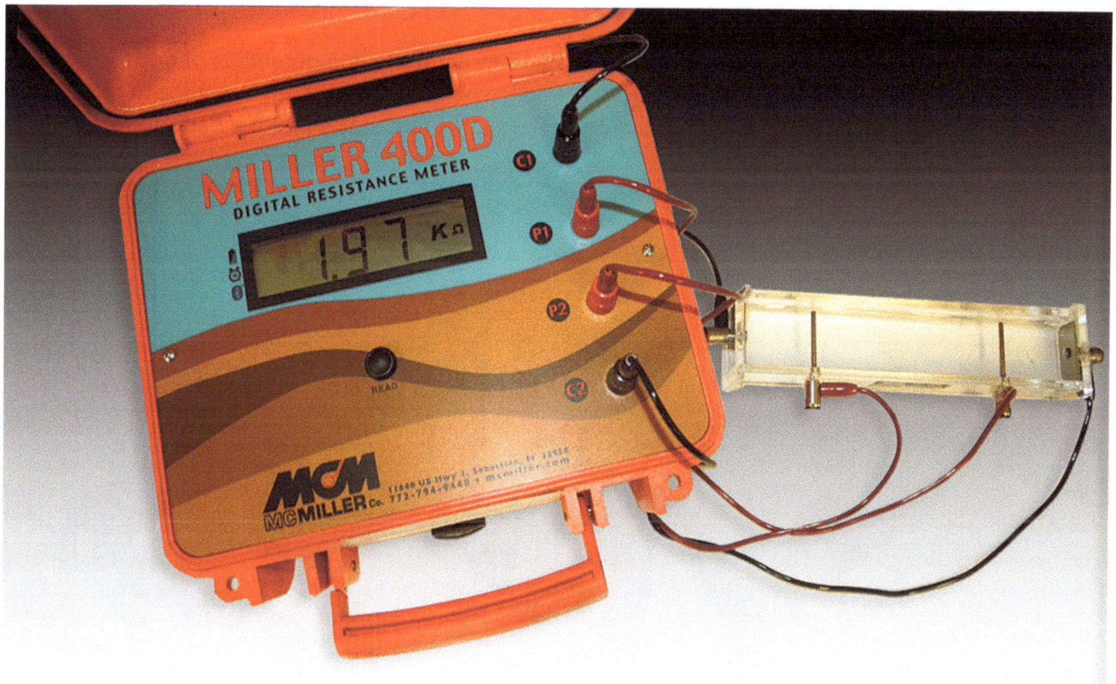

Figure 3 Soil resistance meter.

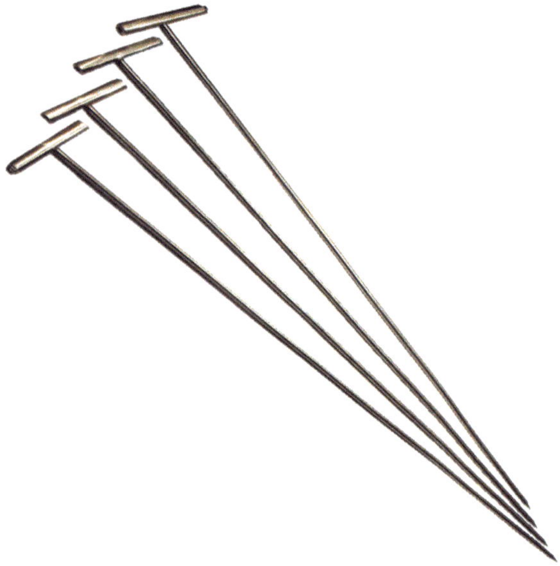

Figure 4 Soil resistance pins.

Figure 5 Soil resistance box.

Cathodic Protection Digital Voltmeters

This DC cathodic protection voltmeter can be used to measure structure-to-soil potentials of buried or submerged metal structures, such as pipelines and storage tanks. It can also be used to measure structure-to-structure and structure-to-anode values. This meter has a connection for the half-cell to screw into the back of the voltmeter to measure structure-to-anode values.

STRUCTURE-TO-ANODE MEASUREMENT

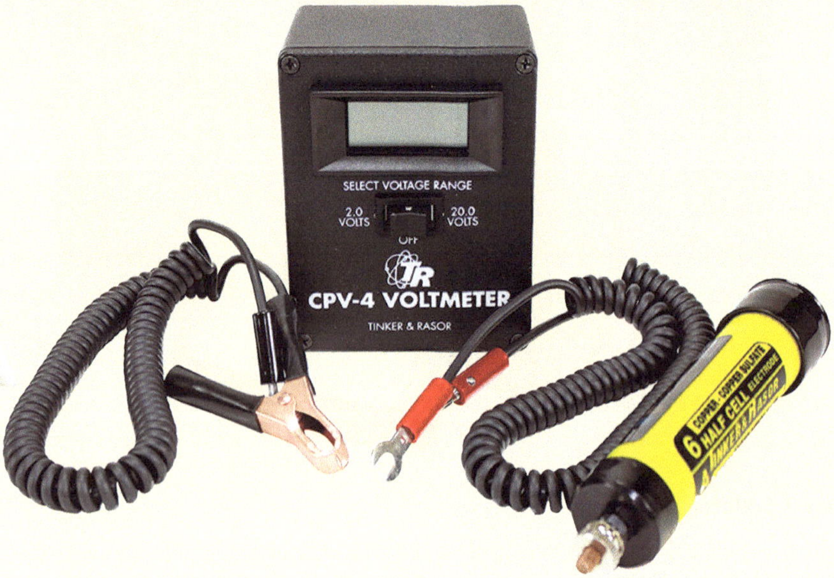

Additional Resources

It is assumed that you will learn to perform these procedures under the direct supervision of your crew leader, foreman, or instructor. For independent study, refer to the recommended Additional Resources at the back of this module.

1.0.0 Section Review

1. When a foreign structure is located near a pipeline that is protected by a cathodic protection system, the structure might _____.
   a. convert any DC stray current it receives to alternating current
   b. receive some cathodic protection by picking up stray current
   c. act as an amplifier and overwhelm the pipeline protection system
   d. store the current like a capacitor before releasing a voltage burst

2. An instrument that is most likely to be used to measure very high resistance values is a(n) _____.
   a. voltmeter
   b. multimeter
   c. ammeter
   d. megohmmeter

SECTION TWO

2.0.0 ELECTRICAL TESTING OF BONDS

Objective

Describe how to inspect and perform electrical testing of bonds (CT1_4-17).

a. Identify the prerequisites, competencies, and abnormal operating conditions (AOCs) associated with inspecting and performing electrical testing of bonds.
b. Describe the procedure for conducting a bond test.

Performance Task

Inspect and perform electrical testing of bonds (CT1_4-17).

- Identify potential abnormal operating conditions that may occur during performance of this CT, and know the appropriate actions to take in response to them.
- Utilize the appropriate personal protective equipment according to relevant company procedures.
- Identify the bond locations where measurements will be taken.
- Conduct a visual inspection of the bond test station for physical damage to the bond station, a damaged shunt, loose connections, disconnected wires, arcing across terminals, etc.
- Select the correct instrumentation (volt-ohmmeter, ammeter, test leads, reference cell) and be prepared to use it properly.
- Make connections with the test equipment to take and record readings.
- Measure the potentials for each of the structures at the bond location.
- Identify the shunt type and size.
- Measure the direction and magnitude of current flow between the structures.
- Field-analyze the readings to ensure they are within a desired range, including a check of the polarity.
- Complete appropriate documentation as required by operator's procedures.

Trade Terms

Abnormal operating conditions (AOCs): Conditions within a pipeline system, identified by the operator, that may indicate (1) a malfunction of a component or (2) a deviation from normal operations that may indicate a condition exceeding design limits or result in a hazard to persons, property, or the environment.

Confined space: A space that is large enough and so configured that an employee can bodily enter and perform assigned work; has limited or restricted means for entry or exit (such as tanks, vessels, vaults, etc.); and is not designed for continuous employee occupancy.

Electrical isolation: The condition of being electrically separated from other metallic structures or the environment.

Polarity: The distinction between the positive and negative potential differences at the terminals of a meter or electrical device.

Reference electrode: An electrode whose open-circuit potential is constant under similar conditions of measurement; used to measure the relative potentials of other electrodes.

Shunt: A calibrated resistor used to determine current flow in the circuit.

Structure-to-soil potential: The potential difference between the surface of a buried or submerged metallic structure (such as a pipe) and the electrolyte that is measured with reference to an electrode in contact with the electrolyte.

Test stations: Specific areas or locations, sometimes called test points, where a connection to a structure is made so that a structure-to-soil potential can be measured.

Personnel inspecting and performing electrical testing of bonds are required to have some basic competencies. They must have the ability to carefully follow the required procedures for conducting bond testing and recording the results. This section describes some of the basic skills needed by pipeline technicians and identifies abnormal operating conditions (AOCs) that personnel may need to respond to when inspecting and testing bonds. It also describes the basic procedures for conducting a bond test.

2.1.0 Prerequisites, Competencies, and Abnormal Operating Conditions

Personnel must possess a variety of skills in order to safely and successfully conduct bond testing. Testing procedures require numerous steps, which a pipeline technician must be competent to perform. Personnel must be aware of abnormal operating conditions (AOCs) that may occur during testing procedures and know how to properly respond to them.

2.1.1 Prerequisites and Competencies

Personnel assigned to conduct bond testing must have a good understanding of hazards that can exist before and during this covered task, as well as what the task entails. Before testing, personnel must be familiar with hazards along the right-of-way (ROW), including those associated with right-of-way encroachments and unsupervised third party construction. They must be familiar with the locations of test stations, and they must be able to properly choose, inspect, calibrate, and operate any tools or instruments to be used during the procedure.

Pipeline work often involves trenched and excavated areas, as well as tank containment areas. For these reasons, personnel must be properly trained for confined space entry and excavation safety.

Personnel must be able to visibly inspect the pipe and determine if there are any unsafe integrity issues, such as evidence of a release or significant mechanical damage (gouges, scrapes, dents), and discontinue the work and make notifications if any are found. If old coating materials that may contain asbestos are encountered, personnel must make the proper notifications so that a qualified person can sample, test, and, if necessary, remove and dispose of any asbestos-containing material (ACM). For these and other safety reasons, personnel must be familiar with local emergency telephone numbers.

This task begins with identifying the location of the bond(s) and ends with the collection of data. Once testing has begun, personnel must know how to position, manipulate, and read any equipment that might be used. If personnel encounter induced AC voltage, which could occur if a pipeline is located near a high voltage AC electrical field, they must be knowledgeable about the effects and hazards associated with the voltage and know how to notify the appropriate company personnel. Personnel must understand that no job is complete until all tasks are documented and the proper notifications are made.

Personnel must have a basic understanding of additional tasks related to bond testing, such as measuring structure-to-soil potential (CT1_1-17), conducting a close interval survey (CT1_2-17), testing to detect interference (CT1_3-17), and inspecting and testing electrical isolation (CT1_5-17).

2.1.2 Abnormal Operating Conditions

Personnel conducting bond testing must be able to identify potential abnormal operating conditions (AOCs) and know the proper response to each, should they occur. The following are some potential AOCs, and the appropriate action personnel should take in response to them:

- *Damage to or unsupervised third party construction on a right-of-way is discovered* – Notify appropriate company personnel and, if the contractor continues to perform unsupervised work on site, discuss the dangers posed by the situation with the contractor.
- *Internal or external corrosion of the pipe is present and to the extent that structure-to-soil potential measurements would be impractical* – Document the condition according to company guidelines and notify personnel responsible for correcting the problem.
- *Broken bond connection is discovered* – Repair or request a repair, and document.
- *Change in the current flow and/or direction across the bond (reverse polarity)* – Ensure that readings were taken correctly, and notify operator personnel as required.
- *Erratic or floating readings occur* – Determine the cause of the erratic readings, and repair test leads or equipment as needed.
- *Induced AC voltage is encountered* – Exercise caution around any equipment or structures and notify the appropriate company personnel.
- *Fire or explosion is caused by the ignition of a hydrocarbon liquid or gas* – Evacuate the area and notify the appropriate firefighting personnel. It may be necessary to call 911 in such an emergency.
- *Unexpected leaks of hazardous liquid or carbon dioxide because of a loss of pipe wall thickness is discovered* – Take steps to eliminate any possible ignition sources and notify the appropriate personnel, possibly including calling 911.

2.2.0 Procedure

When conducting bond testing, always follow company procedures and the manufacturer's instructions for the meter in use. The general steps for a typical procedure for conducting a close interval survey is as follows:

> **WARNING!**
>
> Always obtain and use the appropriate personal protective equipment (PPE) for each task. Stay alert to the possibility of unexpected leaks from the exposed pipeline such as hazardous liquid or carbon dioxide (CO_2). In the event of a leak, safely eliminate all ignition sources and immediately notify the appropriate personnel.

Step 1 Identify potential abnormal operating conditions (AOCs) that may occur, and know the appropriate actions to take in response to them.

Step 2 Utilize the appropriate personal protective equipment according to relevant company procedures.

Step 3 Identify the bond locations where measurements will be taken.

Step 4 Conduct a visual inspection of the bond test station (check for physical damage, a damaged shunt, loose connections, disconnected wires, arcing across terminals, etc.).

Step 5 Select the instrumentation to be used (test leads, ammeter, volt-ohmmeter, data logger, reference electrode, etc.). Inspect the equipment for damage and proper operation prior to taking any measurements.

Step 6 Make connections with the test equipment to take and record readings.

Step 7 Measure the structure-to-soil potentials for each of the structures at the bond location.

Step 8 Identify the shunt type and size in order to calculate the current flow.

Step 9 Measure the direction and magnitude of current flow between the structures.

Step 10 Review the readings as they are taken to ensure that they fall within a desired range. Compare the readings with previous values at the test location.

Step 11 Document all readings as required by operator's procedures.

> **NOTE**
>
> Proper documentation is essential for future analysis and identification of necessary repairs.

Additional Resources

It is assumed that you will learn to perform these procedures under the direct supervision of your crew leader, foreman, or instructor. For independent study, refer to the recommended Additional Resources at the back of this module.

2.0.0 Section Review

1. If extensive internal or external piping corrosion is found prior to conducting a bond test, a pipeline technician should _____.
   a. apply a cathodic patch over the damage and continue with the measurements
   b. make sure the instruments are attached directly to the corroded area
   c. document the condition and notify personnel about correcting the problem
   d. attach an approved anode bank to the structure and adjust the test instrument

2. Before taking any measurements during a bond test, a pipeline technician should first _____.
   a. identify the direction of current flow between structures
   b. check structure-to-soil potentials at the foreign structure
   c. determine the magnitude of current flow involved
   d. identify the bond locations at the structures

SUMMARY

Underground steel piping systems that are located near structures such as high-voltage AC power lines can be affected by the electromagnetic field created by the alternating current. The induced AC current is not only a potential safety risk to personnel, it can also cause excessive pipeline corrosion, especially where there are coating defects. Technicians must be familiar with foreign structures that can cause interference and how a metallic bond is installed to provide corrosion protection. The bond must be tested regularly to ensure that the connection is continuous. This module identified common foreign structures that can cause interference, and instruments that are used to measure electrical resistance. It also explained the basic procedure for conducting a bond test.

Review Questions

1. The cost of troubleshooting cathodic protection on pipeline systems can be significantly reduced when gas utilities and pipeline companies locate contacts with foreign metal structures by using _____.
   a. pipeline mapping systems
   b. satellite-based metering
   c. WAN resistance meters
   d. anodic/cathodic transponders

2. When the Wenner method is used to test soil resistivity, the resistance measured by the test instrument is provided by the _____.
   a. current flow changes when the ohmmeter leads are connected together
   b. voltage rise between the two outer pins that were driven into the earth
   c. grounding rod interference between the pipeline and the foreign structure
   d. voltage drop between the two inner pins that were driven into the earth

3. One of the abnormal operating conditions (AOCs) that a pipeline technician might encounter during a bond test is _____.
   a. having to make test equipment connections
   b. finding and inspecting bond test stations
   c. encountering induced AC current
   d. measuring current flow direction and magnitude

4. The final step in the general procedure for conducting a bond test is to _____.
   a. inspect every piece of tested equipment for damage and proper operation
   b. accurately document required information according to operator's procedures
   c. remove lockout/tagout barriers and restart all holiday detection equipment
   d. conduct a Barcol frequency test to certify the readings to NACE standards

Trade Terms Quiz

Fill in the blank with the correct term that you learned from your study of this module.

1. The distinction between the positive and negative potential differences at the terminals of a meter or electrical device is called ________.
2. The unit of resistance equal to 1,000,000 ohms is a(n) ________.
3. Specific areas or locations, sometimes called test points, where a connection to a structure is made so that a structure-to-soil potential can be measured are called ________.
4. Electrically charged atoms (either positively charged or negatively charged) are ________.
5. The potential difference between the surface of a buried or submerged metallic structure (such as a pipe) and the electrolyte that is measured with reference to an electrode in contact with the electrolyte is known as the ________.
6. A meter that allows measurement of different quantities and whose functions are made available through a selector switch is called a(n) ________.
7. The effect upon cathodic protection levels of a structure due to the interaction between its cathodic protection system and the cathodic protection system of another structure is ________.
8. A method of testing the resistivity of soil that uses four metal probes (pins) and induced current is known as the________.
9. Wires that are connected to a structure being tested that provide easy connection points for a meter are called ________.
10. A meter that measures amperes is called a(n) ________.
11. An electrode whose open-circuit potential is constant under similar conditions of measurement, and which is used to measure the relative potentials of other electrodes is a (n) ________
12. The electrical connection of two structures, which usually have two different cathodic protection levels is called a(n) ________.
13. A pipeline or other metal structure in the vicinity of a pipeline system under cathodic protection is a(n) ________.
14. A calibrated resistor used to determine current flow in the circuit is a(n) ________.
15. Conditions within a pipeline system that may indicate (1) a malfunction of a component or (2) a condition exceeding design limits or resulting in a hazard to persons, property, or the environment are called ________.
16. The condition of being electrically separated from other metallic structures or the environment is known as ________.
17. Resistance of a cube of material (earth/soil) with the sides of the cube being measured in centimeters is called ________.
18. A space that is large enough and so configured that an employee can bodily enter and perform assigned work; has limited or restricted means for entry or exit (such as tanks, vessels, vaults, etc.); and is not designed for continuous employee occupancy is called a(n) ________.
19. The electrode of an electrochemical cell at which oxidation occurs is called a(n) ________.

Trade Terms

Abnormal operating conditions (AOCs)
Ammeter
Anode
Bond
Confined space
Electrical isolation
Foreign structure
Ions
Interference
Megohm
Multimeter
Ohms-cm (ohms per centimeter)
Polarity
Reference electrode
Shunt
Structure-to-soil potential
Test leads
Test stations
Wenner (four-pin) method

Trade Terms Introduced in This Module

Abnormal operating conditions (AOCs): Conditions within a pipeline system, identified by the operator, that may indicate (1) a malfunction of a component or (2) a deviation from normal operations that may indicate a condition exceeding design limits or result in a hazard to persons, property, or the environment.

Ammeter: A meter that measures amperes. It usually comes as direct-connect type and clamp-on type.

Anode: The electrode of an electrochemical cell at which oxidation occurs.

Bond: The electrical connection of two structures, which usually have two different cathodic protection levels.

Confined space: A space that is large enough and so configured that an employee can bodily enter and perform assigned work; has limited or restricted means for entry or exit (such as tanks, vessels, vaults, etc.); and is not designed for continuous employee occupancy.

Electrical isolation: The condition of being electrically separated from other metallic structures or the environment.

Foreign structure: A pipeline or other metal structure in the vicinity of a pipeline system under cathodic protection.

Ions: Electrically charged atoms (either positively charged or negatively charged).

Interference: The effect upon cathodic protection levels of a structure due to the interaction between its cathodic protection system and the cathodic protection system of another structure.

Megohm: A unit of resistance equal to 1,000,000 ohms.

Multimeter: A meter that allows measurement of different quantities and whose functions are made available through a selector switch. Most basic multimeters include an ammeter, voltmeter, and ohmmeter.

Ohms-cm (ohms per centimeter): Resistance of a cube of material (earth/soil) with the sides of the cube being measured in centimeters.

Polarity: The distinction between the positive and negative potential differences at the terminals of a meter or electrical device.

Reference electrode: An electrode whose open-circuit potential is constant under similar conditions of measurement; used to measure the relative potentials of other electrodes.

Shunt: A calibrated resistor used to determine current flow in the circuit.

Structure-to-soil potential: The potential difference between the surface of a buried or submerged metallic structure (such as a pipe) and the electrolyte that is measured with reference to an electrode in contact with the electrolyte.

Test leads: Wires that are connected to a structure being tested that provide easy connection points for a meter.

Test stations: Specific areas or locations, sometimes called test points, where a connection to a structure is made so that a structure-to-soil potential can be measured.

Wenner (four-pin) method: Use of four metal probes (pins) placed equal distances apart in a straight line. A current is induced and current/resistance readings are taken at the appropriate points and recorded. Results are applied into the proper formula and charted to meet company specifications.

Additional Resources

The following recommended resources can provide additional helpful information related to the requirements for this covered task:

API Recommended Practice 1161, Third Edition, © 2014 American Petroleum Institute. API Publishing Services, 1220 L Street, NW, Washington, DC 20005. Distributed by Thompson Reuters (Scientific) LLC, **www.techstreet.com**.

49 *CFR* 192, Subpart I, **www.ecfr.gov**

49 *CFR* 195, Subpart H, **www.ecfr.gov**

NCCER Module AOCCC-17, *Abnormal Operating Conditions Control Center*

NCCER Module AOCFG-17, *Abnormal Operating Conditions Field and Gas*

NCCER Module 26102-14, *Electrical Safety*, from *Electrical*

NCCER Module 26112-14, *Electrical Test Equipment*, from *Electrical*

NCCER Module 62105-14, *Introduction to Pipeline Documents*

NCCER Module 66101-14, *Introduction to the Pipeline Industry*

NCCER Module 75120-13, *Confined Spaces and Excavations*, from *Field Safety*

NCCER Module CT1_1-17, *Measure Structure-to-Soil Potentials*

NCCER Module CT1_2-17, *Conduct Close Interval Survey*

NCCER Module CT1_3-17, *Test and Detect Interference*

NCCER Module CT1_5-17, *Inspect and Test Electrical Isolation*

NCCER Module CT14_1-17, *Locate Line*

API Recommended Practice 651, Cathodic Protection of Aboveground Petroleum Storage Tanks. Fourth Edition, 2014. Available from the American Petroleum Institute at **www.api.org**.

NACE International *SP0169-2013, Control of External Corrosion on Underground or Submerged Metallic Piping Systems*. 2013. Available from NACE International at **www.nace.org**.

Standard Test Method for Field Measurement of Soil Resistivity Using the Wenner Four-Electrode Method. 2012. Available from ASTM International at **www.astm.org**.

Figure Credits

Tinker & Rasor, Module Opener, Figure 5, SA01

Greenlee / A Textron Company, Figure 1

Extech Instruments, a FLIR Company, Figure 2

M. C. Miller Co., Inc., Figures 3, 4

Section Review Answer Key

Answer	Section Reference	Objective
Section One		
1. b	1.1.0	1a
2. d	1.2.1	1b
Section Two		
1. c	2.1.2	2a
2. d	2.2.0	2b

NCCER CURRICULA — USER UPDATE

NCCER makes every effort to keep its textbooks up-to-date and free of technical errors. We appreciate your help in this process. If you find an error, a typographical mistake, or an inaccuracy in NCCER's curricula, please fill out this form (or a photocopy), or complete the online form at **www.nccer.org/olf**. Be sure to include the exact module ID number, page number, a detailed description, and your recommended correction. Your input will be brought to the attention of the Authoring Team. Thank you for your assistance.

Instructors – If you have an idea for improving this textbook, or have found that additional materials were necessary to teach this module effectively, please let us know so that we may present your suggestions to the Authoring Team.

NCCER Product Development and Revision

13614 Progress Blvd., Alachua, FL 32615

Email: curriculum@nccer.org
Online: www.nccer.org/olf

❑ Trainee Guide ❑ Lesson Plans ❑ Exam ❑ PowerPoints Other ______________

Craft / Level: Copyright Date:

Module ID Number / Title:

Section Number(s):

Description:

Recommended Correction:

Your Name:

Address:

Email: Phone:

CT1_5-17

Inspect and Test Electrical Isolation

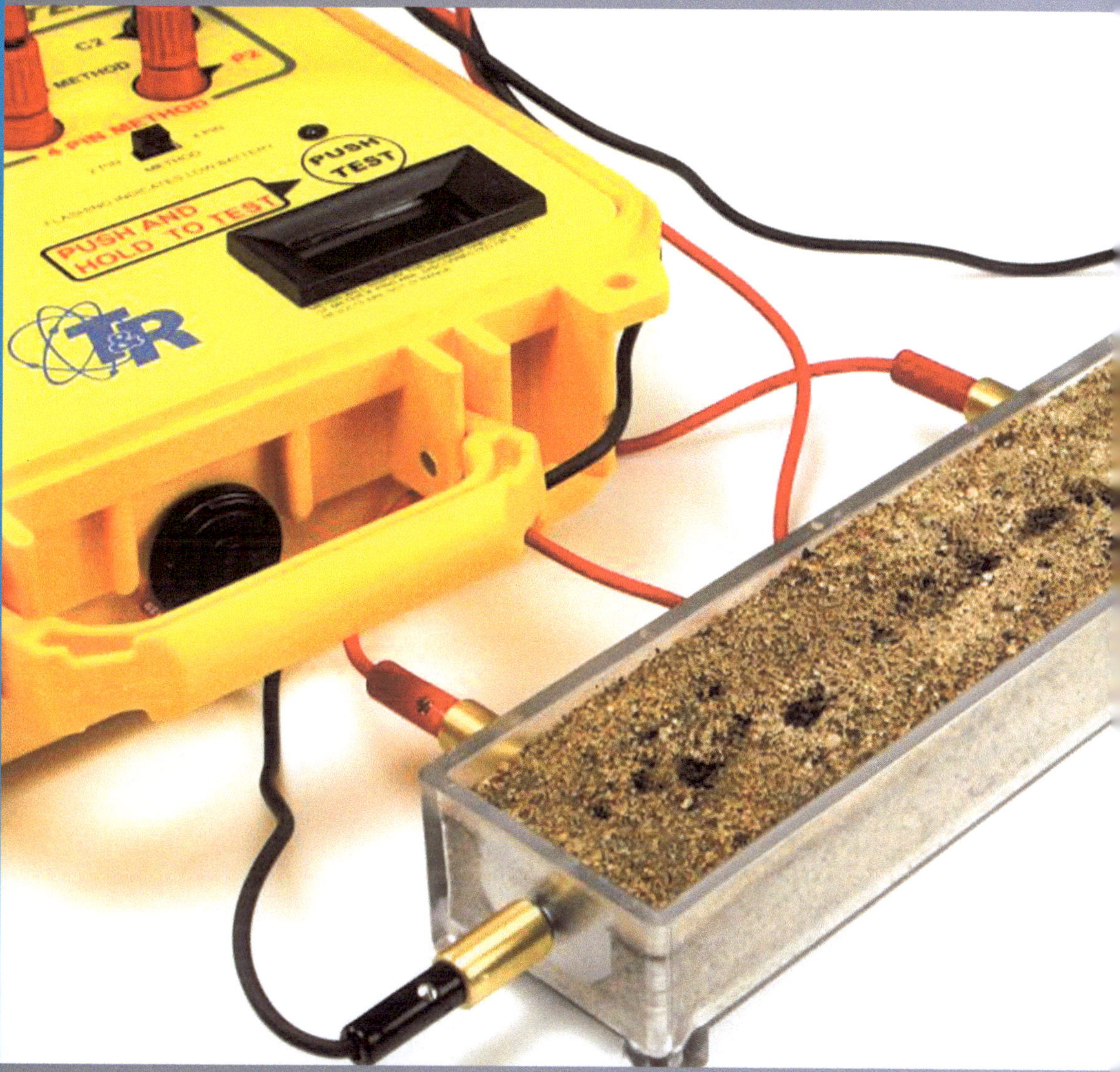

Overview

Insulated flanges are used on transmission lines to electrically isolate the pipeline from terminal facilities or pumping stations. They are also used to divide the pipeline system into sections and enhance the effectiveness of the cathodic protection system. This module identifies components used to isolate and ground pipeline systems as part of corrosion control, and explains the basic procedures for inspecting and testing electrical isolation.

Trainees with successful module completions may be eligible for credentialing through the NCCER Registry. To learn more, go to **www.nccer.org** or contact us at 1.888.622.3720. Our website has information on the latest product releases and training, as well as online versions of our *Cornerstone* magazine and Pearson's product catalog.

Your feedback is welcome. You may email your comments to **curriculum@nccer.org**, send general comments and inquiries to **info@nccer.org**, or fill in the User Update form at the back of this module.

This information is general in nature and intended for training purposes only. Actual performance of activities described in this manual requires compliance with all applicable operating, service, maintenance, and safety procedures under the direction of qualified personnel. References in this manual to patented or proprietary devices do not constitute a recommendation of their use.

CT1_5-17

Inspect and Test Electrical Isolation

Objectives

When you have completed this module, you will be able to do the following:

1. Identify and describe components used to electrically isolate and ground a pipeline system.
   a. Describe insulated flanges and explain how they can be tested and repaired.
   b. Identify devices used to ground and protect insulated flanges.
2. Describe how to inspect and test electrical isolation (CT1_5-17).
   a. Identify the prerequisites, competencies, and abnormal operating conditions associated with inspecting and testing electrical isolation.
   b. Describe the procedure for conducting an electrical isolation test.

Performance Task

Under the supervision of your instructor, you should be able to do the following:

1. Inspect and test electrical isolation (CT1_5-17).
   - Identify potential abnormal operating conditions that may occur during performance of this CT, and know the appropriate actions to take in response to them.
   - Utilize the appropriate personal protective equipment according to relevant company procedures.
   - Identify the isolation locations and isolation device where measurements will be taken.
   - Conduct a visual inspection of the isolation location; check for physical damage to the test station, a damaged shunt, loose connections, disconnected wires, arcing across a terminal, etc.
   - Select the instrumentation (voltmeter, isolation/flange tester, test leads, reference cell).
   - Make connections with the test equipment to take and record readings.
   - If using a reference cell, measure the potential for each of the structures. The reference cell should remain in the same location during the measurements.
   - Check for continuity on flanges using an isolation/flange tester.
   - Document the readings as required by operator's procedures.

Trade Terms

Abnormal operating conditions (AOCs)
Anode
Confined space
Electrical isolation
Foreign structure
Megohm
Polarity
Soil resistivity
Structure-to-soil potential
Test leads
Test stations

Industry Recognized Credentials

If you are training through an NCCER-accredited sponsor, you may be eligible for credentials from NCCER's Registry. The ID number for this module is CT1_5-17. Note that this module may have been used in other NCCER curricula and may apply to other level completions. Contact NCCER's Registry at 888.622.3720 or go to **www.nccer.org** for more information.

Contents

Figures

Section One

1.0.0 Electrical Isolation Components

Objective

Identify and describe components used to electrically isolate and ground a pipeline system.

a. Describe insulated flanges and how they can be tested and repaired.
b. Identify devices used to ground and protect insulated flanges.

Trade Terms

Anode: The electrode of an electrochemical cell at which oxidation occurs.

Foreign structure: A pipeline or other metal structure in the vicinity of a pipeline system under cathodic protection.

Megohm: A unit of resistance equal to 1,000,000 ohms.

Soil resistivity: The ability of the soil to resist current flow to or from a pipeline.

Structure-to-soil potential: The potential difference between the surface of a buried or submerged metallic structure (such as a pipe) and the electrolyte that is measured with reference to an electrode in contact with the electrolyte.

Underground steel piping systems located near a **foreign structure** such as high-voltage AC power lines can be affected by the electromagnetic field created by the alternating current. The induced AC current not only creates a potential safety hazard, it can cause excessive pipeline corrosion—especially where there are defects in the coating. To help improve cathodic protection and minimize corrosion caused by stray current, pipeline sections are joined using insulated flanges. If numerous sources of stray currents are present, these insulated flanges might be closely spaced. The location and spacing of insulated fittings is determined from experience with areas in which stray currents are likely, and on information collected during testing. This section describes insulated flanges and other devices used to isolate the pipeline and provide lightning and surge protection.

1.1.0 Insulated Flanges

Insulated flanges are commonly used on transmission lines to electrically isolate the pipeline from structures such as terminal facilities or pumping stations. There are many different types of flange-insulating kits available for pipelines using corrosion control equipment. The components in a typical kit include a flange gasket that isolates the two flange faces, sleeves that isolate the flange studs or bolts, and washers that isolate the nuts (*Figure 1*).

Bolt holes on standard flanges are 0.125 inch (3.175 millimeters) larger than the standard bolts. This allows full-length insulating sleeves to be used. As long as the sleeves are properly aligned, they will provide adequate insulation.

For new construction, insulated fittings may be assembled and tested before they are welded into the line. Prefabricated insulated flanges can be tested using an ohmmeter. The resistance values should be measured in the **megohm** range to indicate proper insulation. The installed flange has a substantial length of pipe on each side of it. This reduces the resistance of several million ohms measured before installation to a resistance of only a few ohms after installation. The longer the length of the pipe on each side of the flange, the lower the resistance. The resistance measured is also affected by the quality of the coating and the **soil resistivity** in the area.

Measurements of **structure-to-soil potential** on either side of an insulated fitting may be used to identify possible shorts. Use an insulation tester to determine whether the bolts or flanges are shorted. If a defective insulated flange is found, it may be possible to repair it without taking the pipeline out of service. First, identify the cause of the short by measuring the resistance between the flange halves and each bolt. If the short is from a defective insulated bolt, the reading will show a low or zero resistance. The insulation can then be replaced on the shorted bolts individually while the pipeline remains in service (this is only done at reduced pressures in accordance with company policies and procedures). If all possible bypassing routes have been isolated and all of the bolts have a high resistance, the cause is probably a short across the gasket. In this case, the pipeline must be taken out of service so that the flange can be disassembled and the gasket replaced.

In addition to insulation testers, other meters can be used to test the effectiveness of above-ground insulators. *Figure 2* shows a radio frequency tester being used to test a flange.

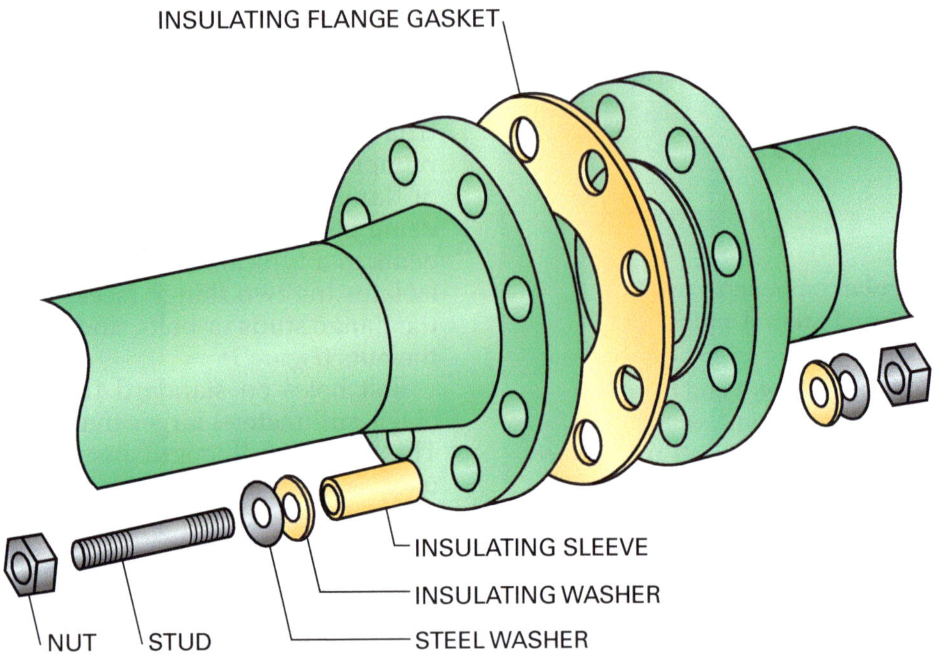

Figure 1 Insulated flange components.

Figure 2 Aboveground insulator tester.

1.2.0 Protective Devices for Insulated Flanges

Insulated flanges can be damaged by abnormal conditions such as lightning strikes and surges. Lightning can cause a very high voltage in the pipeline that can burn through the insulating material if it passes through the flange.

Grounding cells can be used to provide both lightning and surge protection. Grounding cells normally consist of either two or four zinc anodes. (An **anode** is the electrode of an electrochemical cell at which oxidation occurs.) The value of resistance between the anodes will depend on the resistivity of the backfill that is used. Properly installed grounding cells will have a resistance between 0.2 and 0.5 ohm. When a high voltage surge occurs, the grounding cell conducts the voltage across the anodes and lowers the voltage across the insulated flange.

If very high-voltage surges are anticipated, a four-anode ground bed can be used. These ground beds lower the voltage across the insulated flange even more. Also, the zinc anode attached to an unprotected pipeline provides some cathodic protection to the pipe in the flange area.

Other devices that provide grounding and protection for insulated flanges are isolation surge protectors (ISPs), polarization cell replacements (PCRs), solid-state decouplers (SSDs), and polarization cells. ISPs are electronic and block DC current flow while allowing AC current flow to protect the insulated flange. These devices can automatically reset after a fault condition. ISPs provide better protection for the system than grounding cells.

Lightning Strikes

There are millions of lightning strikes each year, and a cloud-to-ground strike can transmit hundreds of millions of volts. Lightning strikes that occur miles away from a pipeline can still induce voltages that are high enough to seriously injure or kill personnel and significantly damage piping and cathodic protection equipment. While devices such as anode beds, surge protectors, and decouplers help protect pipelines, damage from lightning strikes is an ongoing challenge for pipeline companies. With research and experimentation about the fundamental behavior of lightning, more effective ways to control lightning damage to buried pipelines can be developed.

Additional Resources

It is assumed that you will learn to perform these procedures under the direct supervision of your crew leader, foreman, or instructor. For independent study, refer to the recommended Additional Resources at the back of this module.

1.0.0 Section Review

1. To determine the cause of a short on an insulated flange, an insulation tester can be used to check the resistance between _____.
   a. the two pipe sections connected by the flange
   b. any two bolts on the flange
   c. either pipe section and the flange gasket
   d. the flange halves and each bolt

2. Properly installed grounding cells have a resistance between _____.
   a. 0.2 and 0.5 ohm
   b. 2 and 5 ohms
   c. 20 and 50 ohms
   d. 200 and 500 ohms

SECTION TWO

2.0.0 INSPECT AND TEST ELECTRICAL ISOLATION

Objective

Describe how to inspect and test electrical isolation (CT1_5-17).

a. Identify the prerequisites, competencies, and abnormal operating conditions associated with inspecting and testing electrical isolation.
b. Describe the procedure for conducting an electrical isolation test.

Performance Task

Inspect and test electrical isolation (CT1_5-17).

- Identify potential abnormal operating conditions that may occur during performance of this CT, and know the appropriate actions to take in response to them.
- Utilize the appropriate personal protective equipment according to relevant company procedures.
- Identify the isolation locations and isolation device where measurements will be taken.
- Conduct a visual inspection of the isolation location; check for physical damage to the test station, a damaged shunt, loose connections, disconnected wires, arcing across a terminal, etc.
- Select the instrumentation (voltmeter, isolation/flange tester, test leads, reference cell).
- Make connections with the test equipment to take and record readings.
- If using a reference cell, measure the potential for each of the structures. The reference cell should remain in the same location during the measurements.
- Check for continuity on flanges using an isolation/flange tester.
- Document the readings as required by operator's procedures.

Trade Terms

Abnormal operating conditions (AOCs): Conditions within a pipeline system, identified by the operator, that may indicate (1) a malfunction of a component or (2) a deviation from normal operations that may indicate a condition exceeding design limits or result in a hazard to persons, property, or the environment.

Confined space: A space that is large enough and so configured that an employee can bodily enter and perform assigned work; has limited or restricted means for entry or exit (such as tanks, vessels, vaults, etc.); and is not designed for continuous employee occupancy.

Electrical isolation: The electrical separation of two structures.

Polarity: The distinction between the positive and negative potential differences at the terminals of a meter or electrical device.

Test leads: Wires that are connected to a structure being tested that provide easy connection points for a meter.

Test stations: A specific area or location, sometimes called a test point, where a connection to a structure is made so that a structure-to-soil potential can be measured.

Personnel inspecting and testing electrical isolation are required to have some basic competencies. They must have the ability to carefully follow the required procedures for conducting electrical isolation testing and recording the results. This section describes some of the basic skills needed by pipeline technicians and identifies abnormal operating conditions (AOCs) that personnel may need to respond to when inspecting and testing electrical isolation. It also explains the basic procedure used for conducting an electrical isolation test.

2.1.0 Prerequisites, Competencies, and Abnormal Operating Conditions

Personnel must possess a variety of skills in order to safely and successfully conduct electrical isolation testing. Testing procedures require numerous steps, which a pipeline technician must be competent to perform. Personnel must be aware of abnormal operating conditions (AOCs) that may occur during testing procedures and know how to properly respond to them.

2.1.1 Prerequisites and Competencies

Personnel assigned to conduct electrical isolation testing must have a good understanding of hazards that can exist before and during this covered task, as well as what the task entails. Before testing, personnel must be familiar with hazards along the right-of-way (ROW), including those associated with right-of-way encroachments and unsupervised third party construction. They must be familiar with the locations of **test stations**, and they must be able to properly choose, inspect, calibrate, and operate any tools or instruments to be used during the procedure.

Pipeline work often involves trenched and excavated areas, as well as tank containment areas. For these reasons, personnel must be properly trained for **confined space** entry and excavation safety.

Personnel must be able to visibly inspect the pipe and determine if there are any unsafe integrity issues, such as evidence of a release or significant mechanical damage (gouges, scrapes, dents), and discontinue the work and make notifications if any are found. If old coating materials that may contain asbestos are encountered, personnel must make the proper notifications so that a qualified person can sample, test, and, if necessary, remove and dispose of any asbestos-containing material. For these and other safety reasons, personnel must be familiar with local emergency telephone numbers.

This task begins with identification of the isolation device and ends when measurements have been taken and recorded. Once testing has begun, personnel must know how to position, manipulate, and read any equipment that might be used. If personnel encounter induced AC voltage, which could occur if a pipeline is located near a high voltage AC electrical field, they must be knowledgeable about the effects and hazards associated with the voltage and know how to notify the appropriate company personnel. It is also critical for personnel to understand that no job is complete until all tasks are documented and the proper notifications are made.

Personnel must have a basic understanding of additional tasks related to electrical isolation testing, such as measuring structure-to-soil potentials (CT1_1-17), conducting a close interval survey (CT1_2-17), testing to detect interference (CT1_3-17), and inspecting and performing electrical testing of bonds (CT1_4-17).

2.1.2 Abnormal Operating Conditions

Personnel inspecting and testing electrical isolation must be able to identify potential abnormal operating conditions (AOCs) and know the proper response to each, should they occur. The following are some potential AOCs, and the appropriate action personnel should take in response to them:

- *Damage to or unsupervised third party construction on a right-of-way is discovered* – Notify appropriate company personnel and, if the contractor continues to perform unsupervised work on site, discuss the dangers posed by the situation with the contractor.
- *Internal or external corrosion of the pipe is present and to the extent that structure-to-soil potential measurements would be impractical* – Document the condition according to company guidelines and notify personnel responsible for correcting the problem.
- *Erratic or floating readings occur* – Determine the cause of unexpected readings such as reverse **polarity**, broken bond, or change in cathodic system. Repair the **test leads** or equipment, if needed.
- *Induced AC voltage is encountered* – Exercise caution around any equipment or structures and notify the appropriate company personnel.
- *Fire or explosion is caused by the ignition of a hydrocarbon liquid or gas* – Evacuate the area and notify the appropriate firefighting personnel. It may be necessary to call 911 in such an emergency.
- *Unexpected leaks of hazardous liquid or carbon dioxide because of a loss of pipe wall thickness is discovered* – Take steps to eliminate any possible ignition sources and notify the appropriate personnel, possibly including 911.

2.2.0 Procedure

When testing for electrical isolation, always follow company procedures and the manufacturer's instructions for the meter in use. The general steps to perform when testing for electrical isolation are as follows:

WARNING!

Always obtain and use the appropriate personal protective equipment for each task. Stay alert to the possibility of unexpected leaks from the exposed pipeline such as hazardous liquid or carbon dioxide (CO_2). In the event of a leak, safely eliminate all ignition sources and immediately notify the appropriate personnel.

Step 1 Identify potential abnormal operating conditions (AOCs) that may occur, and know the appropriate actions to take in response to them.

Step 2 Utilize the appropriate personal protective equipment according to relevant company procedures.

Step 3 Identify the isolation locations and device where measurements will be taken.

Step 4 Conduct a visual inspection of the isolation location; check for physical damage, a damaged shunt, loose connections, disconnected wires, arcing across terminals, etc.

Step 5 Select the instrumentation to be used (voltmeter, isolation tester, and test leads or reference cell). Inspect the equipment for damage and proper operation prior to taking any measurements.

Step 6 Make connections with the test equipment to take and record readings.

Step 7 If using a reference cell, measure the potentials for each of the structures. The reference cell should remain in the same location during the measurements. This allows for a comparison of structure-to-soil potentials in order to determine if the structures are isolated. If the difference in potential is 100mV or more, the isolation is effective. If it is less then 100mV, further testing may be necessary.

Step 8 Check for continuity on flanges using an isolation tester.

Step 9 Document all readings per company procedures.

NOTE

Proper documentation is essential for future analysis and identification of problem areas.

Additional Resources

It is assumed that you will learn to perform these procedures under the direct supervision of your crew leader, foreman, or instructor. For independent study, refer to the recommended Additional Resources at the back of this module.

2.0.0 Section Review

1. If extensive internal or external piping corrosion is found prior to conducting an electrical isolation test, a pipeline technician should _____.
   a. apply a cathodic patch over the damage and continue with the measurements
   b. make sure that the instruments are attached directly to the corroded area
   c. document the condition and notify personnel responsible for correcting the problem
   d. attach an approved anode bank to the structure and adjust the test instrument

2. If the difference in the structure-to-soil potential for the structures being tested is 100mV or higher, it means _____.
   a. the meter is probably faulty
   b. the isolation is effective
   c. a lightning strike has occurred
   d. further testing is required

Summary

Underground steel piping systems that are located near foreign structures such as high-voltage AC power lines can be affected by the electromagnetic field created by the alternating current. The induced AC current is not only a potential safety hazard; it can also cause excessive pipeline corrosion, especially where there are coating defects. To help improve cathodic protection and minimize corrosion caused by stray current, pipeline sections are joined using insulated flanges.

Insulated flanges used on transmission lines electrically isolate the pipeline from terminal facilities or pumping stations and divide the pipeline system into sections. Technicians must understand how flanges and other components are used to isolate and protect pipelines. This module identified components used to isolate and ground pipeline systems as part of corrosion control, and it explained the basic procedure for inspecting and testing electrical isolation.

Review Questions

1. The sleeves in a typical flange-isolating kit are used to _____.
   a. isolate the flange studs or bolts
   b. insulate between the flange faces
   c. eliminate friction at the nuts
   d. regulate the potential across the gasket

2. If an insulated flange is tested and all of the flange bolts show a high resistance, the most likely cause is a _____.
   a. defective flange stud or bolt
   b. voltage surge in the system
   c. short across the insulating flange gasket
   d. failed grounding cell or decoupler nearby

3. Which of the following devices is used to provide both lightning and surge protection for pipeline fittings?
   a. Ground fault circuit interrupters
   b. Barcol barriers
   c. Wenner pins
   d. Grounding cells

4. One of the abnormal operating conditions (AOCs) that a pipeline technician might encounter while testing electrical isolation is _____.
   a. having to make test equipment connections
   b. finding and inspecting isolation locations
   c. induced AC current
   d. checking for continuity on piping flanges

5. The final step in the general procedure for conducting an electrical isolation test is to _____.
   a. inspect every piece of tested equipment for damage and proper operation
   b. accurately record all required information in accordance with company policy
   c. remove lockout/tagout barriers and restart all holiday detection equipment
   d. conduct a Barcol frequency test to certify the readings to NACE standards

Trade Terms Quiz

Fill in the blank with the correct term that you learned from your study of this module.

1. The potential difference between the surface of a buried or submerged metallic structure (such as a pipe) and the electrolyte that is measured with reference to an electrode in contact with the electrolyte is known as the ________.
2. The distinction between the positive and negative potential differences at the terminals of a meter or electrical device is the ________.
3. Conditions within a pipeline system that may indicate (1) a malfunction of a component or (2) a condition exceeding design limits or resulting in a hazard to persons, property, or the environment are called ________.
4. A unit of resistance equal to 1,000,000 ohms is called a(n) ________.
5. Specific areas or locations, sometimes called test points, where a connection to a structure is made so that a structure-to-soil potential can be measured are called ________.
6. A pipeline or other metal structure in the vicinity of a pipeline system under cathodic protection is a(n) ________.
7. The ability of the soil to resist current flow to or from a pipeline is the ________.
8. Wires that are connected to a structure being tested that provide easy connection points for a meter are called ________.
9. The electrical separation of two structures is called ________.
10. The electrode of an electrochemical cell at which oxidation occurs is called a(n) ________.
11. A space that is large enough and so configured that an employee can bodily enter and perform assigned work; has limited or restricted means for entry or exit (such as tanks, vessels, vaults, etc.); and is not designed for continuous employee occupancy is called a(n) ________.

Trade Terms

Abnormal operating conditions (AOCs)
Anode
Confined space
Electrical isolation
Foreign structure
Megohm
Polarity
Soil resistivity
Structure-to-soil potential
Test leads
Test stations

Trade Terms Introduced in This Module

Abnormal operating conditions (AOCs): Conditions within a pipeline system, identified by the operator, that may indicate (1) a malfunction of a component or (2) a deviation from normal operations that may indicate a condition exceeding design limits or result in a hazard to persons, property, or the environment.

Anode: The electrode of an electrochemical cell at which oxidation occurs.

Confined space: A space that is large enough and so configured that an employee can bodily enter and perform assigned work; has limited or restricted means for entry or exit (such as tanks, vessels, vaults, etc.); and is not designed for continuous employee occupancy.

Electrical isolation: The electrical separation of two structures.

Foreign structure: A pipeline or other metal structure in the vicinity of a pipeline system under cathodic protection.

Megohm: A unit of resistance equal to 1,000,000 ohms.

Polarity: The distinction between the positive and negative potential differences at the terminals of a meter or electrical device.

Soil resistivity: The ability of the soil to resist current flow to or from a pipeline.

Structure-to-soil potential: The potential difference between the surface of a buried or submerged metallic structure (such as a pipe) and the electrolyte that is measured with reference to an electrode in contact with the electrolyte.

Test leads: Wires that are connected to a structure being tested that provide easy connection points for a meter.

Test stations: A specific area or location, sometimes called a test point, where a connection to a structure is made so that a structure-to-soil potential can be measured.

Additional Resources

The following recommended resources can provide additional helpful information related to the requirements for this covered task:

API Recommended Practice 1161, Third Edition, © 2014 American Petroleum Institute. API Publishing Services, 1220 L Street, NW, Washington, DC 20005. Distributed by Thompson Reuters (Scientific) LLC, **www.techstreet.com**.

49 *CFR* 192, Subpart I, **www.ecfr.gov**

49 *CFR* 195, Subpart H, **www.ecfr.gov**

NCCER Module AOCCC-17, *Abnormal Operating Conditions Control Center*

NCCER Module AOCFG-17, *Abnormal Operating Conditions Field and Gas*

NCCER Module 62105-14, *Introduction to Pipeline Documents*

NCCER Module 66101-14, *Introduction to the Pipeline Industry*

NCCER Module CT1_1-17, *Measure Structure-to-Soil Potentials*

NCCER Module CT1_2-17, *Conduct Close Interval Survey*

NCCER Module CT1_3-17, *Test and Detect Interference*

NCCER Module CT1_4-17, *Inspect and Perform Electrical Test of Bonds*

NCCER Module CT14_1-17, *Locate Line*

API Recommended Practice 651, Cathodic Protection of Aboveground Petroleum Storage Tanks. Fourth Edition, 2014. Available from the American Petroleum Institute at **www.api.org**.

NACE International *SP0169-2013, Control of External Corrosion on Underground or Submerged Metallic Piping Systems*. 2013. Available from NACE International at **www.nace.org**.

Standard Test Method for Field Measurement of Soil Resistivity Using the Wenner Four-Electrode Method. 2012. Available from ASTM International at **www.astm.org**.

Figure Credits

Tinker & Rasor, Module Opener, Figure 2

Section Review Answer Key

Answer	Section Reference	Objective
Section One		
1. d	1.1.0	1a
2. a	1.2.0	1b
Section Two		
1. c	2.1.2	2a
2. b	2.2.0	2b

NCCER CURRICULA — USER UPDATE

NCCER makes every effort to keep its textbooks up-to-date and free of technical errors. We appreciate your help in this process. If you find an error, a typographical mistake, or an inaccuracy in NCCER's curricula, please fill out this form (or a photocopy), or complete the online form at **www.nccer.org/olf**. Be sure to include the exact module ID number, page number, a detailed description, and your recommended correction. Your input will be brought to the attention of the Authoring Team. Thank you for your assistance.

Instructors – If you have an idea for improving this textbook, or have found that additional materials were necessary to teach this module effectively, please let us know so that we may present your suggestions to the Authoring Team.

NCCER Product Development and Revision

13614 Progress Blvd., Alachua, FL 32615

Email: curriculum@nccer.org

Online: www.nccer.org/olf

❑ Trainee Guide ❑ Lesson Plans ❑ Exam ❑ PowerPoints Other ______________

Craft / Level: Copyright Date:

Module ID Number / Title:

Section Number(s):

Description:

Recommended Correction:

Your Name:

Address:

Email: Phone:

CT8_1-17

Measure Pit Depth with Pit Gauge

Overview

Pipelines are subject to damage from both general corrosion and localized corrosion in the form of pitting. A single pit in an otherwise healthy system can cause the entire system to fail. It is essential to monitor pipelines for corrosion and perform the necessary testing to prevent failure of the system. One way corrosion testing can be done is by using pit gauges. This module describes different types of pit gauges and explains how to prepare and use these instruments to measure pit depth.

Trainees with successful module completions may be eligible for credentialing through the NCCER Registry. To learn more, go to **www.nccer.org** or contact us at 1.888.622.3720. Our website has information on the latest product releases and training, as well as online versions of our *Cornerstone* magazine and Pearson's product catalog.

Your feedback is welcome. You may email your comments to **curriculum@nccer.org**, send general comments and inquiries to **info@nccer.org**, or fill in the User Update form at the back of this module.

This information is general in nature and intended for training purposes only. Actual performance of activities described in this manual requires compliance with all applicable operating, service, maintenance, and safety procedures under the direction of qualified personnel. References in this manual to patented or proprietary devices do not constitute a recommendation of their use.

CT8_1-17
Measure Pit Depth with Pit Gauge

Objectives

When you have completed this module, you will be able to do the following:

1. Describe types of pit gauges and how to use them, and explain how to gather additional piping wall data after measurements have been taken.
   a. Identify common types of pit gauges and explain how to use them.
   b. Describe how to gather remaining wall strength data.
2. Describe how to measure pit depth with a pit gauge (CT8_1-17).
   a. Identify the prerequisites, competencies, and abnormal operating conditions associated with preparing and using pit gauges.
   b. Describe how to operate a pit gauge.
   c. Describe the procedures used to measure pit depth.

Performance Task

Under the supervision of your instructor, you should be able to do the following:

1. Measure pit depth with a pit gauge (CT8_1-17).
   - Identify potential abnormal operating conditions that may occur during performance of this CT, and know the appropriate actions to take in response to them.
   - Utilize the appropriate personal protective equipment according to relevant company procedures.
   - Perform a visual inspection of the exposed pipe. Look for integrity issues such as evidence of a release or significant metal deformation. If unsafe integrity issues are encountered, discontinue the task and make immediate notifications.
   - Clean and remove debris from the area to be measured.
   - Verify that the gauge is working properly. Calibrate the gauge if necessary.
   - Position the gauge flush and longitudinally across the area to be measured. Hold it firmly against the surface, making sure it is supported on non-corroded surfaces.
   - Move the depth indicator until it contacts the deepest part of the wall loss.
   - Read and record the depth and longitudinal length measurements. Repeat several measurements to verify the deepest area of wall loss.
   - Complete appropriate documentation and make notifications as required by operator's procedures.

Trade Terms

Abnormal operating conditions (AOCs)
Calibrate
Dent
General corrosion
Gouge
Interaction
Localized corrosion
Mil
Pitting
River bottom profile
Wall loss

Industry Recognized Credentials

If you are training through an NCCER-accredited sponsor, you may be eligible for credentials from NCCER's Registry. The ID number for this module is CT8_1-17. Note that this module may have been used in other NCCER curricula and may apply to other level completions. Contact NCCER's Registry at 888.622.3720 or go to **www.nccer.org** for more information.

Contents

Figures

SECTION ONE

1.0.0 PIT GAUGES

Objective

Describe types of pit gauges and how to use them, and explain how to gather additional piping wall data after measurements have been taken.

a. Identify common types of pit gauges and explain how to use them.
b. Describe how to gather remaining wall strength data.

Trade Terms

General corrosion: Corrosion that takes place over the entire surface, resulting in thinning of the material.

Interaction: When two or more areas of localized corrosion are in close proximity and combine to reduce the overall wall thickness.

Localized corrosion: Corrosion in isolated areas.

Mil: Unit of measure equal to 0.001 of an inch.

Pitting: A type of localized corrosion characterized by the formation of small cavities or pits.

River bottom profile: A collection of pit depths depicting peaks and valleys, or the actual profile of the defect.

Wall loss: The loss of pipe metal either as a result of mechanical surface damage or corrosion.

It is essential to monitor pipelines for corrosion to prevent failure of the system. Pipeline pitting is a type of localized corrosion that is often difficult to detect because it can be covered by coatings. Sometimes only a small hole may show on the surface of the piping or equipment. Even when general corrosion rates are low, microbes can accelerate localized pitting and cause it to go through a steel pipeline wall fairly quickly.

Pits can undercut the surface, resulting in wall loss so that only a fragile layer remains on the surface, or they may cut deep fissures in the piping with only tiny cracks visible at the surface. Interaction between areas of localized corrosion can combine to result in a serious loss of wall thickness. The extent of a pitted area is commonly measured using some type of pit gauge.

1.1.0 Types of Pit Gauges

When pipeline corrosion is discovered, the structural integrity or (remaining wall thickness) of the piping must be determined in order to set the priority for repair or replacement of the corroded section of pipe. One part of this process involves measuring the depth of the pits formed by the corrosion. Common tools used for measuring corrosion pitting include mechanical pit gauges, dial-type pit gauges, and micrometer pit gauges. These gauges provide an accurate measurement of pit depth and other anomalies.

1.1.1 Mechanical Pit Gauges

The common mechanical pit gauge (*Figure 1*) is a small, sturdy tool used to determine the extent of corrosion damage. These gauges are usually made of stainless steel and are easily carried in a tool kit. The face of the gauge often includes formulas, tables, and data giving the outside diameter (OD), wall thickness, and weight per foot for various pipe sizes.

The depth measurement scale on a mechanical pit gauge is marked in sixteenths of an inch and in mils (a mil is 0.001 of an inch). Most mechanical pit gauges include formulas and common pipe dimensions such as outside diameter, wall thickness, and weight. Care must be taken to use these gauges properly and to recognize and understand their limitations.

1.1.2 Dial-Type Pit Gauges

When pit depth measurements are critical and accuracy is a priority, dial-type pit gauges may be used. Most dial pit gauges are capable of measuring depth to extremely close tolerances. However, these instruments are more fragile than mechanical gauges and should be handled with care to ensure accuracy. *Figure 2* shows a dial pit gauge positioned on a pipe.

1.1.3 Micrometer Pit Gauges

A micrometer pit gauge provides the most accurate measurement of corrosion pit depth. Some models of manually operated micrometer pit depth indicators measure to an accuracy of +/– 5 microns (0.0002"), and have a digital display so that the precise measurements are easy to read. These measuring tools are somewhat fragile and quite expensive as compared to other indicators. Micrometer pit gauges must be handled with care to ensure accuracy and to safeguard the company's investment.

Figure 1 Mechanical pit gauge.

Figure 2 Dial pit gauge positioned on pipe.

1.2.0 Gather Remaining Wall Strength Data

When pit gauges are used to measure pit depths, it is also necessary to calculate the structural integrity of the pipeline. This process involves gathering remaining wall strength data for the pipe in order to set the priority for repair or replacement of the pipe. Pipeline operators are required by 49 *CFR*192, *Transportation of Natural and Other Gas by Pipeline* and 49 *CFR*195, *Transportation of Hazardous Liquids by Pipeline* to evaluate the remaining strength in any corroded pipeline, using either *ASME/ANSI B31G* or RSTRENG (remaining strength) corrosion assessment methods. The basis for these two calculation approaches has been in existence for more than 25 years. Both methodologies have been validated through burst testing of pipe with corrosion and failures of pipe in service.

The B31G method calculates predicted failure points based on the maximum length and depth of the corrosion defect on the pipe. While acceptable, the B31G method is very conservative, and most strength tests are now done using the RSTRENG method.

RSTRENG is available as software that combines the B31G method with a more detailed profile of the defect by evaluating a greater number of depth measurements (minimum of three with one being the maximum depth in the defect) taken along the horizontal length of the defect. RSTRENG works equally well with single defects or with several interacting corrosion arrays. *Figure 3* shows a typical **river bottom profile** of a corrosion defect as determined through the RSTRENG method.

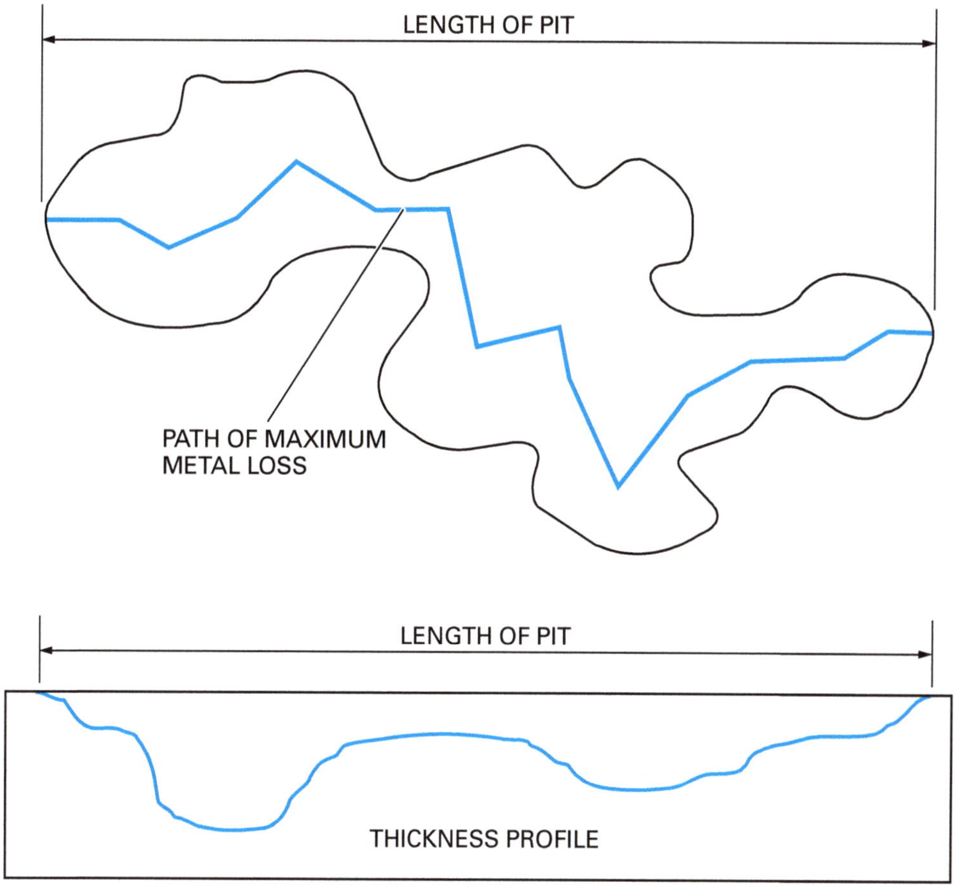

Figure 3 Typical river bottom profile.

Additional Resources

It is assumed that you will learn to perform these procedures under the direct supervision of your crew leader, foreman, or instructor. For independent study, refer to the recommended Additional Resources at the back of this module.

1.0.0 Section Review

1. The type of pit gauge that provides the most accurate measurement of corrosion pit depth is the _____.
   a. dial type
   b. mechanical type
   c. magnetometer type
   d. micrometer type

2. 49 *CFR*195 covers the pipeline transportation of hazardous _____.
   a. gases
   b. solids
   c. liquids
   d. plasma

SECTION TWO

2.0.0 MEASURE PIT DEPTH WITH A PIT GAUGE

Objective

Describe how to measure pit depth with a pit gauge (CT8_1-17).

a. Identify the prerequisites, competencies, and abnormal operating conditions associated with preparing and using pit gauges.
b. Describe how to operate a pit gauge.
c. Describe the procedures used to measure pit depth.

Performance Task

Measure pit depth with a pit gauge (CT8_1-17).

- Identify potential abnormal operating conditions that may occur during performance of this CT, and know the appropriate actions to take in response to them.
- Utilize the appropriate personal protective equipment according to relevant company procedures.
- Perform a visual inspection of the exposed pipe. Look for integrity issues such as evidence of a release or significant metal deformation. If unsafe integrity issues are encountered, discontinue the task and make immediate notifications.
- Clean and remove debris from the area to be measured.
- Verify that the gauge is working properly. Calibrate the gauge if necessary.
- Position the gauge flush and longitudinally across the area to be measured. Hold it firmly against the surface, making sure it is supported on non-corroded surfaces.
- Move the depth indicator until it contacts the deepest part of the wall loss.
- Read and record the depth and longitudinal length measurements. Repeat several measurements to verify the deepest area of wall loss.
- Complete appropriate documentation and make notifications as required by operator's procedures.

Trade Terms

Abnormal operating conditions (AOCs): Conditions within a pipeline system, identified by the operator, that may indicate (1) a malfunction of a component or (2) a deviation from normal operations that may indicate a condition exceeding design limits or result in a hazard to persons, property, or the environment.

Calibrate: A process used to ensure accuracy by comparing an instrument reading against a known standard.

Dent: A depression in the surface of a pipe that affects the curvature of the pipe without reducing the pipe wall thickness.

Gouge: Damage to the surface of a pipe that displaces or removes metal and reduces the pipe wall thickness.

Personnel assigned to take pit depth measurements are required to have some basic competencies. They must be able to follow the required procedures for measuring and recording pit depth. This section describes some of the basic skills needed by pipeline technicians and identifies abnormal operating conditions (AOCs) that personnel may need to respond to during pit depth measurements. It also explains the general procedures used to measure pit depth with a pit gauge.

2.1.0 Prerequisites, Competencies, and Abnormal Operating Conditions

Personnel must possess a variety of skills in order to measure pit depth with a pit gauge. The operation requires a series of steps, which a pipeline technician must be competent to perform. Personnel must be aware of the particular abnormal operating conditions that may occur during the procedure and know the proper response to them.

2.1.1 Prerequisites and Competencies

Personnel assigned to determine the extent of pipeline wall loss by measuring the depth of pits must understand what the task entails. Prior to beginning the task, personnel must be able to properly choose, inspect, calibrate, and operate the mechanical, dial-type, micrometer, or equivalent pit gauging instrument to be used during the procedure.

Personnel must understand that the task begins when the pipe is first exposed, cleaned, and prepared for inspection. They must be able to visibly inspect the pipe and determine if there are any unsafe integrity issues, such as evidence of a release or significant metal deformation, and discontinue the inspection and make notifications if any are found. Once the inspection begins, personnel must know how to position, manipulate, and read the instrument in order to obtain valid measurements of the pit depth. Personnel must understand that the task is not complete until all measurements are documented and the proper notifications are made.

Personnel must have a basic understanding of additional tasks related to measuring pit depth, such as measuring the remaining wall thickness (CT8_2-17) and measuring the area of the corrosion (CT8_3-17). They must also have an understanding of how the *ASME/ANSI B31G* and/or RSTRENG corrosion assessment methods are applied to determine the remaining wall strength for the pipe once pit depth and other data have been gathered.

2.1.2 Abnormal Operating Conditions

Personnel using pit gauges to measure pit depth must be able to identify potential abnormal operating conditions and know the proper response to each, should they occur. The following are some potential AOCs, and the appropriate action personnel should take in response to them:

- *Internal or external corrosion of the pipe is present and to the extent that pit depth measurements would be impractical* – Document the condition according to company policy and notify personnel responsible for correcting the problem.
- *A pipeline has been significantly damaged by line hits or mechanical surface damage* – Document the condition as required and notify personnel responsible for correcting the damage.
- *Fire or explosion caused by the ignition of a hydrocarbon liquid or gas* – Evacuate the area and notify the appropriate firefighting personnel.
- *Unexpected release of hazardous liquid or carbon dioxide due to insufficient pipe wall thickness* – Take steps to eliminate ignition sources and notify the appropriate personnel, possibly including 911.

2.2.0 Using a Pit Gauge

Before using a pit gauge to take pit depth measurements, it is important to thoroughly inspect the gauge for damage or wear and calibrate it (or ensure that it has been calibrated) in accordance with the company's tool calibration procedure. When using a micrometer pit gauge, the gauge must first be zeroed. This will calibrate the gauge. Closely examine the point of the instrument for wear. If a pit gauge has a worn or damaged point, it must be replaced. Only use an instrument that is in excellent condition; otherwise, the measurements taken during the inspection may not be accurate.

Before using the pit gauge, clean the pit in accordance with company policy. Inaccurate readings can result unless the gauge is aligned properly and placed on an even surface. Where severely pitted or partially coated pipe and associated structures are involved, handheld pit gauges provide only an approximation of pit depth.

To use a standard pit gauge, align the gauge lengthwise with the pipe to be inspected. The gauge must rest squarely on the pipe surface to provide an accurate reading of the pits. Move the arm on the gauge so that the pointer is at the bottom of the pit. Read and record the corresponding number on the scale of the gauge. Repeat this process in additional pits to achieve the required level of sampling. *Figure 4* shows the proper technique for measuring pit depth using a mechanical pit gauge.

When measuring pit length, place the pointer of the pit gauge in the pit. Observe the length of the pit on the bottom scale of the mechanical pit gauge. Record all measurements on the appropriate company form.

Most micrometer pit gauges are capable of measuring to extremely close tolerances. To make a measurement, place the gauge squarely on the pipe surface. Push the pointer into the pit to be

Figure 4 Measuring pit depth.

measured using the control button on the side of the gauge. This should register a reading on the gauge. Record this measurement. Repeat this process to achieve the required level of sampling.

2.2.0 Procedures

This section provides general procedures for the use of a mechanical pit gauge to measure pit depth. Always follow the manufacturer's instructions for the instrument in use.

> **WARNING!**
>
> Always obtain and use the appropriate personal protective equipment (PPE) for each task. Stay alert to the possibility of unexpected leaks from the exposed pipeline such as hazardous liquid or carbon dioxide (CO_2). Safely eliminate all ignition sources and immediately notify the appropriate personnel.

Step 1 Ensure that the areas of the pipeline system to be inspected have been carefully excavated to avoid damaging the pipe. Thoroughly inspect the piping for any evidence of leaks or loss of pipe integrity. Look for mechanical surface damage (such as a dent or gouge). Stop if any hazards are present and immediately notify the appropriate personnel.

Step 2 Clean the area to be measured. The presence of dirt and debris will prevent an accurate reading.

Step 3 Inspect the pit gauge for wear and ensure that it has been calibrated.

Step 4 Align the gauge lengthwise with the pipe to be inspected. To provide an accurate reading, the gauge must be positioned flush with the pipe surface and resting on a non-corroded surface. If necessary, a bridging bar can be used for a platform reference where the corroded area is too large to provide a stable surface for pit measurement.

Step 5 Position the arm on the gauge so that the pointer is at the deepest part of the pit.

Step 6 Read and record the corresponding number from the scale of the gauge. With the pointer of the pit gauge still in the pit, record the length of the pit on the bottom scale of the mechanical pit gauge.

> **NOTE**
>
> The operation of a dial-type pit gauge follows a similar procedure. Once the gauge has been calibrated according to the manufacturer's instructions, it should be placed squarely on the pipe surface. Push the pointer into the pit so that the control button is on the side of the gauge. This should register a reading on the gauge.

Step 7 Repeat this process until an adequate sample of readings has been recorded to establish a pit profile and verify the deepest pit measurement. This also helps to obtain the average pit depth.

> **CAUTION**
>
> If pit depth is equal to 80 percent or more of the pipe company specification, the affected area of the pipe must be replaced immediately.

Step 8 Once the readings have been completed, document the findings on the appropriate company inspection form and make notifications according to company policy. When measuring multiple defects, always follow company guidelines.

Additional Resources

It is assumed that you will learn to perform these procedures under the direct supervision of your crew leader, foreman, or instructor. For independent study, refer to the recommended Additional Resources at the back of this module.

2.0.0 Section Review

1. If an unexpected leak of hazardous liquid or carbon dioxide is encountered during a pit depth measurement, the technician should _____.
   a. document the problem and attempt to repair the leak
   b. eliminate ignition sources and notify the appropriate personnel
   c. put on HAZMAT gear and continue with the procedure
   d. evacuate the area and call 911

2. If the corroded area on a pipe is too large to provide a stable surface for a pit gauge, _____.
   a. a bridging bar can be used for a platform reference
   b. the entire measuring procedure must be stopped
   c. flexible filler material can be added under the gauge
   d. only perimeter depth measurements should be taken

SUMMARY

Pitting from general and/or localized corrosion can cause an entire pipeline system to fail. To prevent such failures, pipelines are routinely inspected and tested. A common test performed on a corroded pipeline is measuring the pit depth using a pit gauge.

Personnel must understand how to prepare pit gauges, accurately take and record pit depth measurements, and follow approved techniques to determine the structural integrity of the pipe. This enables pipeline technicians to determine the extent of the damage and the urgency of pipe repair or replacement.

This module provided an overview of pit gauges and how they are used, identified factors associated with measuring pit depth, and explained the general procedure for using a pit gauge to measure pit depth.

Review Questions

1. A common tool used to check the extent of corrosion defects is a(n) _____.
   a. RSTRENG calculator
   b. optical parallax gauge
   c. magnetic nanometer
   d. mechanical pit gauge

2. Dial-type pit gauges are designed for _____.
   a. use where common pit gauges will not fit
   b. more accurate measurement of pit depth
   c. use in areas where access is limited
   d. training use only

3. When using a pit gauge, it is important to first _____.
   a. take four readings to get a valid sample
   b. measure the depth of the largest pit
   c. prepare the corroded area with RSTRENG
   d. inspect and calibrate the gauge

4. The B31G corrosion assessment method _____.
   a. is acceptable, but conservative in nature
   b. is the preferred method for evaluating corrosion
   c. is no longer approved for use by the API
   d. was developed to correct faults with the RSTRENG method

5. When a pipe is exposed for inspection, the pipeline technician must be able to _____.
   a. perform a B31G test before measuring pit depth
   b. complete a flame test at any potential leak sites
   c. visually identify any unsafe integrity issues
   d. demagnetize the pipe to prevent sparking

6. To provide an accurate reading, the pit gauge must be _____.
   a. resting on the corroded pipe surface
   b. positioned flush with the pipe surface
   c. aligned crosswise with the pipe
   d. positioned directly under the bridging bar

Trade Terms Quiz

Fill in the blank with the correct term that you learned from your study of this module.

1. To ensure test instrument accuracy by comparing the reading to a known standard is to ________.
2. The peaks and valleys of a corroded area can be depicted using a(n) ________.
3. Corrosion that takes place over an entire surface resulting in thinning of the material is called ________.
4. The loss of pipe metal either as a result of mechanical surface damage or corrosion is called ________.
5. When two or more areas of localized corrosion are in close proximity and combine to reduce the overall wall thickness a process takes place known as ________.
6. A unit of measure equal to 0.001 of an inch is called a(n) ________.
7. A type of localized corrosion is referred to as ________.
8. Conditions within a pipeline system, identified by the operator, that may indicate (1) a malfunction of a component or (2) a deviation from normal operations that may indicate a condition exceeding design limits or result in a hazard to persons, property, or the environment are called ________.
9. Corrosion in isolated areas is known as ________.
10. A depression in the surface of a pipe that affects the curvature of the pipe without reducing the pipe wall thickness is a(n) ________.
11. Damage to the surface of a pipe that displaces or removes metal and reduces the pipe wall thickness is a(n) ________.

Trade Terms

Abnormal operating conditions (AOCs)
Calibrate
Dent
General corrosion
Gouge
Interaction
Localized corrosion
Mil
Pitting
River bottom profile
Wall loss

Trade Terms Introduced in This Module

Abnormal operating conditions (AOCs): Conditions within a pipeline system, identified by the operator, that may indicate (1) a malfunction of a component or (2) a deviation from normal operations that may indicate a condition exceeding design limits or result in a hazard to persons, property, or the environment.

Calibrate: A process used to ensure accuracy by comparing an instrument reading against a known standard.

Dent: A depression in the surface of a pipe that affects the curvature of the pipe without reducing the pipe wall thickness.

General corrosion: Corrosion that takes place over the entire surface, resulting in thinning of the material.

Gouge: Damage to the surface of a pipe that displaces or removes metal and reduces the pipe wall thickness.

Interaction: When two or more areas of localized corrosion are in close proximity and combine to reduce the overall wall thickness.

Localized corrosion: Corrosion in isolated areas.

Mil: Unit of measure equal to 0.001 of an inch.

Pitting: A type of localized corrosion characterized by the formation of small cavities or pits.

River bottom profile: A collection of pit depths depicting peaks and valleys, or the actual profile of the defect.

Wall loss: The loss of pipe metal either as a result of surface damage or corrosion.

Additional Resources

The following recommended resources can provide additional helpful information related to the requirements for this covered task:

API Recommended Practice 1161, Third Edition, © 2014 American Petroleum Institute. API Publishing Services, 1220 L Street NW, Washington, DC 20005. Distributed by Thompson Reuters (Scientific) LLC, **www.techstreet.com**.

NCCER Module AOCCC-17, *Abnormal Operating Conditions Control Center*

NCCER Module AOCFG-17, *Abnormal Operating Conditions Field and Gas*

NCCER Module 62105-14, *Pipeline Documentation*

NCCER Module 66101-14, *Introduction to the Pipeline Industry*

NCCER Module CT14_1-17, *Locate Line*

NCCER Module CT8_2-17, *Measure Wall Thickness with Ultrasonic Meter*

NCCER Module CT8_3-17, *Measure Corroded Area*

49 *CFR*, Section 192, *Transportation of Natural and Other Gas by Pipeline, Subpart I, Requirements for Corrosion Control*, available from the US Government Printing Office at **www.gpo.org**.

49 *CFR*, Section 195, *Transportation of Hazardous Liquids by Pipeline, Subpart H, Corrosion Control*, available from the US Government Printing Office at **www.gpo.org**.

ASTM G46-94 (2013), *Standard Guide for Examination and Evaluation of Pitting Corrosion*, available at **www.astm.org**.

Figure Credits

Photo courtesy of Alan Kehr, Alan Kehr Anti-Corrosion, LLC, Module Opener, Figure 4

Western Instruments, Figures 1, 2

Section Review Answer Key

Answer	Section Reference	Objective
Section One		
1. d	1.1.3	1a
2. c	1.2.0	1b
Section Two		
1. b	2.1.2	2a
2. a	2.3.0	2b

NCCER CURRICULA — USER UPDATE

NCCER makes every effort to keep its textbooks up-to-date and free of technical errors. We appreciate your help in this process. If you find an error, a typographical mistake, or an inaccuracy in NCCER's curricula, please fill out this form (or a photocopy), or complete the online form at **www.nccer.org/olf**. Be sure to include the exact module ID number, page number, a detailed description, and your recommended correction. Your input will be brought to the attention of the Authoring Team. Thank you for your assistance.

Instructors – If you have an idea for improving this textbook, or have found that additional materials were necessary to teach this module effectively, please let us know so that we may present your suggestions to the Authoring Team.

NCCER Product Development and Revision

13614 Progress Blvd., Alachua, FL 32615

Email: curriculum@nccer.org

Online: www.nccer.org/olf

❑ Trainee Guide ❑ Lesson Plans ❑ Exam ❑ PowerPoints Other ____________________

Craft / Level: Copyright Date:

Module ID Number / Title:

Section Number(s):

Description:

Recommended Correction:

Your Name:

Address:

Email: Phone:

CT8_2-17

Measure Wall Thickness with Ultrasonic Meter

Overview

Testing performed on in-service equipment is commonly done by pipeline companies to avoid unnecessary downtime and loss of production. This testing is especially useful when internal access to equipment is impractical. One common example of such testing is measuring the remaining wall thickness of a corroded pipeline using digital ultrasonic thickness gauging. This module describes ultrasonic thickness (UT) meter operation and explains how a UT meter is used to measure pipe wall thickness.

Trainees with successful module completions may be eligible for credentialing through the NCCER Registry. To learn more, go to **www.nccer.org** or contact us at 1.888.622.3720. Our website has information on the latest product releases and training, as well as online versions of our *Cornerstone* magazine and Pearson's product catalog.

Your feedback is welcome. You may email your comments to **curriculum@nccer.org**, send general comments and inquiries to **info@nccer.org**, or fill in the User Update form at the back of this module.

This information is general in nature and intended for training purposes only. Actual performance of activities described in this manual requires compliance with all applicable operating, service, maintenance, and safety procedures under the direction of qualified personnel. References in this manual to patented or proprietary devices do not constitute a recommendation of their use.

CT8_2-17
Measure Wall Thickness with Ultrasonic Meter

Objectives

When you have completed this module, you will be able to do the following:

1. Describe the basic operation and calibration of an ultrasonic thickness (UT) meter.
   a. Describe the basic operation of a UT meter.
   b. Describe the calibration of a UT meter.
2. Describe how to measure pipe wall thickness using a UT meter (CT8_2-17).
   a. Identify the prerequisites, competencies, and abnormal operating conditions associated with measuring pipe wall thickness using a UT meter.
   b. Describe the procedures used to measure pipe wall thickness using a UT meter.

Performance Task

Under the supervision of your instructor, you should be able to do the following:

1. Measure wall thickness using an ultrasonic meter (CT8_2-17).
   - Identify potential abnormal operating conditions that may occur during performance of this CT, and know the appropriate actions to take in response to them.
   - Utilize the appropriate personal protective equipment according to relevant company procedures.
   - Perform a visual inspection of the exposed pipe. Look for integrity issues such as evidence of a release or significant metal deformation. If unsafe integrity issues are encountered, discontinue the task and make immediate notifications.
   - Assemble, check, and calibrate an ultrasonic thickness (UT) meter. Ensure it is ready for operation.
   - Clean and remove debris from the area to be measured.
   - Apply an acceptable couplant.
   - Place the transducer in contact with the couplant/surface, ensuring it is oriented to the pipe surface according to the manufacturer's instructions.
   - Observe the wall thickness measurement on the meter display.
   - Repeat several measurements to confirm the nominal wall thickness.
   - Complete appropriate documentation and make notifications as required by operator's procedures.

Trade Terms

Abnormal operating conditions (AOCs)
Backwall
Calibrate
Couplant
Cross-talk barrier
Dent
Doubling
Gouge
Laminations
Nominal wall thickness
Pulse-echo
Sound waves
Transducer
Velocity of sound

Industry Recognized Credentials

If you are training through an NCCER-accredited sponsor, you may be eligible for credentials from NCCER's Registry. The ID number for this module is CT8_2-17. Note that this module may have been used in other NCCER curricula and may apply to other level completions. Contact NCCER's Registry at 888.622.3720 or go to **www.nccer.org** for more information.

Contents

Figures and Tables

Section One

1.0.0 Ultrasonic Thickness Meters

Objective

Describe the basic operation and calibration of an ultrasonic thickness (UT) meter.

a. Describe the basic operation of a UT meter.
b. Describe the calibration of a UT meter.

Trade Terms

Backwall: The sidewall of the material being tested opposite from the spot where the ultrasonic probe emits the thickness measurement signal.

Calibrate: A process used to ensure accuracy by comparing an instrument reading against a known standard.

Couplant: A liquid, grease, paste, or gel placed between the transducer of an ultrasonic meter and the surface of the test piece to enhance the transmission of ultrasonic energy (sound waves).

Cross-talk barrier: A thin metal barrier inside a transducer probe that prevents the signals from the transmitting and receiving elements in the probe from interfering with one another.

Doubling: An effect that can occur when a UT meter is used to measure a material that is below the minimum thickness range of the probe which results in an indicated value that is twice the actual material thickness.

Laminations: Defects or imperfections in piping material that can result in metal separation and wall weakness.

Nominal wall thickness: The expected wall thickness of a pipe as determined by alignment sheets or other records.

Pulse-echo: A non-destructive test in which pulses of energy are sent into the material and the time for return echo from one or more surfaces is measured.

Sound waves: Particle displacement that propagates through a medium. Sound waves above the audible range are called ultrasonic.

Transducer: A device that converts an input signal into an output signal of a different form.

Velocity of sound: The rate at which sound travels through different materials.

The remaining wall thickness of a corroded pipe is a major indicator of the overall mechanical integrity of the pipe. Digital ultrasonic thickness gauging is a testing method commonly used to determine the remaining wall thickness of a corroded pipe. This test uses high-frequency sound waves to measure the pipe wall thickness and compares it against the **nominal wall thickness**, which is the expected wall thickness as determined by alignment sheets or other records. The general use and calibration of ultrasonic meters are discussed in the following sections.

1.1.0 Basic UT Meter Operation

The operation of an ultrasonic thickness (UT) meter is based on the non-destructive qualities of **pulse-echo** testing. During operation, a UT meter generates an electrical energy pulse. A **transducer** probe converts this pulse into high-frequency **sound waves**. When the transducer probe is placed on the surface of a pipe, the sound waves travel through the pipe wall and are reflected from the interior surface (**backwall**) of the pipe. The reflected wave (echo) is picked up by the transducer probe and converted back into electrical energy that is measured by the gauge. A thin metal barrier inside the transducer probe, called a **cross-talk barrier**, prevents the signals from the transmitting and receiving elements from interfering with one another. *Figure 1* shows an ultrasonic wall thickness meter in use.

Charts are available that show the **velocity of sound** through a given material. Using the value for the given pipe material, the time required for

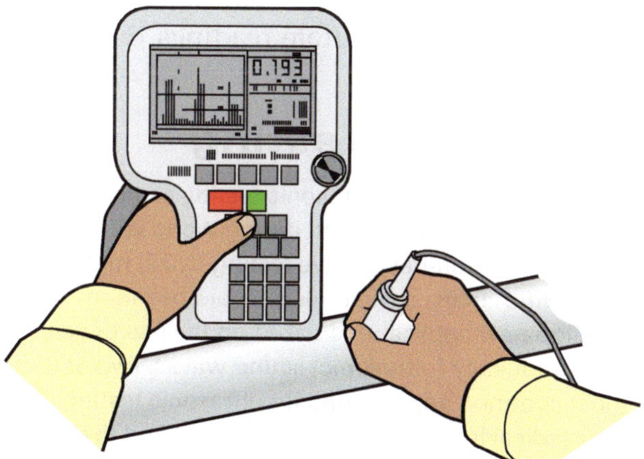

Figure 1 Typical ultrasonic wall thickness meter with transducer probe.

the emitted sound wave to travel to the interior wall and back can be converted into a thickness value. It is critical to know the correct velocity of sound for the pipe material being measured in order to achieve accurate results.

UT meters are reliable tools for measuring equipment subject to internal corrosion. Where corrosion damage exists, the damaged area has a different molecular structure and therefore a different velocity of sound. The transmitted signal reflects off the line between the good and damaged material and can be used to determine the thickness of the good material.

1.1.1 Through-Coating Measurement

Typically, aboveground equipment is protected by a thick layer of paint, or one of many other coating materials. This poses a problem for ultrasonic measurement if the operator does not know the exact thickness of the coating. Another problem is caused by the extremely slow velocity of sound values for most coating materials. Because of this difference in velocity of sound values between the coating and the underlying metal, the UT meter measures the typical coating as being two to three times thicker than the actual thickness, which results in false data.

In the past, accurate thickness readings could only be obtained by removing the coating and measuring only the pipe material itself. The paint-stripped area was then cleaned and recoated, a costly process in terms of material and labor hours.

An improved method for measuring through-coating thickness is available with the use of dual-element transducer probes. These probes are designed to emit a more focused signal beam that is capable of more accurate readings than single beam probes when measuring areas of varying thicknesses. *Figure 2* shows how the dual-element transducer probe is placed on the pipe to generate a readout of wall thickness.

UT meters can be equipped with a multiple echo readout capability. This type of thickness gauge is used to identify and measure metal thickness apart from coating thickness and display the results in a single readout. *Figure 3* shows how dual-element transducer signals are converted into digital values for both the coating and the pipe material. Operators may need to adjust the gauge for starting point and gate width for accurate measurement of pipe or equipment material thickness.

Ultrasonic thickness gauge software can interpret massive amounts of data to simplify decision-making. By viewing the data in color levels related to thickness, it is easy to see if part of a component is below the minimum thickness value. *Figure 4* shows a typical thickness level displayed in grid block view.

The presence of pipe laminations may produce false readings and must be considered when the wall thickness exceeds the expected value. Some ultrasonic thickness detectors have the ability to detect and measure laminations. Refer to the operator's manual for the meter in use.

Software programs are often used in refineries, chemical processing plants, and power generation plants to predict the remaining service life of equipment or components, and schedule when each item must be inspected. *Figure 5* shows a typical thickness measurement location line graph showing corrosion rates and remaining life.

1.1.2 Using Couplants

An ultrasonic couplant consists of a liquid, paste, or gel placed between the transducer of an ultrasonic meter and the surface of the test piece to eliminate the air gap and improve probe coupling for the transmission of ultrasonic energy (sound waves). Thicker couplants can be used in vertical applications to ensure that the couplant adheres to the surface being measured. Always refer to the manufacturer's couplant recommendations for the meter in use.

Couplants must be applied in a uniform layer, as variations in couplant thickness may produce errors in the test results. *Figure 6* shows an ultrasonic thickness gauge and couplant.

Ultrasonic Testing

The inspiration for ultrasonic testing came from the use of sonar before and during World War II and the need for non-destructive testing (NDT) methods. During the 1940s, civilian researchers began to develop new applications based on the use of very high-frequency (ultrasonic) sound waves. Two of the most common applications for ultrasonic testing are medical imaging (ultrasonography) and the non-destructive inspection and testing of materials.

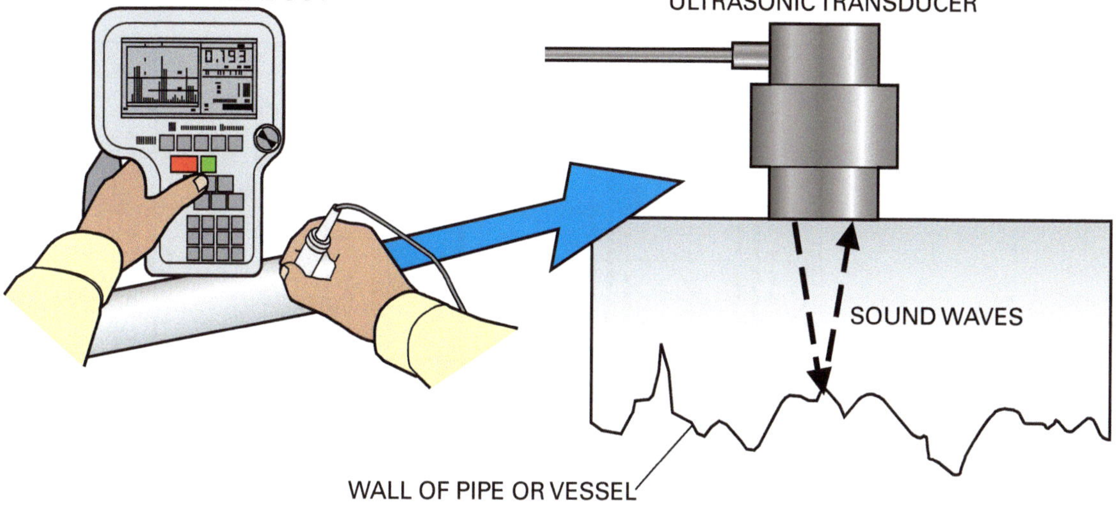

Figure 2 Placement of a dual element transducer.

Couplants improve probe coupling and acoustic transmission properties by focusing the high-frequency sound waves onto the target pipeline. The resulting penetration of the pipe wall with higher amplitude signals provides stronger echoes back to the transducer and yields more accurate test results.

1.1.3 Data Analysis

Depending on the size of the system being tested, the number of thickness readings required may number in the hundreds or even thousands. Software programs are used to analyze large amounts of thickness data, schedule future inspections, and predict the service life of the piping or equipment.

Data management software can interpret massive amounts of data to simplify decision-making. By viewing the data in color levels related to thickness, it is easy to recognize areas that are below the minimum thickness value.

High-level software programs are typically used in large refineries, chemical processing plants, or power generation plants. Database history and complex analysis calculations within the program can predict the remaining service life of an in-service equipment item or component and calculate when the item needs re-inspection.

Other benefits of these programs include documentation of test results, record retention, and the ability to compare historical data for performance trends and code compliance (where required).

1.2.0 Calibrating a UT Meter

Before any measurement is taken, a UT meter must be calibrated to ensure it will provide accurate data. It is important to calibrate UT meters because of variations in factors such as the sound velocity of the material being tested, the temperature, and the material thickness. Best results are obtained if the instrument is calibrated on a velocity reference block made of the same material as the test piece. This block should be flat, smooth, and as thick as the maximum thickness of the test piece. The calibration procedures vary depending on the manufacturer of the test instrument.

General calibration requirements are discussed in the following sections. For specific procedures, refer to the operating manual for the instrument in use.

1.2.1 Calibration of Sound Velocity

Sound velocity may not be constant in the material being tested. For example, heat treating can cause significant changes in sound velocity. This must be considered when evaluating the accuracy of the thickness indicated by an instrument. To minimize testing errors, calibrate instruments before testing, and check the calibration again after testing is completed.

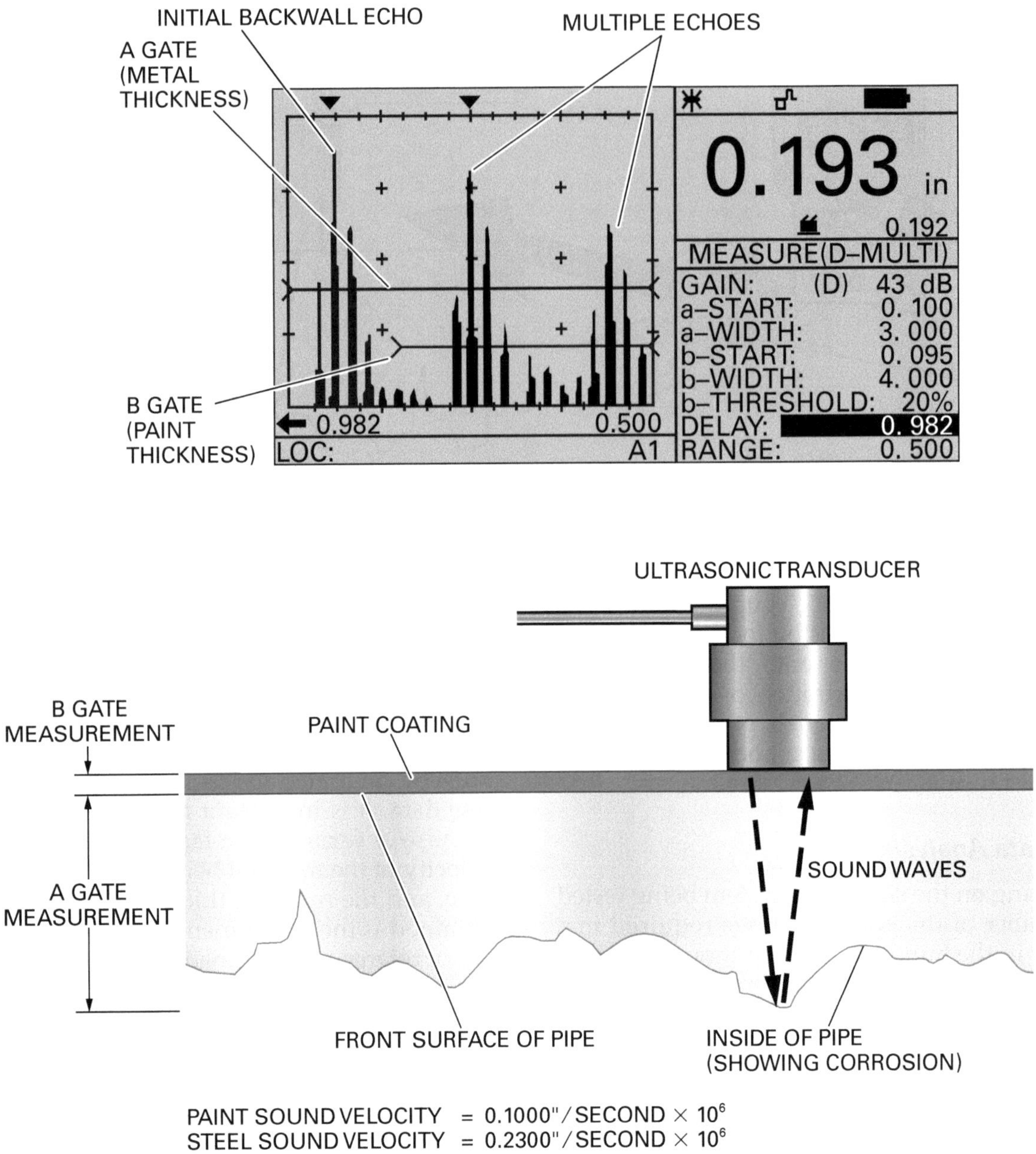

Figure 3 Ultrasonic transducer placement and sample reading.

1.2.2 Effects of Temperature on Calibration

Temperature variations will change the sound velocity of materials and transducer delay times (the time delay between sending and receiving). This also changes calibration requirements. To minimize errors due to temperature variations, perform all calibrations on site, using test blocks of the same temperature as the test piece.

When calibrating a new transducer, or when ultrasonically measuring a test piece for the first time, calibrate the instrument/transducer combination on reference blocks that represent the complete range of possible thicknesses that may be encountered during the test activity. This is particularly important when the thickness history of the item to be tested is unknown and the measurements will serve as a baseline.

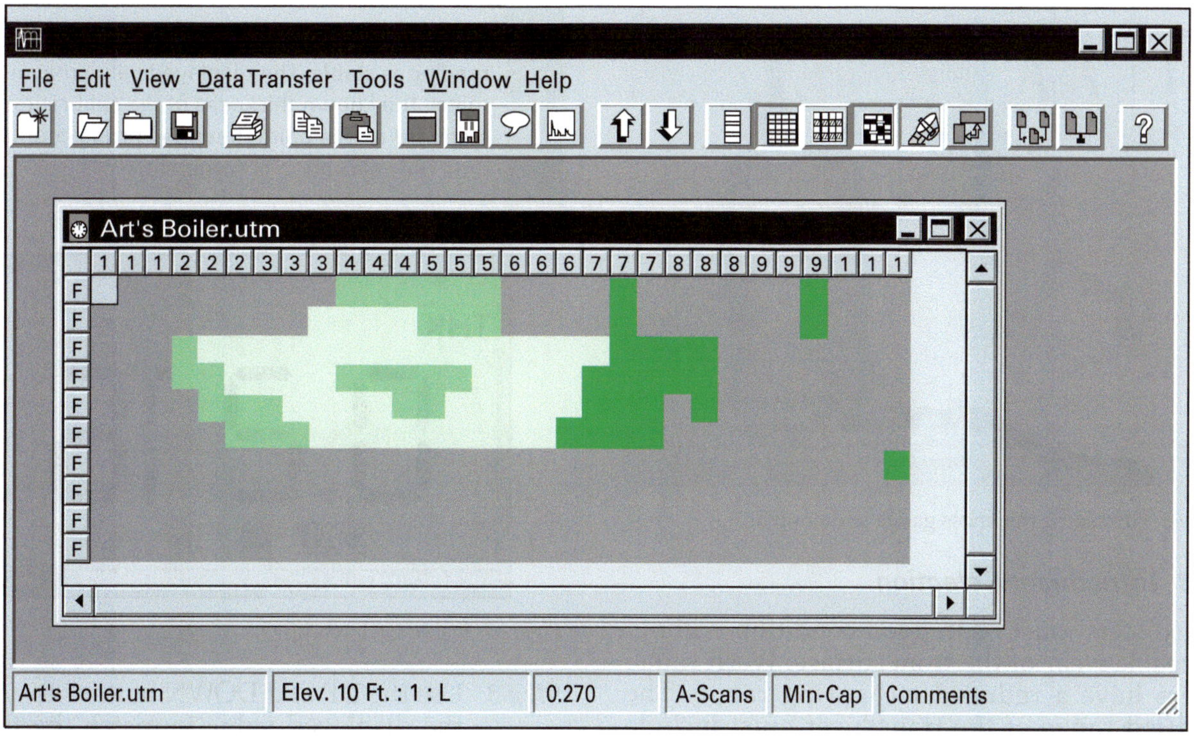

Figure 4 Thickness level grid block readout.

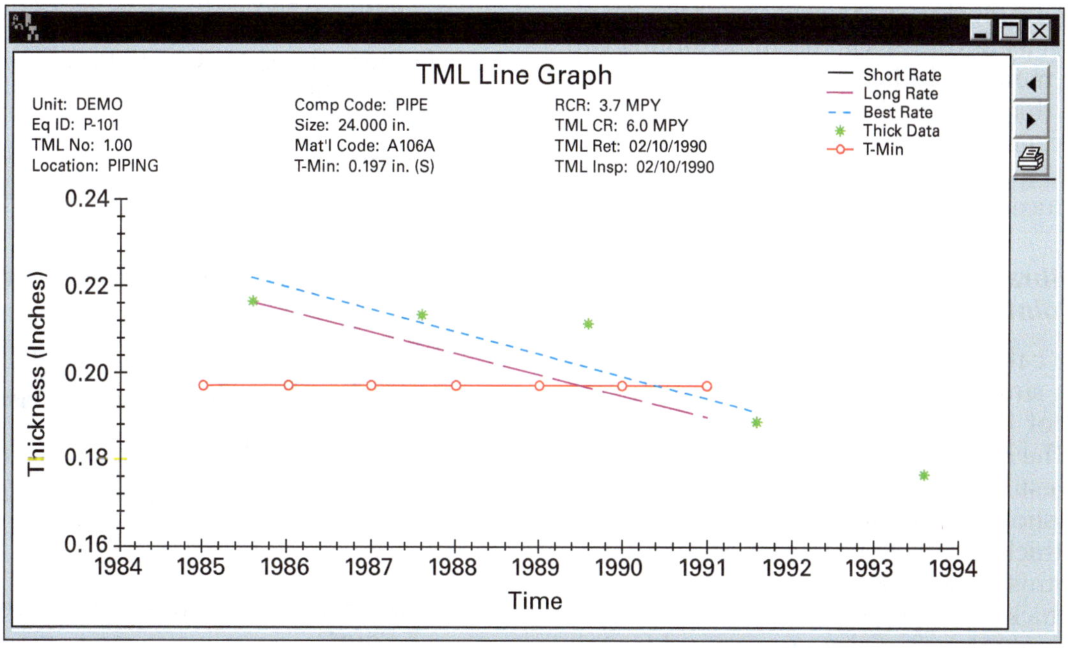

Figure 5 Thickness measurement line graph.

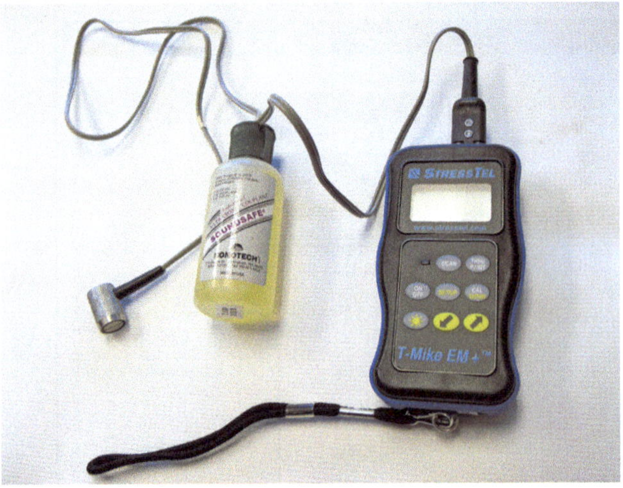

Figure 6 Ultrasonic thickness gauge and couplant.

1.2.3 Transducer Selection

Transducers must be in good condition without noticeable wear of the front surface. Worn transducers have a reduced measuring range. The specified range of the transducer must include the complete range of thicknesses to be tested. If a UT meter is used on materials that are below the minimum thickness range of the probe, an effect known as **doubling** can occur. Doubling can cause the meter to indicate a value that is twice the actual thickness of the material. Since temperature affects measurement accuracy, the temperature of the material to be tested must be within the transducer's temperature range.

1.2.4 Calibration to a Known Thickness, 1-Point

A 1-point calibration is used to calibrate a UT meter to a single known thickness. A calibration standard of the same material and velocity of sound as the material to be measured is required. For best results, the thickness of the calibration standard should be equal to, or slightly greater than, the thickest area to be measured.

Follow this basic procedure to calibrate an ultrasonic thickness gauge to a known thickness. These are only the general steps required for this calibration; always refer to the operating instructions for the instrument in use.

Step 1 Press the CAL/ON button.

Step 2 With the CAL flashing, couple the probe to the calibration standard. *Figure 7* shows a stable CAL reading.

> **NOTE**
> Be sure the coupling indicator lights up and the reading is stable. The displayed value may not match the known thickness of the calibration standard at this point. Keep the probe coupled until the calibration is complete.

Figure 7 Stable CAL reading.

Step 3 Use the UP and DOWN buttons to adjust the displayed value to match the thickness of the calibration standard. *Figure 8* shows how the display should appear with a 0.200" calibration standard.

Step 4 Press the CAL/ON button again to end the calibration procedure.

Figure 8 Display with a 0.200" calibration standard.

1.2.5 Calibration to a Known Thickness, 2-Point

A 2-point calibration is used to calibrate a UT meter using two known thicknesses. This procedure requires two calibration standards (or one standard with two different thicknesses) of the same material, velocity, and curvature as the material under test. For best results, the thicker standard should be equal to, or slightly greater than, the maximum thickness to be measured. The thickness of the thinner standard should be as close as possible to the lower end of the expected measurement range. Calibrate the gauge to both standards.

These are the general steps required for this calibration; always refer to the operating instructions for the instrument in use.

Step 1 Press the CAL/ON button to display the current status.

Step 2 Use the UP and DOWN buttons to enable 2-PT (*Figure 9*).

Figure 9 Enable 2-PT.

Step 3 Press the CAL/ON button again. The 2-PT indicator now flashes and the prompt to couple the LO (thin) calibration standard should appear (*Figure 10*).

Figure 10 LO display.

Step 4 Couple the probe to the thin calibration standard. Ensure that the coupling indicator light comes on and the reading is stable. Keep the probe coupled, or uncouple the probe and wipe the excess couplant from the probe face.

Step 5 Use the UP and DOWN buttons to adjust the displayed value to match the thickness of the thin calibration standard.

Step 6 Press the CAL/ON button again. The prompt to couple the HI (thick) calibration standard should be displayed.

Step 7 Couple the probe to the thick calibration standard. Ensure that the coupling indicator light comes on and the reading is stable.

Step 8 Use the UP and DOWN buttons to adjust the displayed value to match the thick calibration standard.

Step 9 Press the CAL/ON button to exit 2-PT calibration mode (*Figure 11*). The 2-PT indicator remains illuminated to show that the calibration mode is 2-PT and that the automatic probe is set at zero.

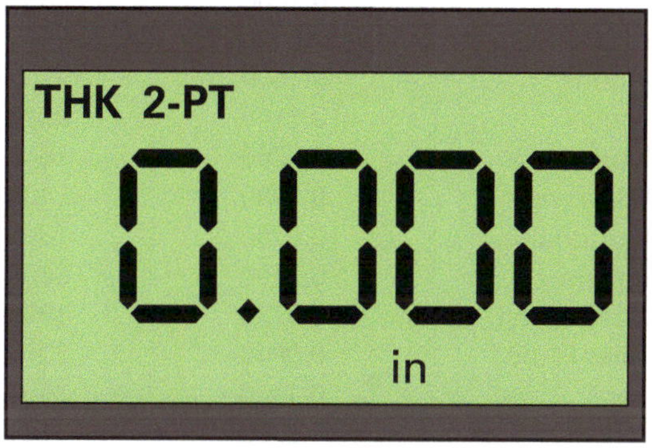

Figure 11 2-PT calibration readout.

The 2-PT calibration procedure is complete and the instrument is ready for thickness measurement.

1.2.6 Calibration to a Known Velocity

If the velocity of sound of the material to be measured is known, the instrument can be calibrated to the known velocity without using a calibration standard. However, the calibration should be always be verified by measuring a known thickness (or calibration standard) of the same material and velocity as the material to be measured.

Refer to *Table 1*, which is a chart of calculated sound velocities for different materials. Using the listed value for the pipe material under test, the time required for the emitted sound wave to travel to the interior wall of the pipe and return to the probe can be converted into a thickness value. Using the correct velocity of sound for the pipe

Table 1. Typical Sound Velocities in Various Metals

Material	in/µs	m/s
Aluminum	0.2500	6300
Brass	0.1700	4300
Cadmium	0.1100	2800
Cast Iron	0.1800	4600
Copper	0.1800	4600
Epoxy Resin	0.1100	2800
Glass (Crown)	0.2200	5600
Glass (Window)	0.2700	6800
Gold	0.1300	3300
Inconel	0.2200	5600
Lead	0.0850	2200
Magnesium	0.2300	5800
Manganese	0.1800	4600
Molybdenum	0.2500	6300
Monel	0.2100	5300
Neoprene	0.0630	1600
Nickel	0.2200	5600
Nylon	0.1000	2500
Phenolic	0.0560	1400
Platinum	0.1600	4100
Plexiglas	0.1100	2800
Polyethylene	0.0700	1800
Polystyrene	0.0930	2400
Polyurethane	0.0700	1800
Porcelain	0.2200	5600
Rubber (Butyl)	0.0730	1900
Rubber (Vulc.)	0.0900	2300
Silver	0.1400	3600
Steel	0.2300	5800
Tin	0.1300	3300
Titanium	0.2400	6100
Tungsten	0.2100	5300
Zinc	0.1600	4100
Zircaloy 2	0.1900	4700

material being measured is critical for accurate results. Softer materials, such as rubber, tend to have a lower velocity of sound than harder materials, such as steel.

Use the following procedure to calibrate the ultrasonic thickness gauge to a known velocity. When performing this procedure for velocity, the 2-PT calibration mode must be set to the OFF position. These are the general steps required for this calibration; always refer to the operating instructions for the instrument in use.

Step 1 Press the MODE button until VEL illuminates. The current velocity setting is shown in *Figure 12*.

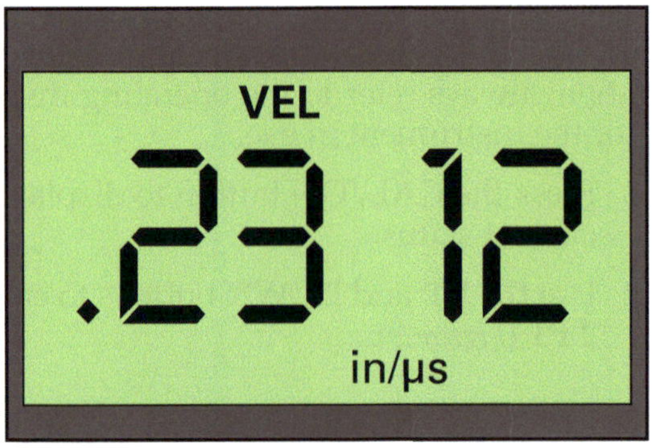

Figure 12 Current velocity setting.

Step 2 Press the CAL/ON button to enable calibration mode.

Step 3 With the CAL indicator flashing, use the UP and DOWN buttons to set the displayed velocity to that of the material under test. *Figure 13* shows the velocity setting of 0.2328" per microsecond.

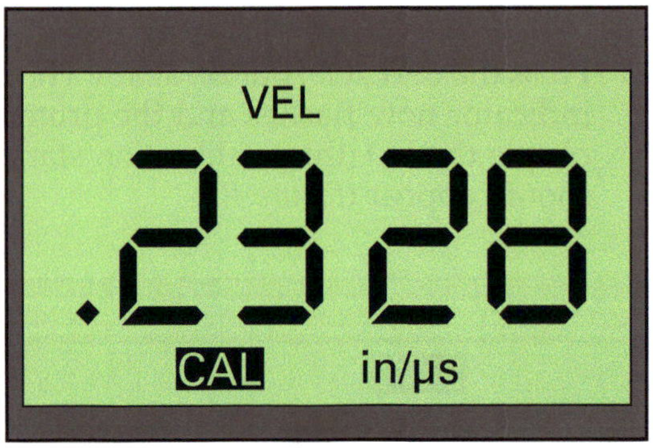

Figure 13 Velocity setting of 0.2328" per microsecond.

Step 4 Press the CAL/ON button again to exit the velocity calibration mode.

1.2.7 Data Loggers/Recorders

Some UT meters have internal memory that allows the printing of reports and storage of thickness readings. Consult the manufacturer's literature for the use of these functions.

Additional Resources

It is assumed that you will learn to perform these procedures under the direct supervision of your crew leader, foreman, or instructor. For independent study, refer to the recommended Additional Resources at the back of this module.

1.0.0 Section Review

1. The part of a UT meter that converts an electrical energy pulse to high-frequency sound waves is the _____.
   a. sonic coupler
   b. optical pulsar
   c. transducer probe
   d. echo inductor

2. Technicians using UT meters must take into consideration that piping heat treating can cause significant changes in _____.
   a. sound velocity
   b. light transmission
   c. magnetic properties
   d. laminar echoes

Section Two

2.0.0 Measuring Wall Thickness with a UT Meter

Objective

Describe how to measure pipe wall thickness using a UT meter (CT8_2-17).

a. Identify the prerequisites, competencies, and abnormal operating conditions associated with measuring pipe wall thickness using a UT meter.
b. Describe the procedures used to measure pipe wall thickness using a UT meter.

Performance Task

Measure wall thickness using an ultrasonic meter (CT8_2-17).

- Identify potential abnormal operating conditions that may occur, and know the appropriate actions to take in response to them.
- Utilize the appropriate personal protective equipment according to relevant company procedures.
- Perform a visual inspection of the exposed pipe. Look for integrity issues such as evidence of a release or significant metal deformation. If unsafe integrity issues are encountered, discontinue the task and make immediate notifications.
- Assemble, check, and calibrate an ultrasonic thickness (UT) meter. Ensure it is ready for operation.
- Clean and remove debris from the area to be measured.
- Apply an acceptable couplant.
- Place the transducer in contact with the couplant/surface, ensuring it is oriented to the pipe surface according to the manufacturer's instructions.
- Observe the wall thickness measurement on the meter display.
- Repeat several measurements to confirm the nominal wall thickness.
- Complete appropriate documentation and make notifications as required by operator's procedures.

Trade Terms

Abnormal operating conditions (AOCs): Conditions within a pipeline system, identified by the operator, that may indicate (1) a malfunction of a component or (2) a deviation from normal operations that may indicate a condition exceeding design limits or result in a hazard to persons, property, or the environment.

Dent: A depression in the surface of a pipe that affects the curvature of the pipe without reducing the pipe wall thickness.

Gouge: Damage to the surface of a pipe that displaces or removes metal and reduces the pipe wall thickness.

Accurately determining the extent of pipeline corrosion and the urgency of repairs needed to correct a problem requires pipeline technicians to have some basic competencies as well as the ability to carefully follow the procedures required to measure and record wall thickness. This section describes some of the basic skills needed by pipeline technicians and identifies **abnormal operating conditions (AOCs)** that personnel may need to respond to during wall thickness measurements. It also describes the procedures used to measure wall thickness with a UT meter.

2.1.0 Prerequisites, Competencies, and Abnormal Operating Conditions

Personnel must possess a variety of skills in order to measure the remaining wall thickness of a corroded pipeline. The operation requires a series of steps, which a pipeline technician must be competent to perform. Personnel must be aware of the particular abnormal operating conditions that may occur during the procedure and know the proper response to them.

2.1.1 Prerequisites and Competencies

Personnel assigned to determine the structural integrity of a pipeline by measuring the remaining wall thickness must understand what the task entails. Prior to beginning the task, personnel must be able to properly select, calibrate, and operate the UT meter to be used during the procedure.

Personnel must understand that the task begins when the pipe is first exposed, cleaned, and prepared for inspection. They must be able to visibly inspect the pipe and determine if there are any unsafe integrity issues, such as evidence of a release or significant metal deformation, and discontinue the task and make immediate notifications if any are found. Once the inspection begins, personnel must know how to position, manipulate, and read the instrument in order to obtain valid measurements of the wall thickness. Personnel must understand that the task is not complete until all measurements are documented and the proper notifications are made.

Personnel must have a basic understanding of additional tasks related to measuring wall thickness depth, such as measuring pit depth (covered in NCCER Module CT8_1-17) and measuring the area of the corrosion (covered in NCCER Module CT8_3-17). They must also have an understanding of how the *ASME/ANSI B31G* and/or RSTRENG corrosion assessment methods are applied to determine the remaining wall strength for the pipe once wall thickness and other data have been gathered.

2.1.2 Abnormal Operating Conditions

Prior to, or during, wall thickness measurements, pipeline technicians must be able to recognize the following potential AOCs and know the proper response to each, should they occur:

- *Internal or external corrosion of the pipe is present* – Document the condition according to company guidelines and notify personnel responsible for correcting the problem.
- *A pipeline has been significantly deformed by line hits or mechanical surface damage* – Document the condition accordingly and notify personnel responsible for correcting the problem.
- *Fire or explosion caused by the ignition of a hydrocarbon liquid or gas* – Immediately evacuate the area and notify the appropriate firefighting personnel. It may be necessary to call 911 in such an emergency.
- *Unexpected leaks of hazardous liquid or carbon dioxide occurs because of a loss of pipe wall thickness* – Take steps to eliminate any possible ignition sources and notify the appropriate personnel, possibly including 911.

2.2.0 Procedures

This section describes the general procedures for using a UT meter to measure wall thickness. Always follow the manufacturer's instructions for the instrument in use.

WARNING!

Always obtain and use the appropriate personal protective equipment for each task. Stay alert to the possibility of unexpected leaks from the exposed pipeline such as hazardous liquid or carbon dioxide (CO_2). Safely eliminate all ignition sources and immediately notify the appropriate personnel.

Step 1 Identify potential abnormal operating conditions that may occur, and know the appropriate actions to take in response to them.

Step 2 Utilize the appropriate personal protective equipment according to relevant company procedures.

Step 3 Visually inspect the pipe for any evidence of leaks or loss of pipe integrity. Look for mechanical surface damage, such as a dent or a gouge. Dents alter the external shape of the pipe without reducing the wall thickness, while gouges displace or remove surface metal and reduce the wall thickness. If any unsafe integrity issues are present, stop the task and immediately notify the appropriate personnel.

Step 4 Properly assemble, check, and calibrate the meter according to the manufacturer's instructions.

Step 5 Prepare and clean the area of the pipe to be measured. The presence of dirt and debris will prevent an accurate reading.

Step 6 Apply the appropriate couplant to the area being measured.

Step 7 Orient the transducer on the pipe surface according to the manufacturer's instructions (typically perpendicular). Make sure it is placed firmly within the couplant and in contact with the surface to be measured.

Step 8 Read the measurement on the display. Be aware of conditions that can cause false thickness readings. For example, doubling can occur if the pipe wall thickness is below the minimum thickness range of the transducer probe. Laminations, which are defects in piping material that can result in metal separation and wall weakness, can also cause false thickness readings.

CAUTION

A major drop in the transducer reading may indicate lamination. Laminated pipe cannot be repaired and must be replaced. Laminations are a mill defect. If laminations are found, locate and test all pipe with the same lot number whenever possible.

Step 9 Repeat this process to verify the nominal wall thickness. Note any discrepancies between the measured values and the expected (nominal) value. Some UT meters have a differential mode that displays the difference (+ or –) from a set nominal value.

Step 10 Document the findings on the appropriate company inspection form and make notifications according to company policy.

Additional Resources

It is assumed that you will learn to perform these procedures under the direct supervision of your crew leader, foreman, or instructor. For independent study, refer to the recommended Additional Resources at the back of this module.

2.0.0 Section Review

1. If a hazardous liquid or carbon dioxide leak occurs during a wall thickness measuring activity, the pipeline technician should _____.
   a. put on HAZMAT gear and use couplant to try to stop the leak
   b. quickly finish the thickness testing and evacuate the area
   c. use the ultrasonic meter to determine if the liquid or gas is combustible
   d. take steps to eliminate ignition sources, and notify the appropriate personnel

2. A false thickness reading can occur when there are defects in the piping material caused by _____.
   a. sagging
   b. laminations
   c. discoloration
   d. alligatoring

SUMMARY

Performing non-destructive tests on in-service equipment enables companies to avoid unnecessary downtime and loss of production. This type of testing is especially useful where internal access to equipment is impractical. One common example of such testing is measuring the remaining wall thickness of a corroded pipeline using an ultrasonic thickness (UT) meter.

Understanding how to prepare a UT meter, accurately take and record wall thickness measurements, and follow approved techniques to determine the structural integrity of the pipe enables pipeline technicians to ascertain the extent of the damage and the urgency of pipe repair or replacement.

This module described how to calibrate and use an ultrasonic thickness meter to measure the remaining wall thickness of a corroded pipe, and it presented the basic procedures for safely performing this task.

Review Questions

1. Digital ultrasonic thickness gauging measures the pipe wall thickness and compares it against the _____.
   a. maximum wall thickness
   b. median wall thickness
   c. minimum wall thickness
   d. nominal wall thickness

2. During operation, a UT meter generates an electrical energy pulse that a transducer probe converts into _____.
   a. low-pressure pneumatic waves
   b. high-frequency sound waves
   c. high-voltage current waves
   d. low-frequency sound waves

3. Using a UT meter on a pipe which has a thickness below the minimum thickness range of the transducer probe can lead to an inaccurate reading effect called _____.
   a. doubling
   b. localizing
   c. cross-talking
   d. echoing

4. If a fire or explosion occurs while performing a wall thickness measuring procedure, personnel should _____.
   a. make every effort to extinguish the fire before calling 911
   b. lie flat under a polyester cover and wait for help to arrive
   c. immediately evacuate the area and notify the appropriate personnel
   d. spray hydrocarbon liquid on the fire and evacuate the area

5. Surface metal that has been displaced or removed with reduced pipe wall thickness may be the result of _____.
   a. backwalls
   b. gouges
   c. coatings
   d. dents

6. Before placing a UT meter's transducer probe onto the surface of the piping, the technician should first _____.
   a. apply an appropriate couplant to the area being measured
   b. enter median thickness values into the data collector
   c. locate the grooved mounting laminations on the pipe
   d. remove any couplant material from the excavated pipe

Trade Terms Quiz

Fill in the blank with the correct term that you learned from your study of this module.

1. To ensure accuracy by comparing an instrument reading against a known standard is to ________.
2. Damage to the surface of a pipe that displaces or removes metal and reduces the pipe wall thickness is a(n) ________.
3. A liquid, grease, paste, or gel placed between the transducer of an ultrasonic meter and the surface of the test piece to enhance the transmission of ultrasonic energy (sound waves) is a(n) ________.
4. Defects or imperfections in piping material that can result in metal separation and wall weakness are ________.
5. Conditions within a pipeline system, identified by the operator, that may indicate (1) a malfunction of a component or (2) a deviation from normal operations that may indicate a condition exceeding design limits or result in a hazard to persons, property, or the environment are called ________.
6. A non-destructive test in which pulses of energy are sent into the material and the time for return echo from one or more surfaces is measured is called ________.
7. The rate at which sound travels through different materials is the ________.
8. An effect that can occur when a UT meter is used to measure a material that is below the minimum thickness range of the probe which results in an indicated value that is twice the actual material thickness is called ________.
9. Particle displacement that propagates through a medium are in the form of ________.
10. The sidewall of the material being tested opposite from the spot where the ultrasonic probe emits the thickness measurement signal is the ________.
11. A depression in the surface of a pipe that affects the curvature of the pipe without reducing the pipe wall thickness is a(n) ________.
12. A device that converts an input signal into an output signal of a different form is a(n) ________.
13. A thin metal barrier inside a transducer probe that prevents the signals from the transmitting and receiving elements in the probe from interfering with one another is a(n) ________.
14. The expected wall thickness of a pipe as determined by alignment sheets or other records is known as the ________.

Trade Terms

Abnormal operating conditions (AOCs)
Backwall
Calibrate
Cross-talk barrier
Couplant
Dent
Doubling
Gouge
Laminations
Nominal wall thickness
Pulse-echo
Sound waves
Transducer
Velocity of sound

Trade Terms Introduced in This Module

Abnormal operating conditions (AOCs): Conditions within a pipeline system, identified by the operator, that may indicate (1) a malfunction of a component or (2) a deviation from normal operations that may indicate a condition exceeding design limits or result in a hazard to persons, property, or the environment.

Backwall: The sidewall of the material being tested opposite from the spot where the ultrasonic probe emits the thickness measurement signal.

Calibrate: A process used to ensure accuracy by comparing an instrument reading against a known standard.

Couplant: A liquid, grease, paste, or gel placed between the transducer of an ultrasonic meter and the surface of the test piece to enhance the transmission of ultrasonic energy (sound waves).

Cross-talk barrier: A thin metal barrier inside a transducer probe that prevents the signals from the transmitting and receiving elements in the probe from interfering with one another.

Dent: A depression in the surface of a pipe that affects the curvature of the pipe without reducing the pipe wall thickness.

Doubling: An effect that can occur when a UT meter is used to measure a material that is below the minimum thickness range of the probe which results in an indicated value that is twice the actual material thickness.

Gouge: Damage to the surface of a pipe that displaces or removes metal and reduces the pipe wall thickness.

Laminations: Defects or imperfections in pipeline material that can result in metal separation and wall weakness.

Nominal wall thickness: The expected wall thickness of a pipe as determined by alignment sheets or other records.

Pulse-echo: A non-destructive test in which pulses of energy are sent into the material and the time for return echo from one or more surfaces is measured.

Sound waves: Particle displacement that propagates through a medium. Sound waves above the audible range are called ultrasonic.

Transducer: A device that converts an input signal into an output signal of a different form.

Velocity of sound: The rate at which sound travels through different materials.

Additional Resources

The following recommended resources can provide additional helpful information related to the requirements for this covered task:

API Recommended Practice 1161, Third Edition, © 2014 American Petroleum Institute. API Publishing Services, 1220 L Street NW, Washington, DC 20005. Distributed by Thompson Reuters (Scientific) LLC, **www.techstreet.com**.

49 *CFR* 192, *Subpart I*,**www.ecfr.gov.**

49 *CFR* 195, *Subpart H*,**www.ecfr.gov.**

NCCER Module AOCCC-17, *Abnormal Operating Conditions Control Center*

NCCER Module AOCFG-17, *Abnormal Operating Conditions Field and Gas*

NCCER Module 62105-14, *Pipeline Documentation*

NCCER Module 66101-14, *Introduction to the Pipeline Industry*

NCCER Module CT8_1-17, *Measure Pit Depth with Pit Gauge*

NCCER Module CT8_3-17, *Measure Corroded Area*

NCCER Module CT14_1-17, *Locate Line*

ASTM G46-94 (2013), *Standard Guide for Examination and Evaluation of Pitting Corrosion*, **www.astm.org**.

Figure Credits

Photo courtesy of Alan Kehr, Alan Kehr Anti-Corrosion, LLC, Module Opener

GE Energy Measurement & Control, Figures 1–4

Section Review Answer Key

Answer	Section Reference	Objective
Section One		
1. c	1.1.0	1a
2. a	1.2.1	1b
Section Two		
1. d	2.1.2	2a
2. b	2.2.0	2b

NCCER CURRICULA — USER UPDATE

NCCER makes every effort to keep its textbooks up-to-date and free of technical errors. We appreciate your help in this process. If you find an error, a typographical mistake, or an inaccuracy in NCCER's curricula, please fill out this form (or a photocopy), or complete the online form at **www.nccer.org/olf**. Be sure to include the exact module ID number, page number, a detailed description, and your recommended correction. Your input will be brought to the attention of the Authoring Team. Thank you for your assistance.

Instructors – If you have an idea for improving this textbook, or have found that additional materials were necessary to teach this module effectively, please let us know so that we may present your suggestions to the Authoring Team.

NCCER Product Development and Revision

13614 Progress Blvd., Alachua, FL 32615

Email: curriculum@nccer.org

Online: www.nccer.org/olf

❑ Trainee Guide ❑ Lesson Plans ❑ Exam ❑ PowerPoints Other ____________

Craft / Level: Copyright Date:

Module ID Number / Title:

Section Number(s):

Description:

Recommended Correction:

Your Name:

Address:

Email: Phone:

CT8_3-17
Measure Corroded Area

Overview

Determining the extent of pipeline corrosion requires various testing procedures to collect data on factors such as corrosion pit depth, pipe wall thickness, and remaining wall strength. If it is determined that the corrosion loss is localized, profile measurements of the corroded area must be obtained. This module describes procedures involved in measuring corroded areas of piping and documenting the extent and location of corrosion damage.

Trainees with successful module completions may be eligible for credentialing through the NCCER Registry. To learn more, go to **www.nccer.org** or contact us at 1.888.622.3720. Our website has information on the latest product releases and training, as well as online versions of our *Cornerstone* magazine and Pearson's product catalog.

Your feedback is welcome. You may email your comments to **curriculum@nccer.org**, send general comments and inquiries to **info@nccer.org**, or fill in the User Update form at the back of this module.

This information is general in nature and intended for training purposes only. Actual performance of activities described in this manual requires compliance with all applicable operating, service, maintenance, and safety procedures under the direction of qualified personnel. References in this manual to patented or proprietary devices do not constitute a recommendation of their use.

CT8_3-17
Measure Corroded Area

Objectives

When you have completed this module, you will be able to do the following:

1. Explain the need to determine defect length and the extent and location of damage.
   a. Describe how the interaction rule is used to determine defect length.
   b. Describe how to document the extent and location of damage.
2. Describe how to measure a corroded area (CT8_3-17).
   a. Identify the prerequisites, competencies, and abnormal operating conditions associated with measuring a corroded area.
   b. Describe the procedures used to measure a corroded area.

Performance Task

Under the supervision of your instructor, you should be able to do the following:

1. Measure a corroded area (CT8_3-17).
   - Identify potential abnormal operating conditions that may occur during performance of this CT, and know the appropriate actions to take in response to them.
   - Utilize the appropriate personal protective equipment according to relevant company procedures.
   - Perform a visual inspection of the exposed pipe. Look for integrity issues such as evidence of a release or significant metal deformation. If unsafe integrity issues are encountered, discontinue the task and make immediate notifications.
   - Gather the necessary tools and materials.
   - Clean and remove debris from the area to be measured.
   - Create a representation of the pipe surface to be inspected so that corroded areas on the pipeline can be accurately represented. Identify the long seam and the nearest girth weld as reference points and identify corrosion in proximity.
   - Overlay a grid on each area of localized corrosion.
   - Measure the longitudinal length and circumferential width of each area of localized corrosion.
   - Measure the distance between each area of localized corrosion.
   - Obtain profile measurements of the corrosion region (CT8_1-17 and/or CT8_2-17).
   - Obtain and determine the nominal wall thickness (CT8_2-17).
   - Identify the areas of greatest wall loss within each area of localized corrosion and obtain pit depth measurements (CT8_1-17 and/or CT8_2-17).
   - Denote all areas of localized corrosion, distances between those areas and pit depth readings on the representation.
   - Complete appropriate documentation and make notifications as required by operator's procedures.

Trade Terms

Abnormal operating conditions (AOCs)
Calibrate
Circumferential
Dent
General corrosion
Gouge
Interaction
Localized corrosion
Longitudinal
Nominal wall thickness
Pitting
Profile
Wall loss

Industry Recognized Credentials

If you are training through an NCCER-accredited sponsor, you may be eligible for credentials from NCCER's Registry. The ID number for this module is CT8_3-17. Note that this module may have been used in other NCCER curricula and may apply to other level completions. Contact NCCER's Registry at 888.622.3720 or go to **www.nccer.org** for more information.

Contents

Figures

Section One

1.0.0 Determine Defect Length and Extent of Damage

Objective

Explain the need to determine defect length and the extent and location of damage.

a. Describe how the interaction rule is used to determine defect length.
b. Describe how to document the extent and location of damage.

Trade Terms

General corrosion: Corrosion that takes place over the entire surface, resulting in thinning of the material.

Interaction: When two or more areas of localized corrosion are in close proximity and combine to reduce the overall wall thickness.

Localized corrosion: Corrosion in isolated areas.

Longitudinal: Pertaining to measurements taken along the length of a pipe.

Pitting: A type of localized corrosion characterized by the formation of small cavities or pits.

Wall loss: The loss of pipe metal either as a result of mechanical surface damage or corrosion.

Pipelines are subject to damage from both general corrosion and localized corrosion in the form of pitting. Pitting can undercut the surface of the piping, resulting in wall loss so that only a fragile layer remains on the surface. Interaction between areas of localized corrosion can combine to result in a serious loss of wall thickness.

The remaining wall thickness of a corroded pipe is a major factor in the overall mechanical integrity of the pipe. Technicians use specific procedures when measuring a corroded area to determine if two or more metal loss defects are close enough together to interact as a single defect. They must then accurately document the extent and location of the damage.

1.1.0 Using the Interaction Rule to Determine Defect Length

The interaction rule is the most widely used method of determining defect length. As shown in *Figure 1*, the interaction rule is a guide to help determine if two or more defects found during corrosion measurements are close enough together to interact as a single defect. The rule states that two or more defects may interact as one if the distance between them is less than the longitudinal length of the smallest defect. Procedures may vary; always follow company guidelines when measuring multiple defects.

1.1.1 Example Corrosion Array #1

The longitudinal length of the shortest defect (B) in corrosion array #1 (*Figure 2*) is ³⁄₄". The distance between the two defects (G) is shown as ¹⁄₂". The two defects interact because G is less than the longitudinal length of the shortest defect. The total length of array #1 is 2¹⁄₄".

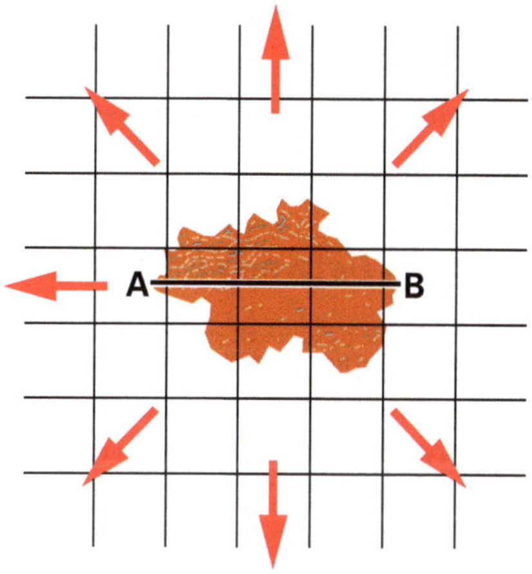

Figure 1 Interaction rule.

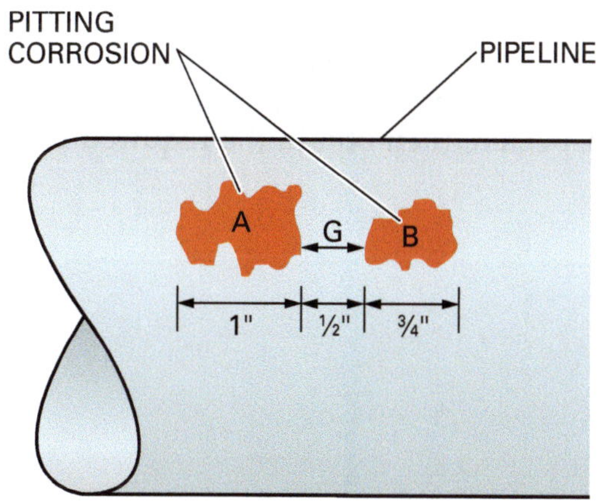

Figure 2 Corrosion array #1.

1.1.2 Example Corrosion Array #2

The longitudinal length of the shortest defect (F) in corrosion array #2 (*Figure 3*) is ½". The distance between the two defects (G) is shown as ¼". The two defects interact because G is less than the longitudinal length of the shortest defect. The total longitudinal length of array #2 is 1¼".

1.1.3 Example Corrosion Array #3

The longitudinal length of the shortest defect (D) in corrosion array #3 (*Figure 4*) is ¼". The distance between the two defects (G) is shown as ¼". The two defects interact because G is less than the longitudinal length of the shortest defect. The total longitudinal length of array #3 is ½".

> **NOTE**
>
> In corrosion array #3, the longitudinal length of D is equal to G. Although G is not less than the length of the shortest defect, as spelled out in the rule, the most conservative approach is to assume that the two corrosion defects, C and D, interact.

1.1.4 Interaction Between Multiple Defect Arrays

The interaction rule addresses interaction between defects, not defect arrays. *Figure 5* shows multiple corrosion defect arrays on a pipeline. These arrays, as shown, do not interact. The distances between the separate arrays are greater than the longitudinal length of the shortest defects in each array.

1.2.0 Documenting the Extent and Location of Damage

When determining the thickness of pipe, the alignment sheet for the pipeline being inspected must be obtained. This shows the location and features of the pipeline when it was originally built. Determine the exact location of the area to be inspected, then obtain the required information.

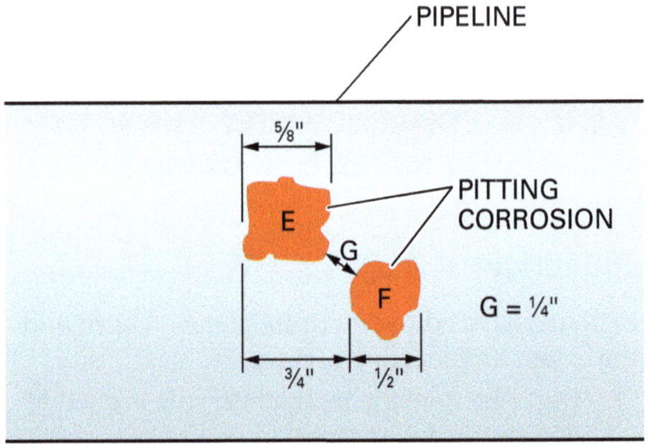

Figure 3 Corrosion array #2.

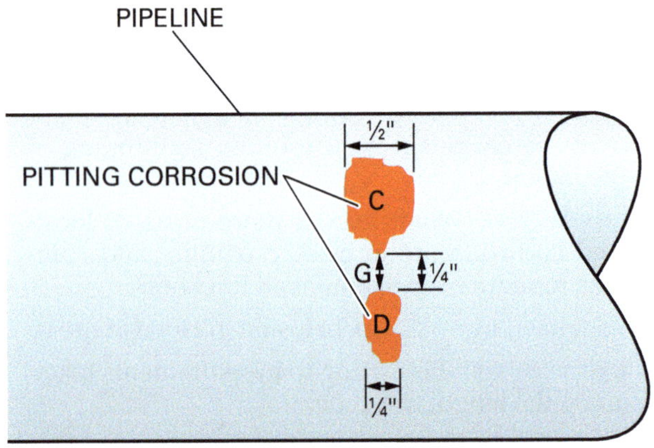

Figure 4 Corrosion array #3.

Document this information on the company pipeline inspection report form. The form for each company will differ, but most will require the same information. At a minimum, the form must contain the exact location of the inspection (such as the wheel count and clock position) and the results of the inspection, along with the inspector's contact information.

Always follow company procedures when completing an inspection report. This ensures that all the information needed to evaluate the extent of the corrosion and schedule any necessary repairs is provided. These reports can be compared to previous readings to determine the rate of corrosion. *Figure 6A*, *Figure 6B*, and *Figure 6C* show a typical pipeline inspection form.

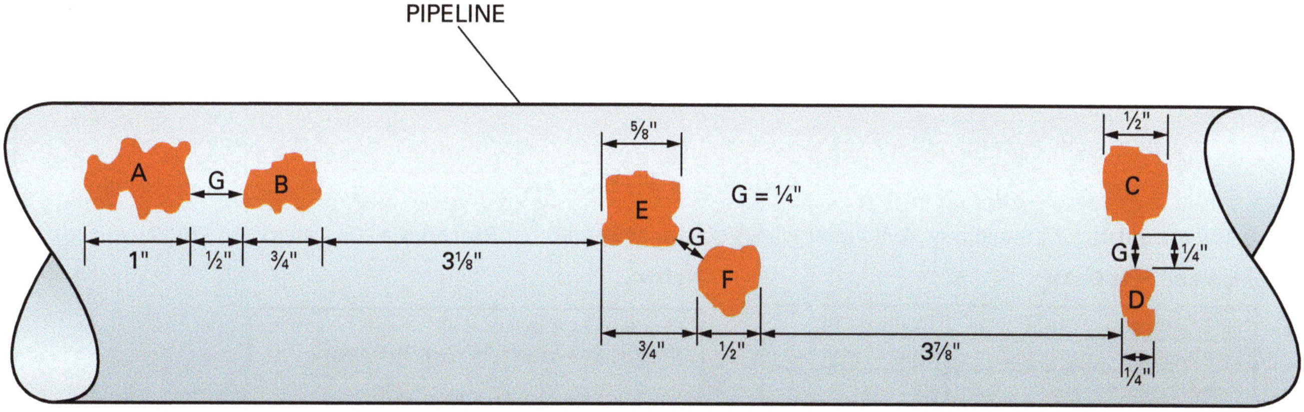

Figure 5 Multiple corrosion defect arrays.

Finite Element Analysis (FEA)

Different corrosion assessment standards have been developed and modified over the years to help technicians determine the effect of corrosion on piping. Technological advances continue to play a crucial role in new pipeline design and installation, as well as in determining the structural integrity of corroded piping. One example of a modern alternative assessment technique is based on finite element analysis (FEA). FEA software packages enable users to create accurate computer models of pipeline defects. Users can then estimate the structural integrity and burst strength of a corroded pipeline found to have interacting defects.

ANOMALY SITE DATA

Below is the data for the anomaly identified at wheel count **48340.60** In: **Rincon, GA** On Date: **7/11/2012**

Pipe and Site Data:

Line Segment	McIntosh PL
Line Number	N/A
Pipe Diameter:	24″
Nominal Wall Thickness:	.375″
SMYS:	65,000
Seam Type:	ERW
Seam Clock Position:	11:40
Actual Wall Thickness:	.375
Valve Section:	N/A
Class Location:	2
MAOP:	1200
In HCA:	No
Anomaly Type:	Metal Loss
Anomaly Rating from ILI/DCVG:	36%

Notes:

Data from I.L.I. Run:
36% Depth, 4.92"Lx7.96"W, at 7:15 o'clock

Effective Area Analysis

CAOP EAM – 1106.9
Burst Pressure (Pf) – 1844.9

External Corrosion *(Pitting)*:

Target Defect:	
Wheel Count	48340.60
Clock Position	7:00 to 7:55
Length:	4.60″
Width:	6.50″
Depth:	.219″
Interacting with:	Girth Weld .10″ Penetration
MPI Results: *(Lengths)*	No MPI
CAOP Mod B31G -	1042.6
Pf Mod B31G -	1737.7
Verified By	

Defect 2:	
Wheel Count	
Clock Position	
Length:	
Width:	
Depth:	
Interacting with	
MPI Results: *(Lengths)*	
CAOP Mod B31G -	
Pf Mod B31G -	
Verified By	

Defect 3:	
Wheel Count	
Clock Position	
Length:	
Width:	
Depth:	
Interacting with	
MPI Results: *(Lengths)*	
CAOP Mod B31G -	
EAM -	

Notes:

Corrective Action:
Installed **StrongBack Composite Wrap** centered over target feature at WC 48340.60

U/S Edge of Wrap GPS:
32.39810805, -86.5880206

D/S Edge of Wrap GPS:
32.39810817, -86.5880209

Recoat Information:
Recoated with SPC 2888

Page 1 of 3

Figure 6A Typical pipeline inspection form (1 of 3).

ANOMALY SITE DATA

Verified By	

Comments Regarding Repair Decisions:

Pipe Inspection Data:

Coating Type:	FBE
Bond To Pipe:	Good
Coating Thickness:	16mils
P/S At Surface:	-1.075V
P/S At Spring Line:	-1.050V
Anchor Profile:	N/A
Pipe Temperature:	75.3F
Ambient Temperature:	83.1F
Soil Type:	Clay
Soil PH:	6
Soil Resistivity/Ohms Cm :	4,300Ω
Cover to Top of Pipe:	67″
Length of Expose Pipe:	60″
Surface Drainage:	Good
Deposits under Coating:	Yes, Liquid pH 9
Calcareous Deposits:	None
MPI Results:	No MPI Indications
Any Other Interactions:	N/A

Notes:

Coating Thickness: (12:00 – 17mils) (03:00 – 15mils) (06:00 – 15mils) (09:00 – 16mils)

Notch Test: Tight

13.86VAC

Shrink Sleeve Evaluation:

Wrinkles located from 2:10o'clock to 3:00o'clock.

Gap in shrink sleeve at 7:15o'clock. Dirt and liquids found behind shrink sleeve, with a strong smell of Sulfide present.

Additional Threats *(Mechanical Damage/Gouges)*:

Type Threat:	
Length:	
Width:	
Depth:	
Interacting with	
CAOP – B31G	
Verified By	

UT Around Threat:	
3:00	0.376″
6:00	0.375″
9:00	0.380″
12:00	0.377″
Apex	

Type Threat:	
Length:	
Width:	
Depth:	
Interacting with	
CAOP – B31G	
Verified By	

Notes

Page 2 of 3

Figure 6B Typical pipeline inspection form (2 of 3).

ANOMALY SITE DATA

Dent Data:

Dent Length:	
Dent Width:	
Dent Depth at Apex:	
Dent % Depth	
Sharpness (L/d)	
Distance from U/S GW:	
Distance from D/S GW:	
Apex Distance from Edge of Dent U/S:	
Apex Distance from Edge of Dent D/S:	
Clock Position:	
Longitudinal Strain:	
Circumferential Strain:	
Interacting with Seam or Girth Weld:	
Depth at Weld	
Longitudinal Strain at Weld:	
Circumferential Strain at Weld:	
MPI Results: *(Get Linear Indication Lengths)*	
Any Other Interactions:	

Notes:

Additional Threats Associated With Dent *(Corrosion/Gouges)*:

Type Threat:	
Length:	
Width:	
Depth:	
Interacting with	
CAOP – B31G	
Verified By	

UT Around Pipe:	
3:00	
6:00	
9:00	
12:00	

Type Threat:	
Length:	
Width:	
Depth:	
Interacting with	
CAOP – B31G	
Verified By	

Notes

Comments Regarding Dent Repair Decisions:

Page 3 of 3

Figure 6C Typical pipeline inspection form (3 of 3).

Additional Resources

It is assumed that you will learn to perform these procedures under the direct supervision of your crew leader, foreman, or instructor. For independent study, refer to the recommended Additional Resources at the back of this module.

1.0.0 Section Review

1. The interaction rule states that two or more defects may interact as one if the distance between them is less than the _____.
   a. circumferential length of the smallest defect
   b. circumferential length of the largest defect
   c. longitudinal length of the smallest defect
   d. longitudinal length of the largest defect

2. To determine the exact location of pipeline corrosion, technicians use a(n) _____.
   a. alignment sheet
   b. AOC array
   c. flame grid
   d. interaction sheet

SECTION TWO

2.0.0 MEASURE A CORRODED AREA

Objective

Describe how to measure a corroded area (CT8_3-17).

a. Identify the prerequisites, competencies, and abnormal operating conditions associated with measuring a corroded area.
b. Describe the procedures used to measure a corroded area.

Performance Task

Measure a corroded area (CT8_3-17).

- Identify potential abnormal operating conditions that may occur, and know the appropriate actions to take in response to them.
- Utilize the appropriate personal protective equipment according to relevant company procedures.
- Perform a visual inspection of the exposed pipe. Look for integrity issues such as evidence of a release or significant metal deformation. If unsafe integrity issues are encountered, discontinue the task and make immediate notifications.
- Gather the necessary tools and materials.
- Clean and remove debris from the area to be measured.
- Create a representation of the pipe surface to be inspected so that corroded areas on the pipeline can be accurately represented. Identify the long seam and the nearest girth weld as reference points and identify corrosion in proximity.
- Overlay a grid on each area of localized corrosion.
- Measure the longitudinal length and circumferential width of each area of localized corrosion.
- Measure the distance between each area of localized corrosion.
- Obtain profile measurements of the corrosion region (CT8_1-17) and/or CT8_2-17).
- Obtain and determine the nominal wall thickness (CT8_2-17).
- Identify the areas of greatest wall loss within each area of localized corrosion and obtain pit depth measurements (CT8_1-17) and/or CT8_2-17).
- Denote all areas of localized corrosion, distances between those areas and pit depth readings on the representation.
- Complete appropriate documentation and make notifications as required by operator's procedures.

Trade Terms

Abnormal operating conditions (AOCs): Conditions within a pipeline system, identified by the operator, that may indicate (1) a malfunction of a component or (2) a deviation from normal operations that may indicate a condition exceeding design limits or result in a hazard to persons, property, or the environment.

Calibrate: A process used to ensure accuracy by comparing an instrument reading against a known standard.

Circumferential: Pertaining to measurements taken in the direction of the curvature of a pipe.

Dent: A depression in the surface of a pipe that affects the curvature of the pipe without reducing the pipe wall thickness.

Gouge: Damage to the surface of a pipe that displaces or removes metal and reduces the pipe wall thickness.

Nominal wall thickness: The expected wall thickness of a pipe as determined by alignment sheets or other records.

Profile: A graphic representation of a corroded pipe surface that shows depth and length detail.

Pipeline technicians assigned to determine the extent of pipeline corrosion and the urgency of any needed repairs must have some basic competencies as well as the ability to follow the procedures for measuring a corroded area and recording the results. This section describes some of the basic skills needed by pipeline technicians, and identifies abnormal operating conditions (AOCs) that personnel may need to respond to when measuring corroded areas. It also describes the basic procedures used to measure corroded areas.

2.1.0 Prerequisites, Competencies, and Abnormal Operating Conditions

Personnel must possess a variety of skills in order to assess and measure the corroded area of a pipeline. The operation requires a series of steps, which a pipeline technician must be competent to perform. Personnel must be aware of the particular abnormal operating conditions that may occur during the procedure and know the proper response to them.

2.1.1 Prerequisites and Competencies

Personnel assigned to determine the structural integrity of a pipeline by measuring corroded areas must understand what the task entails. Prior to beginning the task, personnel must be able to properly select, calibrate, and operate any tools or instruments to be used during the procedure.

Personnel must understand that the task begins when the pipe is first exposed, cleaned, and prepared for inspection. They must be able to visibly inspect the pipe and determine if there are any unsafe integrity issues, such as evidence of a release or significant metal deformation, and discontinue the task and make immediate notifications if any are found. Once the inspection begins, personnel must know how to position, manipulate, and read graphical representations of the corroded area as well as any equipment used to obtain valid measurements. Personnel must understand that the task is not complete until all measurements are documented and the proper notifications are made.

Personnel must have a basic understanding of additional tasks related to measuring a corroded area, such as measuring pit depth (CT8_1-17) and measuring wall thickness (CT8_2-17). They must also understand how the *ASME/ANSI B31G* and/or RSTRENG corrosion assessment methods are applied to determine the remaining wall strength for the pipe once pit depth, wall thickness, and other data have been gathered.

2.1.2 Abnormal Operating Conditions

Prior to measuring a corroded area, pipeline technicians must be able to recognize potential AOCs and know the proper response to each, should they occur. The following are some potential AOCs, and the appropriate action personnel should take in response to them:

- *Internal or external corrosion of the pipe is present* – Document the condition according to company guidelines and notify personnel responsible for correcting the problem.
- *A pipeline has been significantly deformed by line hits or mechanical surface damage* – Document the condition accordingly and notify personnel responsible for correcting the problem.
- *Fire or explosion caused by the ignition of a hydrocarbon liquid or gas* – Immediately evacuate the area and notify the appropriate firefighting personnel. It may be necessary to call 911 in such an emergency.
- *Unexpected leaks of hazardous liquid or carbon dioxide occurs because of a loss of pipe wall thickness* – Take steps to eliminate any possible ignition sources and notify the appropriate personnel, possibly including calling 911.

2.2.0 Procedures

If prior tests and measurements indicate that the pipeline damage is from localized corrosion, profile measurements of the corroded area must be obtained. A measurement grid of the corroded area can be used to help determine the appropriate number of measurements to profile (*Figure 7*). Use the following basic procedure to measure a corroded area (always follow company procedures when doing so):

WARNING!

Always obtain and use the appropriate personal protective equipment for each task. Stay alert to the possibility of unexpected leaks from the exposed pipeline such as hazardous liquid or carbon dioxide (CO_2). Safely eliminate all ignition sources and immediately notify the appropriate personnel.

Step 1 Identify potential abnormal operating conditions that may occur, and know the appropriate actions to take in response to them.

Step 2 Utilize the appropriate personal protective equipment according to relevant company procedures.

Step 3 Visually inspect the piping for any evidence of leaks or loss of pipe integrity. Look for mechanical surface damage such as a dent or a gouge. Dents alter the external shape of the pipe without reducing the wall thickness, while gouges displace or remove surface metal and reduce the wall thickness. If any unsafe integrity issues are present, stop the task and immediately notify the appropriate personnel.

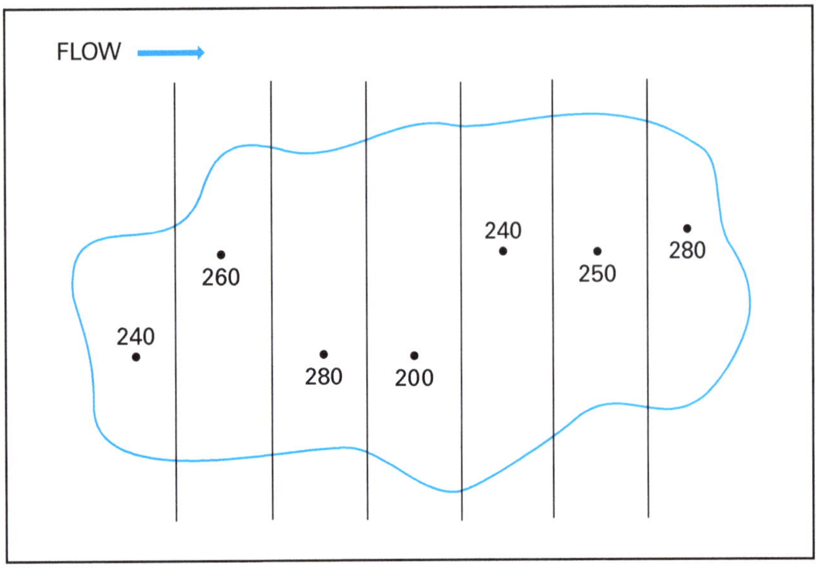

Drawing No.: XCW-123674
Direction to Reference Weld: Upstream
Distance to Reference Weld: 14 ft.
Length Increment: 1/2 inch
Depth Units: mils

Prepared by: F. Weasley
Date: 9/05/14

Figure 7 Corrosion measurement grid.

Step 4 Obtain all necessary equipment and tools (for example, a tape measure, pit gauge, ultrasonic thickness meter, metallic ruler, bridging bar, graph paper, carbon paper, carpenter's pencil, etc.).

Step 5 Prepare, clean, and remove debris from the surface to be measured. Coatings, primer, and surface deposits may interfere with accurate readings and need to be removed.

Step 6 Create a representation of the pipe surface to be inspected. This is normally done by visually splitting the pipe from the top or bottom and drawing it as a flat rectangle. Always follow company policy for pipe representation. Identify the nearest girth weld and the long seam as reference points. Note areas of corrosion on the drawing.

Step 7 Overlay a grid on each area of localized corrosion. Grids are used to represent areas of general corrosion to provide additional detail for assessment.

Step 8 Measure the longitudinal length of each area of localized corrosion. This measurement is required for assessment of pipeline integrity.

Step 9 Measure the **circumferential** width of each area of localized corrosion. This is used to properly identify the localized corrosion on the overall representation.

Step 10 Measure the distance between each area of localized corrosion. This is used to properly identify the localized corrosion on the overall representation and to determine the interaction between discrete areas of localized corrosion.

Step 11 Obtain profile measurements of the corrosion region. The profile measurements can be represented as the remaining wall thickness or actual pit / general corrosion depth.

NOTE

This must be performed by an individual qualified on CT8_1-17 and/or CT8_2-17.

Step 12 Obtain and determine the **nominal wall thickness**.

NOTE

This must be performed by an individual qualified on CT8_2-17.

Step 13 Identify areas of greatest wall loss within each area of localized corrosion and obtain pit depth measurements.

> **NOTE**
>
> This must be performed by an individual qualified on CT8_1-17 and/or CT8_2-17.

Step 14 Using the graph paper lines as a guide, measure and record the deepest pits (greatest wall loss) in each segment. Include flow direction, orientation, distance to the reference weld, date, and other company-required data on the measurement grid.

> **CAUTION**
>
> If pit depth is equal to 80 percent or more of the pipe company specification, the affected area of the pipe must be replaced immediately.

Step 15 Document all findings on the appropriate company inspection form and make notifications according to company policy.

Additional Resources

It is assumed that you will learn to perform these procedures under the direct supervision of your crew leader, foreman, or instructor. For independent study, refer to the recommended Additional Resources at the back of this module.

2.0.0 Section Review

1. A pipeline technician should eliminate ignition sources and notify the appropriate personnel if there is _____.
   a. any visible denting on the pipeline surface
   b. a hazardous liquid or carbon dioxide leak
   c. evidence that the pipeline is still in service
   d. any indication of internal piping corrosion

2. When measuring each area of localized corrosion on a pipeline, a pipeline technician must measure the _____.
   a. latitudinal width and the circumferential depth
   b. horizontal length and latitudinal height
   c. diametrical width and vertical depth
   d. longitudinal length and the circumferential width

Summary

Determining the extent of pipeline corrosion requires various testing procedures to collect and analyze data. If it is determined that the corrosion loss is localized, profile measurements of the corroded area must also be obtained. This module described procedures involved in measuring corroded areas of piping and documenting the extent and location of corrosion damage. Understanding how to measure, assess, and record corroded areas of piping enables technicians to determine the extent of the damage and the urgency of pipe repair or replacement.

Review Questions

1. Two corrosion defects are identified in a piping section. If the longitudinal length of the shortest defect is 1" and the distance between the two defects is $^3/_4$", which of these statements is true?
   a. The two defects interact because they both have longitudinal lengths greater than 1".
   b. There is no interaction between the defects because the longitudinal length between them is less than the circumferential width of the pipe.
   c. The two defects interact because the distance between them is less than the longitudinal length of the shortest defect.
   d. There is no interaction because the defects are separated by at least $^1/_2$".

2. If the longitudinal length of the shortest defect in a corrosion array is the same as the distance between the two defects, it is best to assume that _____.
   a. the defects interact
   b. there is ample separation between the defects
   c. the defects do not interact
   d. there have been measurement errors during testing

3. At a minimum, a company pipeline inspection report form must contain the inspector's contact information along with _____.
   a. atmospheric conditions that were present during the inspection
   b. the approximate time and place of the inspection
   c. a basic summary of the testing equipment and methods
   d. the exact location and results of the inspection

4. Personnel assigned to determine the structural integrity of a pipeline by measuring the corroded area must understand that the task begins _____.
   a. once RSTRENG personnel have approved the inspection
   b. when the pipe is first exposed, cleaned, and prepared for inspection
   c. after detailed damage estimates have been established
   d. as soon as the local governing body completes its testing

5. When creating a representation of the pipe surface to be inspected, a technician should visually split the pipe from the top or bottom and draw it as a flat _____.
   a. rectangle
   b. circle
   c. cylinder
   d. sphere

Trade Terms Quiz

Fill in the blank with the correct term that you learned from your study of this module.

1. To ensure accuracy by comparing an instrument reading against a known standard is to _________.
2. Damage to the surface of a pipe that displaces or removes metal and reduces the pipe wall thickness is a(n) _________.
3. A graphic representation of a corroded pipe surface that shows depth and length detail is a(n) _________.
4. When two or more areas of localized corrosion in close proximity combine to reduce the overall wall thickness there is said to be a(n) _________.
5. Measurements taken along the length of a pipe are _________.
6. Corrosion in adjacent or isolated areas is known as _________.
7. Measurements taken in the direction of the curvature of a pipe are _________.
8. A type of localized corrosion characterized by the formation of small cavities or pits is called _________.
9. The loss of pipe metal either as a result of mechanical surface damage or corrosion is called _________.
10. A depression in the surface of a pipe that affects the curvature of the pipe without reducing the pipe wall thickness is known as a(n) _________.
11. Corrosion that takes place over the entire surface and results in thinning of the material is known as _________.
12. The expected wall thickness of a pipe as determined by alignment sheets or other records is known as the _________.
13. Conditions within a pipeline system, identified by the operator, that may indicate (1) a malfunction of a component or (2) a deviation from normal operations that may indicate a condition exceeding design limits or result in a hazard to persons, property, or the environment are called _________.

Trade Terms

Abnormal operating conditions (AOCs)
Calibrate
Circumferential
Dent
General corrosion
Gouge
Interaction
Localized corrosion
Longitudinal
Nominal wall thickness
Pitting
Profile
Wall loss

Trade Terms Introduced in This Module

Abnormal operating conditions (AOCs): Conditions within a pipeline system, identified by the operator, that may indicate (1) a malfunction of a component or (2) a deviation from normal operations that may indicate a condition exceeding design limits or result in a hazard to persons, property, or the environment.

Calibrate: A process used to ensure accuracy by comparing an instrument reading against a known standard.

Circumferential: Pertaining to measurements taken in the direction of the curvature of a pipe.

Dent: A depression in the surface of a pipe that affects the curvature of the pipe without reducing the pipe wall thickness.

General corrosion: Corrosion that takes place over the entire surface, resulting in thinning of the material.

Gouge: Damage to the surface of a pipe that displaces or removes metal and reduces the pipe wall thickness.

Interaction: When two or more areas of localized corrosion are in close proximity and combine to reduce the overall wall thickness.

Localized corrosion: Corrosion in isolated areas.

Longitudinal: Pertaining to measurements taken along the length of a pipe.

Nominal wall thickness: The expected wall thickness of a pipe as determined by alignment sheets or other records.

Pitting: A type of localized corrosion characterized by the formation of small cavities or pits.

Profile: A graphic representation of a corroded pipe surface that shows depth and length detail.

Wall loss: The loss of pipe metal either as a result of mechanical surface damage or corrosion.

Additional Resources

The following recommended resources can provide additional helpful information related to the requirements for this covered task:

API Recommended Practice 1161, Third Edition, © 2014 American Petroleum Institute. API Publishing Services, 1220 L Street NW, Washington, DC 20005. Distributed by Thompson Reuters (Scientific) LLC, **www.techstreet.com**.

49 *CFR* 192, *Subpart I*,**www.ecfr.gov**.

49 *CFR* 195,*Subpart H*,**www.ecfr.gov**.

NCCER Module AOCCC-17, *Abnormal Operating Conditions Control Center.*

NCCER Module AOCFG-17, *Abnormal Operating Conditions Field and Gas.*

NCCER Module 62105-14, *Introduction to Pipeline Documents.*

NCCER Module 66101-14, *Introduction to the Pipeline Industry.*

NCCER Module CT8_1-17, *Measure Pit Depth with Pit Gauge.*

NCCER Module CT8_2-17, *Measure Wall Thickness with Ultrasonic Meter.*

NCCER Module CT14_1-17, *Locate Line.*

ASTM G46-94 (2013),*Standard Guide for Examination and Evaluation of Pitting Corrosion*, **www.astm.org**.

Figure Credits

Photo courtesy of Alan Kehr, Alan Kehr Anti-Corrosion, LLC, Module Opener

Energy Maintenance Services, EMS USA, Figures 6A–C

Section Review Answer Key

Answer	Section Reference	Objective
Section One		
1. c	1.1.0	1a
2. a	1.2.0	1b
Section Two		
1. b	2.1.2	2a
2. d	2.2.0	2b

NCCER CURRICULA — USER UPDATE

NCCER makes every effort to keep its textbooks up-to-date and free of technical errors. We appreciate your help in this process. If you find an error, a typographical mistake, or an inaccuracy in NCCER's curricula, please fill out this form (or a photocopy), or complete the online form at **www.nccer.org/olf**. Be sure to include the exact module ID number, page number, a detailed description, and your recommended correction. Your input will be brought to the attention of the Authoring Team. Thank you for your assistance.

Instructors – If you have an idea for improving this textbook, or have found that additional materials were necessary to teach this module effectively, please let us know so that we may present your suggestions to the Authoring Team.

NCCER Product Development and Revision

13614 Progress Blvd., Alachua, FL 32615

Email: curriculum@nccer.org
Online: www.nccer.org/olf

❑ Trainee Guide ❑ Lesson Plans ❑ Exam ❑ PowerPoints Other ______________

Craft / Level: Copyright Date:

Module ID Number / Title:

Section Number(s):

Description:

Recommended Correction:

Your Name:

Address:

Email: Phone:

CT9_1-17
Install Bonds

Overview

Various cathodic protection systems—which include the installation of bonds, galvanic anodes, rectifiers, and impressed current groundbeds—are used to mitigate stray current interference and prevent corrosion in pipelines. The purpose of a cathodic protection bond is to connect two or more structures electrically to better use cathodic protection systems and prevent possible structure damage caused by interference. This module describes the general procedures for the installation of bonds.

Trainees with successful module completions may be eligible for credentialing through the NCCER Registry. To learn more, go to **www.nccer.org** or contact us at 1.888.622.3720. Our website has information on the latest product releases and training, as well as online versions of our *Cornerstone* magazine and Pearson's product catalog.

Your feedback is welcome. You may email your comments to **curriculum@nccer.org**, send general comments and inquiries to **info@nccer.org**, or fill in the User Update form at the back of this module.

This information is general in nature and intended for training purposes only. Actual performance of activities described in this manual requires compliance with all applicable operating, service, maintenance, and safety procedures under the direction of qualified personnel. References in this manual to patented or proprietary devices do not constitute a recommendation of their use.

CT9_1-17
Install Bonds

Objectives

1. Describe stray current interference and methods for reducing or eliminating it.
   a. Identify types of stray current and describe how they affect cathodic protection.
   b. Identify and describe methods used to reduce or eliminate stray current interference.
2. Describe how to install bonds (CT9_1-17).
   a. Identify the prerequisites, competencies, and abnormal operating conditions associated with installing bonds.
   b. Describe the procedure for installing bonds.

Performance Task

Under the supervision of the instructor, you should be able to do the following:

1. Install bonds (CT9_1-17).
   - Identify potential abnormal operating conditions that may occur during performance of this CT, and know the appropriate actions to take in response to them.
   - Utilize the appropriate personal protective equipment according to relevant company procedures.
   - Identify the structures to be connected.
   - Install the test leads and bond cables/wires on both structures at the location of current discharge.
   - Attach the test leads and bond cables/wires by exothermic (thermite) weld, pin brazing, or other method which will yield a permanent, low-resistance connection.
   - Terminate the test leads and bond cables/wires inside of the test box/station that is accessible to both structures.
   - Install shunts for the measurement of current flow and resistance inside of the test box/station.
   - Install blocking diodes as required.
   - Conduct tests to determine the effectiveness of the installed interference bond.
   - Complete appropriate documentation as required by operator's procedures.

Trade Terms

Abnormal operating conditions (AOCs)
Anode banks
Bonds
Diodes
Dynamic stray current
Electrolyte
Interference
Potential gradients
Rectifier
Resistivity
Shunt
Static stray current
Stray currents
Structure-to-soil potential
Test leads

Industry Recognized Credentials

If you are training through an NCCER-accredited sponsor, you may be eligible for credentials from NCCER's Registry. The ID number for this module is CT9_1-17. Note that this module may have been used in other NCCER curricula and may apply to other level completions. Contact NCCER's Registry at 888.622.3720 or go to **www.nccer.org** for more information.

Contents

Figures

Section One

1.0.0 Stray Current Interference

Objective

Describe stray current interference and methods for reducing or eliminating it.

a. Identify types of stray current and describe how they affect cathodic protection.
b. Identify and describe methods used to reduce or eliminate stray current interference.

Trade Terms

Anode banks: Sets of two or more anodes installed in a specific configuration where the anodes are connected together by a header cable/wire; also referred to as groundbeds.

Bonds: Electrical connections of two structures that usually have two different cathodic protection levels.

Diodes: Devices used to allow current to flow in only one direction.

Dynamic stray current: A portion of a current that flows over a path other than the intended path in an on/off or intermittent fashion and may cause electrochemical corrosion of metals in contact with electrolytes.

Electrolyte: A substance that ionizes (releases free electrons) in an electric field.

Interference: The effect on the cathodic protection level of a structure due to the interaction between its cathodic protection system and the cathodic protection system of another structure.

Potential gradients: Changes in potential with respect to distance, expressed in millivolts per unit of distance.

Rectifier: A device used to convert AC current to DC current. In an impressed cathodic protection system, it supplies current to the structure in the soil.

Resistivity: The ability of a material to resist the flow of electric current.

Shunt: A conductor that joins two points in an electrical circuit to form a parallel or alternative path through which a portion of the current may pass.

Static stray current: A portion of a current that flows over a path other than the intended path at a constant, unbroken rate and that may cause electrochemical corrosion of metals in contact with electrolytes.

Stray currents: Current flows through a path other than the intended circuit.

Structure-to-soil potential: A measurement of the voltage or potential difference between a buried or submerged metal structure and an electrolyte, such as soil; also referred to as structure-to-electrolyte potential.

Corrosion on a pipeline can be caused by **stray currents**, in which electrical currents flow through a path other than the intended circuit. The current flow results in the problem called **interference**, or *stray current interference*. This section describes the types and sources of stray currents, explains how to test for interference current, and reviews some of the methods that can be used to eliminate stray current interference.

1.1.0 Stray Current

Stray currents are direct currents flowing in the earth that are from a source other than the corrosion protection source. **Potential gradients** cause natural occurrences of stray current in the earth that are transferred to the protected pipeline. *Figure 1* shows potential gradients around a protected pipeline.

Normally, this outside current enters the pipeline at one point, exits the pipeline at another point, and returns to the earth. The area where the current returns to earth, completing the circuit, is the point at which the corrosion may occur if the material is in contact with an **electrolyte**. The measurement of the voltage or potential difference between a buried or submerged metal structure and an electrolyte, such as soil, is the **structure-to-soil potential**, also referred to as the *structure-to-electrolyte potential.*

1.1.1 Types of Stray Current

Stray currents are either static (constant) or dynamic (intermittent). Sources of stray current include cathodic protection (CP) systems on other pipelines, DC transit systems, DC welding op-

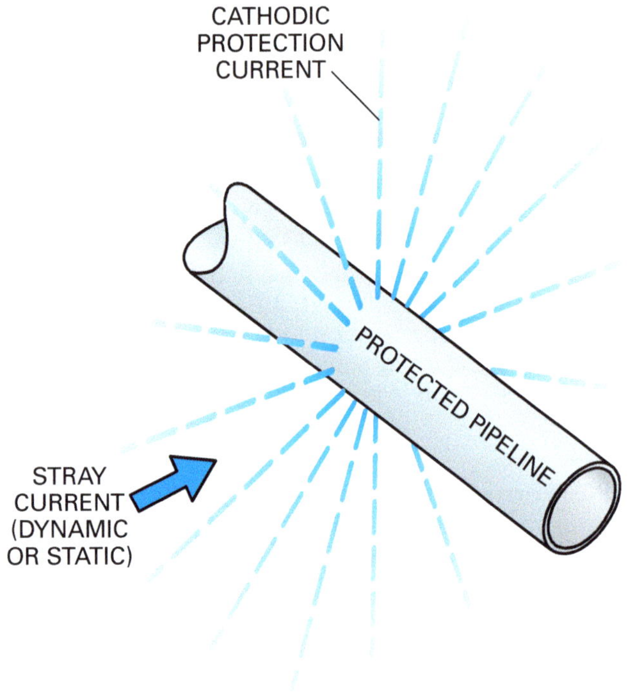

Figure 1 Potential gradients around a protected pipeline.

erations, DC mining operations, high-voltage DC transmission systems, or disturbances in the earth's magnetic field. If environmental conditions allow rectifying junctions such as copper oxide films, AC current flow can be rectified. This results in a direct current that can be mitigated.

There are many different sources of interference current in a cathodic protection system. For example, stray current interference on adjacent pipelines can be caused by impressed current cathodic protection systems. This depends on the locations of the anode banks and the operating characteristics of the cathodic protection system.

Figure 2 shows stray current to a crossover foreign pipeline. In this example, the foreign pipeline picks up current from the anode bank and transfers the current to the protected pipeline when it crosses under it. The potential for corrosion is greatly increased at the crossing point of the pipelines.

Figure 3 shows stray current to a parallel foreign pipeline. In this example, the parallel foreign pipeline is affected by the anode bank of the protected pipe. This is because the foreign pipeline travels through the anode bank and picks up current. When this occurs, the effectiveness of the CP system will be reduced in this area of the pipeline. In the affected area of the foreign pipeline, it becomes positive to the soil and discharges current to the earth. This increases the potential for corrosion on the foreign pipeline in the affected area.

1.1.2 Testing for Interference Current

Static stray current interference is easy to test for as long as there is no presence of dynamic stray current. Dynamic stray current can be caused by sources such as DC-powered transit systems, underground mining, welding, and pipeline interference. In *Figure 4*, a section of coated pipeline has an impressed current CP system applied to it and has two foreign pipelines crossing it. In this example, "B" is a strong indicator of interference because it has the largest difference in voltage. Always look for large changes in voltage to identify interference. Follow company guidelines for determining the need to mitigate interference.

To distinguish the current in the foreign pipeline from the current in the protected line, an automatic current interrupter is installed in the output of the rectifier. The current interrupter is a device that automatically opens and closes an electrical circuit at a preset time. With the current interrupter operating, measure the potential of the lines under both on and off conditions. *Figure 5* shows how to measure this potential at the test station. As shown, a shunt is installed for measurement of current flow and resistance inside of the test station.

These tests are performed using a copper sulfate electrode placed over the crossing point of the lines. Pipeline locators can be used to determine the exact location of the crossing.

WARNING!

Always be alert when testing for interference current. A release could be found while testing.

The data obtained at each pipeline crossing must be properly recorded. *Figure 6* shows typical interference testing potentials. These records are also called field sheets. Field data sheets should also include any information about the protected line, the date, and the current output of the rectifier.

Telluric Currents

Telluric currents are caused by sunspots and solar flares and can result in inaccurate readings during intense periods of solar activity.

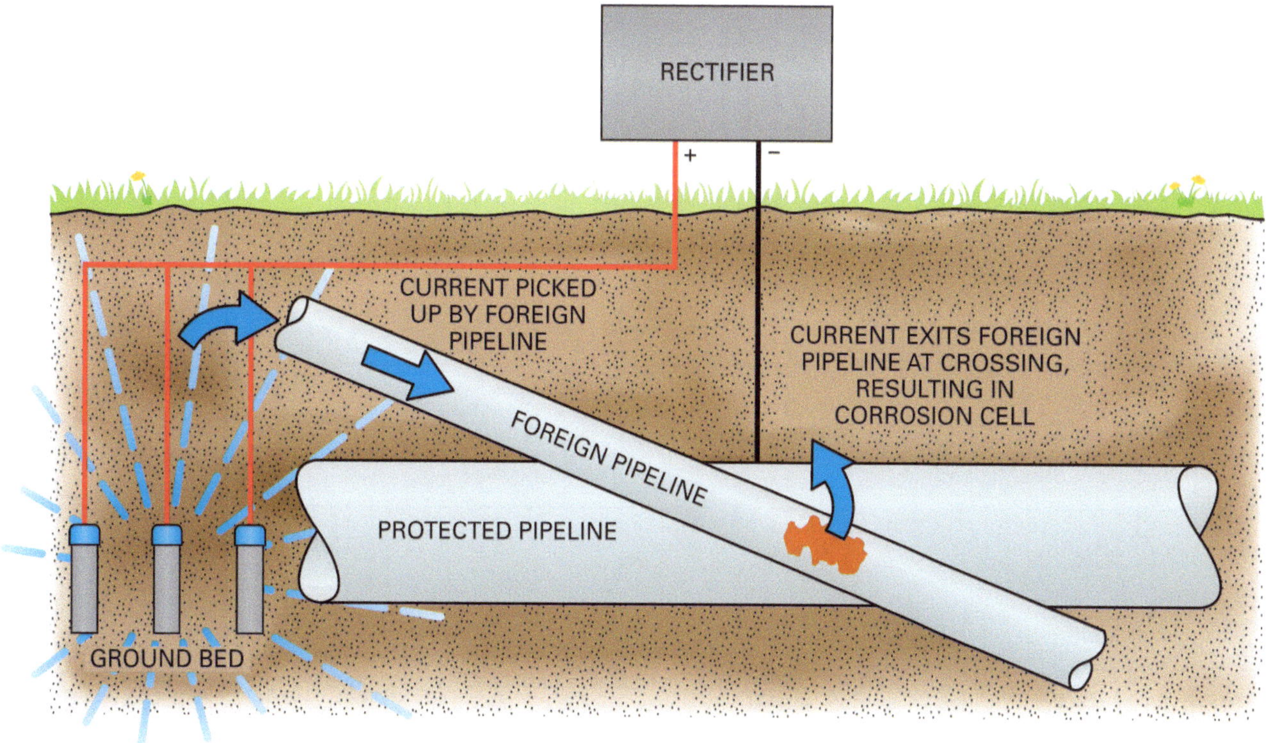

Figure 2 Stray current to a crossover foreign pipeline.

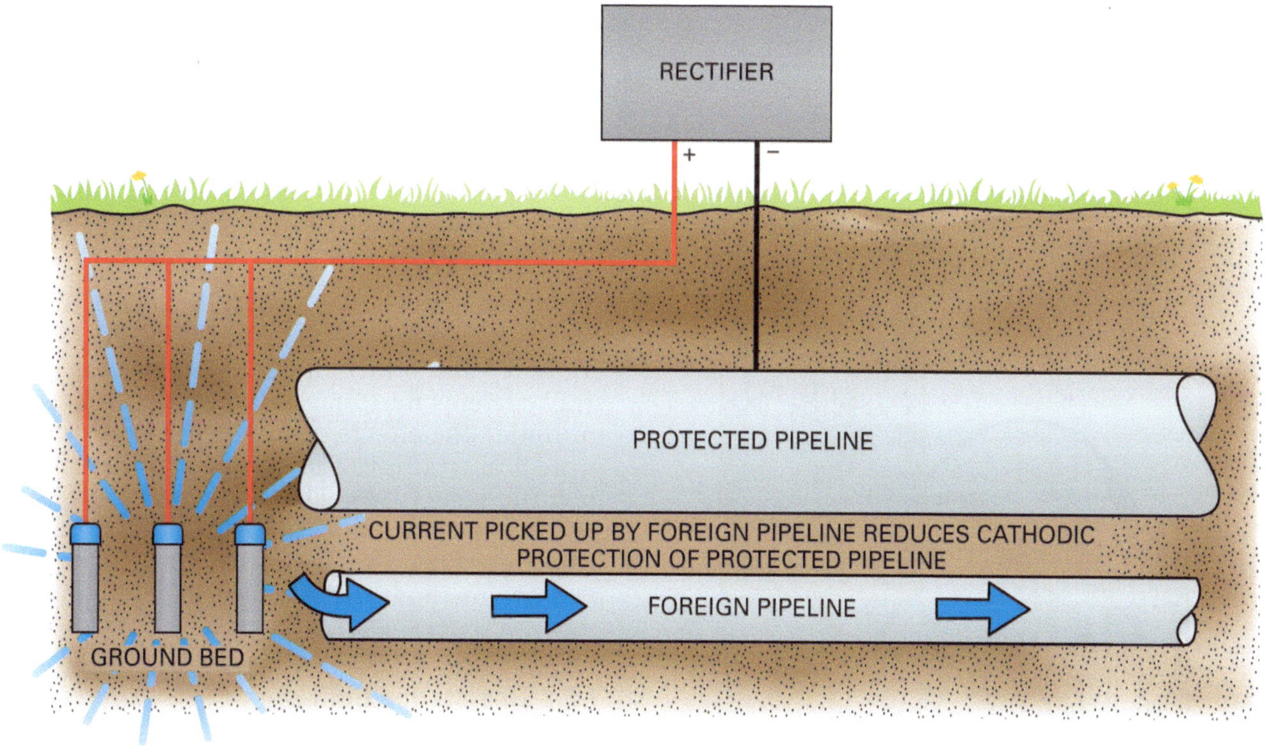

Figure 3 Stray current to a parallel foreign pipeline.

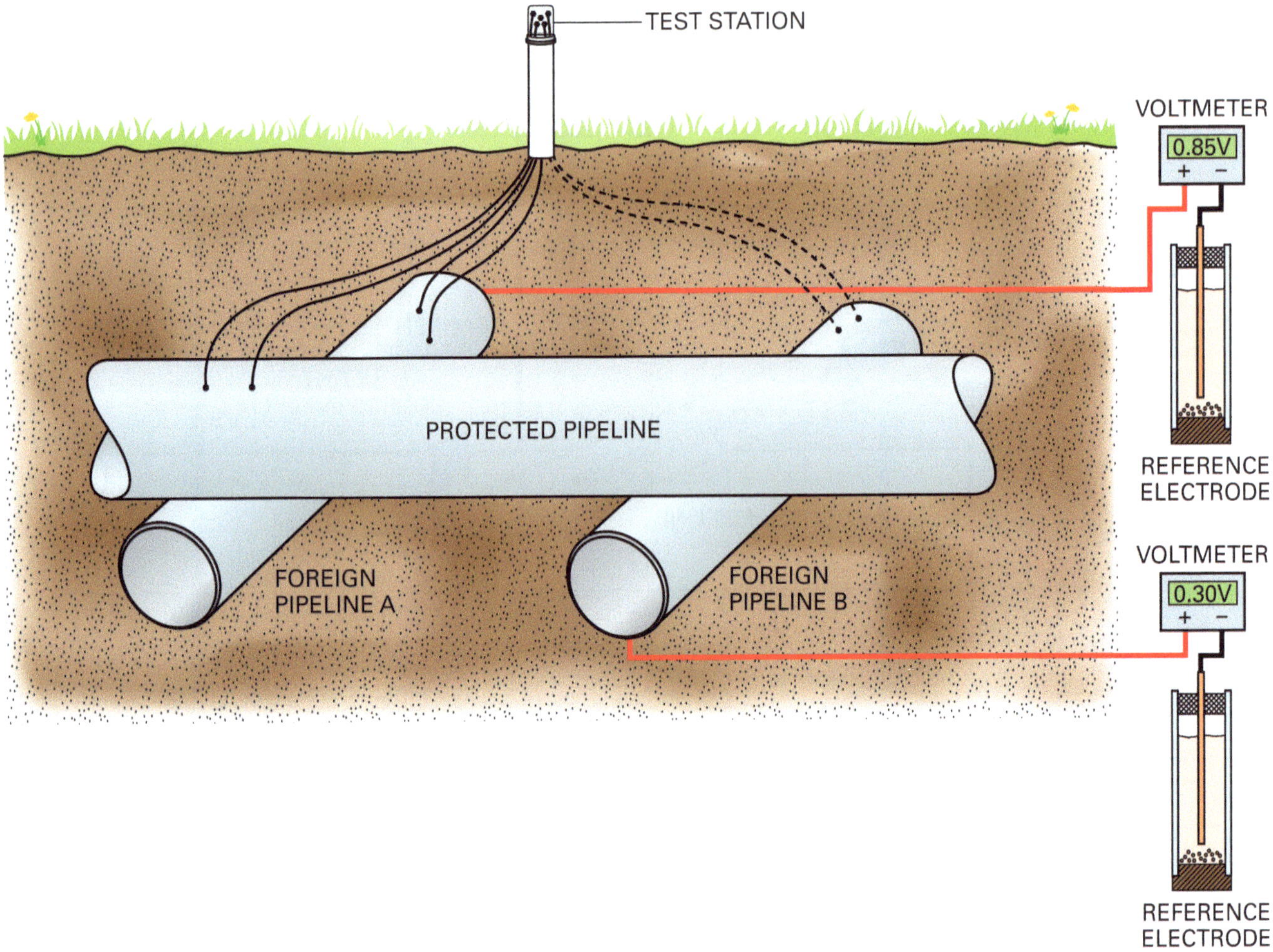

Figure 4 Testing foreign pipeline crossings for stray current interference from CP installations.

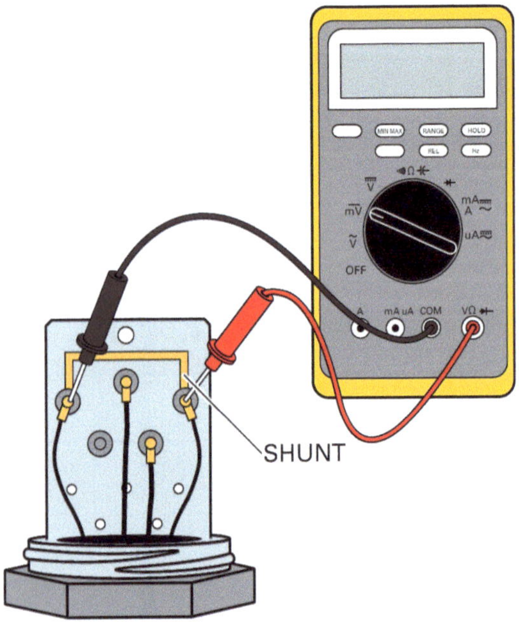

Figure 5 Measuring potential at the test station.

1.2.0 Methods for Reducing or Eliminating Stray Current Interference

According to the *Code of Federal Regulations* 49 *CFR* 192.473 (for gas pipelines) and 49*CFR* 195.577 (for liquid pipelines), any pipeline system subjected to stray currents must have an ongoing program to minimize the effects of these currents and help protect surrounding or adjacent underground metallic structures. There are many different methods used to reduce or eliminate stray current interference from a CP system. Some of the most common methods include the use of bonds, coatings, electrical shields, and galvanic anodes.

1.2.1 Resistance Bonds

A bond is simply a metallic connection between structures. There are two types of bonds: critical and non-critical. Critical bonds are identified by the operator's procedures and include those bonds that may cause damage to the structure if disconnected. Like test leads, bond leads are terminated in a test station.

FOREIGN LINE		PROTECTED PIPELINE (volts)			INTERFERING PIPELINE (volts)		
NAME	PIPELINE STATION	VOLTAGE ON	OFF	Δ volts	VOLTAGE ON	OFF	Δ volts
A	140 + 00	-0.91	-0.90	-0.01	-0.85	-0.88	+0.03
B	1210 + 00	-1.94	-1.00	-0.94	-0.30	-0.68	+0.38

Figure 6 Sample of recording survey notes.

Foreign pipelines can be subject to stray current damage from the impressed current cathodic protection system on another pipeline. To correct this problem, a resistance bond is connected between the two pipelines. The amount of resistance in the bond is adjusted to drain enough current from the affected line to eliminate the corrosion potential.

CAUTION

For liability reasons, the representative of the foreign pipeline company should be on site before procedures to correct the problem on the affected line begin.

For a resistance bond, **diodes** are installed as basic components of a rectifier circuit. Single-phase rectifiers are probably the most common units used for cathodic protection. The diode is installed to allow current to flow in one direction only. This characteristic of the diode causes all the current waves to be more aligned when entering into the cables. Along with other components, the diodes reduce ripple effects in the power supply and filter out interference.

1.2.2 Coatings

If the testing of a foreign line indicates that the interference is caused by a foreign pipeline that is bare or poorly coated, the bond base may not be able to correct the condition. If a high current density is flowing from a well-coated line through the soil and onto the bare foreign line, localized corrosion could occur. To correct this problem, reduce the density of the current flow to the crossing area of the foreign line. The best way to accomplish this is to apply a coating to the line. *Figure 7* shows the use of a coating to correct interference at foreign pipeline crossings.

With proper coating, the voltage drops in the earth around the foreign line are reduced or eliminated. Over-the-line potential measurements are used to identify the portion of the foreign line subject to interference. This information is then used to determine the area on the foreign line that must be coated to mitigate the interference.

CAUTION

Take special precautions to avoid bare spots in the coating (holidays) to prevent localized corrosion.

1.2.3 Electrical Shields

Electrical shields are defined as barriers to stray current flow. Electrical shielding can be provided by nonmetallic insulating barriers or by diverting the current from the affected pipeline to another structure. Shields are particularly useful for protecting the pipeline in congested areas such as pumping stations and tank farms. *Figure 8* shows bare pipe used as electrical shields around a pipeline.

1.2.4 Galvanic Anodes

Galvanic anodes can provide cathodic protection for buried or submerged structures. They can also be used to correct interference, as spot protection, and for alternating current mitigation. When used to correct interference, the anodes are attached to the affected line (see *Figure 9*). They are located in the area that is subject to interference from the foreign line. Anodes are buried in a special backfill that must be wet in order for the anodes to operate.

This method uses the anodic potential fields that are around the galvanic anode to offset

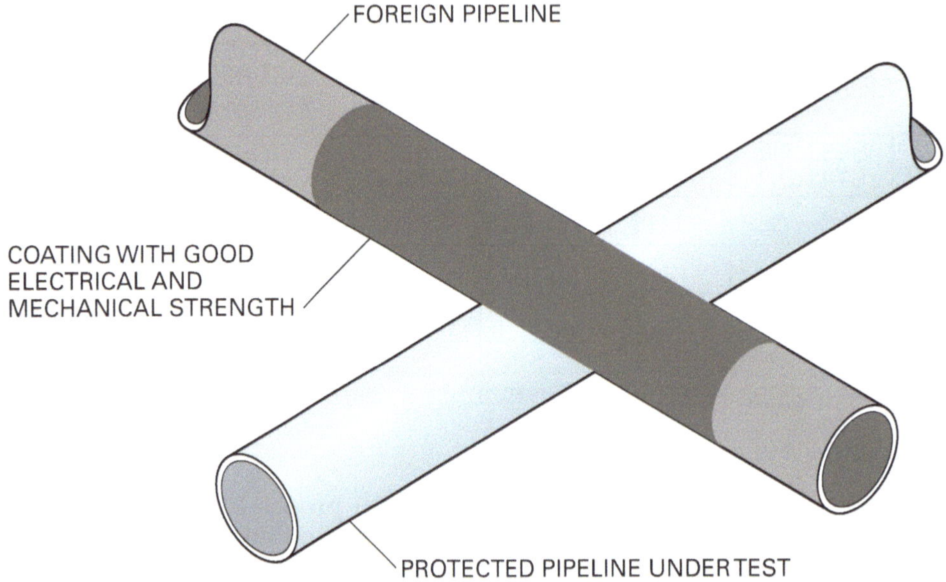

Figure 7 Using a coating to correct interference at foreign pipeline crossings.

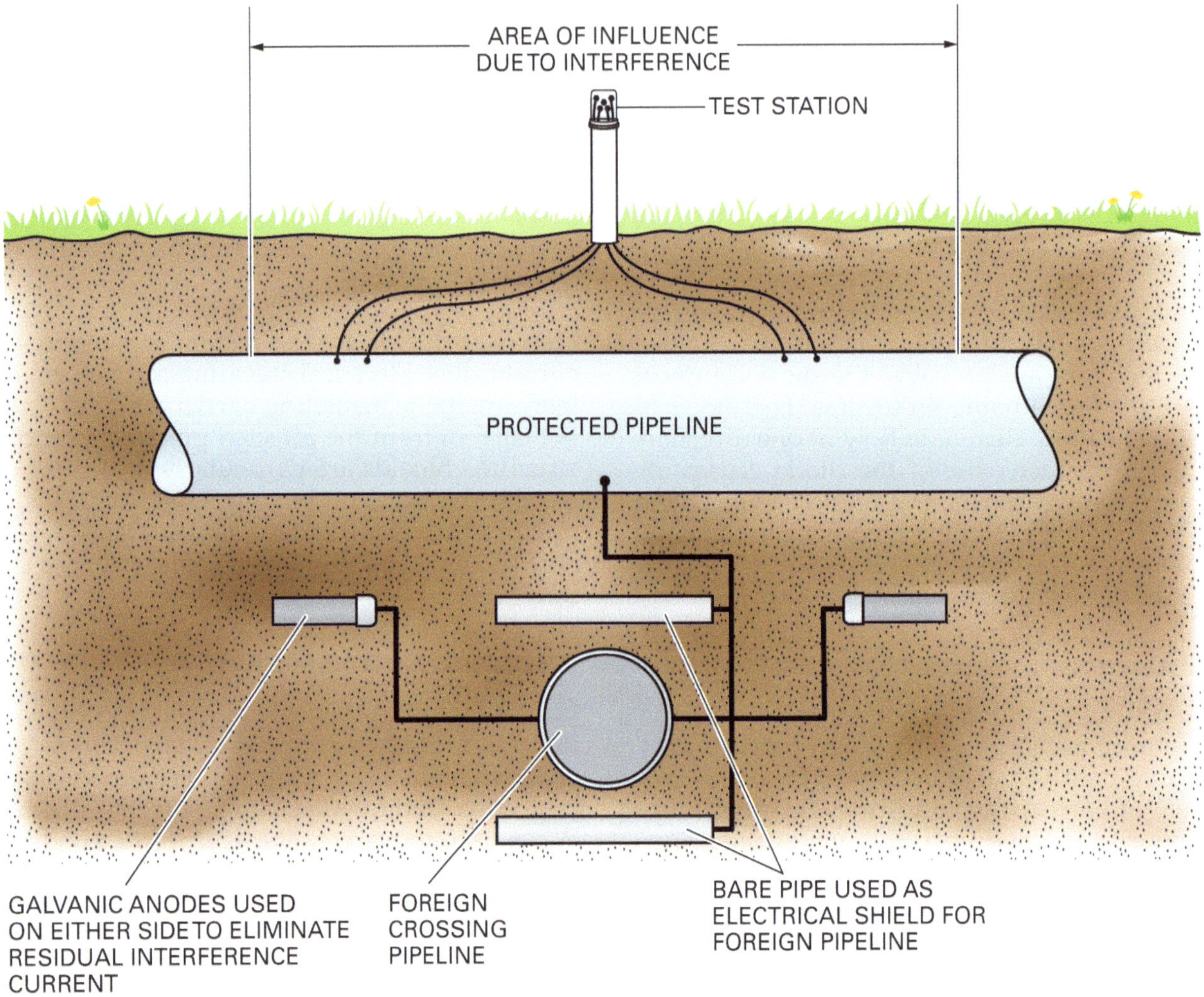

Figure 8 Use of electrical shields to reduce stray current interference.

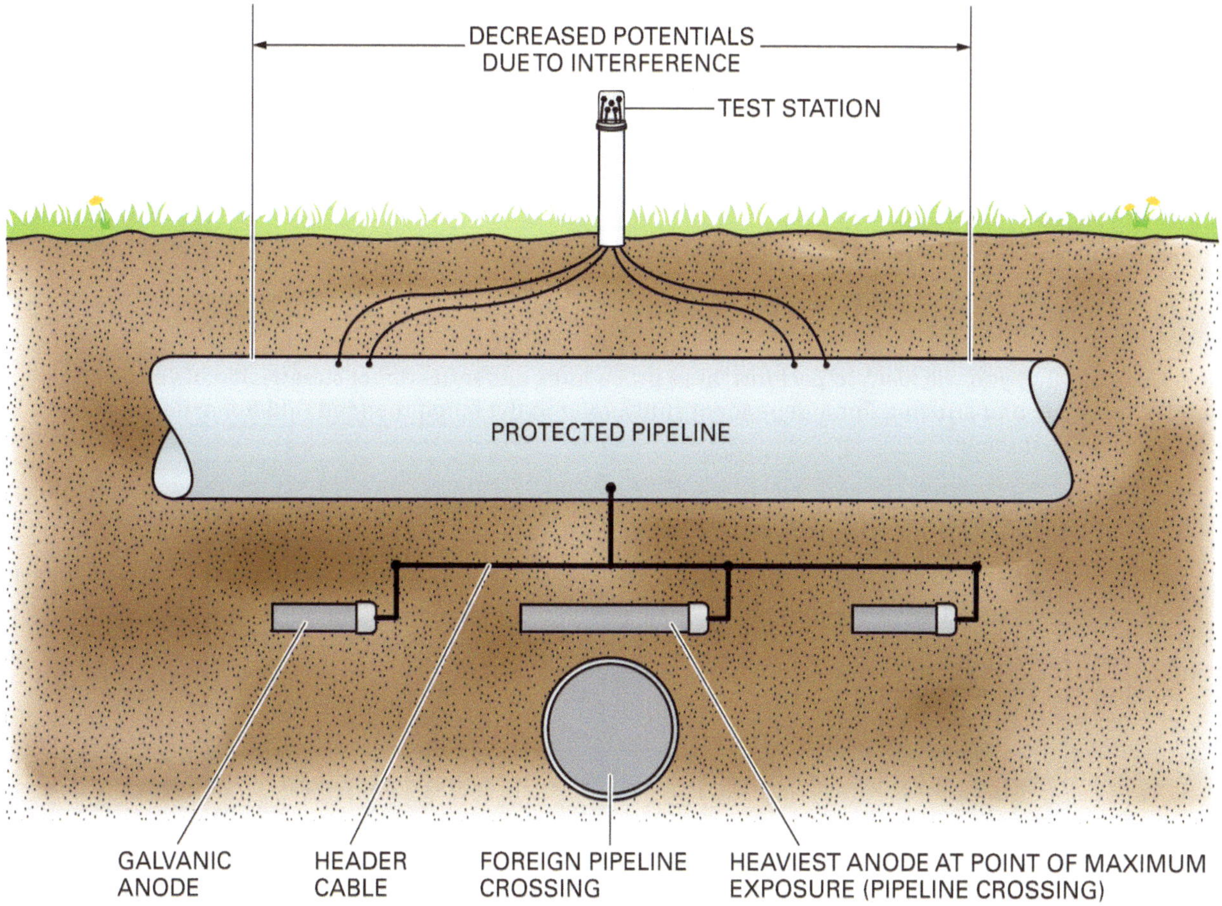

Figure 9 Galvanic anodes attached to a line under test.

the potential gradient field around the protected pipeline. Usually an anode bank consisting of a single line of anodes between the pipeline and the foreign pipeline causing the interference will mitigate the effects of the interference. When properly designed, the remaining current will discharge from the anodes instead of the pipeline. Several strings of anodes may be required when potential differences are very high. The anode life using this arrangement is short unless the surrounding soil has a high **resistivity**.

Anode banks work to ensure that the electrical potential of the pipeline is negative with respect to the surrounding earth. Remote anodes, also referred to as *remote groundbeds*, can be used to protect sections of pipeline located at a distance, while distributed anodes are more suited for protecting smaller, nearby areas of pipeline.

A buried pipeline extending across an area influenced by an anode bank may be subject to influence from stray electrical current surrounding the bed. Specific conditions in the area where this takes place determine how serious the effect will be on the foreign pipeline. In cases where the electrical potential of the stray current is high enough, protective measures must be put in place to prevent corrosion damage from current flowing away from anodic areas on the pipe.

One approach is to employ bare pipe shields on either side of the foreign pipeline where it lies within the potential gradient of the anode bank and picks up stray current. If the bare pipe shields are electrically connected to the rectifier for the anode bank, they will emit a negative cathodic field that opposes the positive field of the anodes.

High installation and operating costs must be weighed against other alternatives. It may be more cost effective to relocate the anode bank than to consider the use of a shield.

1.2.5 Notifications

Pipeline operators must notify owners of any nearby structures when planning any changes to existing CP systems or any new installations. The courtesy should be reciprocated. Stray currents may weaken or completely eliminate existing CP on structures within the same field of influence. Open communications between all owners and operators in an area should be encouraged. Owners should provide strip maps to each other, with each owner responsible for marking applicable structures. The exact location, size, and scope of any new projects can have serious effects on others. Test points in the same area may be shared. To avoid future problems, put all communications in writing.

Additional Resources

It is assumed that you will learn to perform these procedures under the direct supervision of your crew leader, foreman, or instructor. For independent study, refer to the recommended Additional Resources at the back of this module.

1.0.0 Section Review

1. An impressed current cathodic protection system can subject an adjacent pipeline to _____.
   a. endwise current
   b. stray current interference
   c. current discharge
   d. potential gradients

2. Galvanic anode banks work to ensure that the electrical potential of the pipeline with respect to the surrounding earth is _____.
   e. positive
   f. negative
   g. neutral
   h. at the same polarity

SECTION TWO

2.0.0 INSTALLING BONDS

Objective

Describe how to install bonds (CT9_1-17).

a. Identify the prerequisites, competencies, and abnormal operating conditions associated with installing bonds.
b. Describe the procedure for installing bonds.

Performance Tasks

Install bonds (CT9_1-17).

- Identify potential abnormal operating conditions that may occur, and know the appropriate actions to take in response to them.
- Utilize the appropriate personal protective equipment according to relevant company procedures.
- Identify the structures to be connected.
- Install the test leads and bond cables/wires on both structures at the location of current discharge.
- Attach the test leads and bond cables/wires by exothermic (thermite) weld, pin brazing, or other method which will yield a permanent, low-resistance connection.
- Terminate the test leads and bond cables/wires inside of the test box/station that is accessible to both structures.
- Install shunts for the measurement of current flow and resistance inside of the test box/station.
- Install blocking diodes as required.
- Conduct tests to determine the effectiveness of the installed interference bond.
- Complete appropriate documentation as required by operator's procedures.

Trade Terms

Abnormal operating conditions (AOCs): Conditions within a pipeline system, identified by the operator, that may indicate (1) a malfunction of a component or (2) a deviation from normal operations that may indicate a condition exceeding design limits or result in a hazard to persons, property, or the environment.

Test leads: Connections to the structure being tested, usually wires in a supporting stand with easy connection points for the meter.

Installing bonds to connect structures electrically promotes better use of cathodic protection systems and prevents possible damage of the structures that could result from interference. The task of bond installation begins with the defined need of a bond on a cathodic protection system and ends with the installation of the bond and verification of its effectiveness.

This section describes some of the basic skills required to install bonds and identifies abnormal operating conditions (AOCs) that personnel may need to respond to when performing this task. It also describes the general procedure for installing bonds to eliminate stray current interference.

2.1.0 Prerequisites, Competencies, and Abnormal Operating Conditions

Personnel must possess a variety of skills in order to install bonds safely and efficiently. The various operations require a series of steps, which personnel must be competent to perform. Personnel must also be aware of the particular abnormal operating conditions that may occur during the procedure and know the proper responses to them.

2.1.1 Prerequisites and Competencies

Personnel assigned to install bonds must possess the following basic knowledge and skills prior to performing this task:

- Familiarity with types of bonds
- Familiarity with the requirements for attaching test leads
- Experience using multimeters
- Experience in exothermic (thermite) welding
- Experience in terminating bond leads

Personnel must have knowledge of cathodic protection systems and components comparable to NACE Certification Level CP 2. This includes, but is not limited to, the following:

- Differentiating between critical and non-critical types of bonds. (Critical bonds are bonds that, if disconnected, may be detrimental to one of the structures. Critical bonds are determined by an operator's procedures.)
- Installing interference bond facilities at the location of current discharge, if possible.
- Using exothermic welding, pin brazing, or bolt-on connections to connect bond wires/cables and test leads to structures.
- Terminating bond leads and test leads in a test station to allow inspection.
- Using shunts between structures to determine the current amplitude and polarity between the structures.

- Using blocking diodes to prevent current flow in the opposite direction, as necessary, such as when bonding to DC traction systems.

2.1.2 Abnormal Operating Conditions

Before installing bonds to eliminate stray current interference, a pipeline technician must be able to identify and properly respond to potential AOCs, should they occur. The following are some potential AOCs, and the appropriate action personnel should take in response to them:

- *CP levels are low* – Make proper notifications.
- *Test lead or bond cable is broken or damaged* – Notify operator personnel for repair or replacement of the damaged lead or cable, as required.
- *Component failure occurs* – Evaluate the extent of the failure/damage and make company notifications as applicable.
- *Pipeline system is damaged due to exothermic welding burn-through, excessive corrosion, dents and/or gouges* – Anticipate the proper response for unexpected hazardous liquid or carbon dioxide encounter or fire/explosion, if appropriate. Determine the extent of the damage and make proper company notifications.
- *Fire or explosion occurs from burn-through during exothermic welding* – Limit access to the area as much as possible and make company notifications.
- *Unexpected leak of hazardous liquid or carbon dioxide occurs within or outside casings* – Make immediate notifications per company procedure. Try to identify and isolate the source of the leak if possible and if properly trained to do so.

2.2.0 Procedure

When it has been determined that a damaging stray current exists, the source of the current must be found. The following steps present the general procedure for installing a bond at the source of a stray current:

WARNING!

Always obtain and use the appropriate personal protective equipment for each task. Only qualified individuals may make electrical connections and operate equipment. Always verify the absence of voltage and current before making connections. Follow the manufacturer's instructions for the equipment being installed.

Step 1 Identify potential AOCs that may occur, and know the appropriate actions to take in response to them.

Step 2 Utilize the appropriate personal protective equipment according to relevant company procedures.

Step 3 Identify the structures to be bonded. They should be properly identified and marked for connections. Test for the presence of DC current interference. If possible, conduct cooperative interference tests with a representative of the foreign company to determine the location of current discharge and magnitude of interference current.

NOTE

Potential sources include, but are not limited to, other pipeline cathodic protection facilities, DC traction systems (e.g., subways, light rail systems, and trolleys), high-voltage DC transmission, welding, or other facilities.

Step 4 Install the test leads and bond cables/wires on both structures at the location of the current discharge.

NOTE

Test leads are not used for current-carrying connections. In addition to the bond cables/wires being installed, test lead wires may also be installed on both structures so that personnel avoid taking potentials on a current-carrying connection.

Step 5 Attach the test leads and bond cables/wires to the structure using exothermic welding, pin brazing, or other accepted method to yield a permanent, low-resistance connection. This provides a very low resistance path for current return and promotes optimal current transfer.

Step 6 Terminate the test leads inside of the appropriate test station that is accessible to both structures. This enables representatives of both companies to monitor the bond.

Step 7 Install shunts for measurement of current flow and resistance (as required to limit current interchange) inside of the test station. It is important to monitor the magnitude and direction of current flow.

Step 8 Install blocking diodes, as required.

> **NOTE**
> In some situations, such as when bonding to DC transit systems, current must be prevented from flowing in the opposite direction. In such cases, blocking diodes may be required.

Step 9 Conduct tests to determine the effectiveness of the installed interference bond. Determining when the stray current interference has been eliminated is critical. To determine when the bond has been properly adjusted, set up a voltmeter to measure the potential of the foreign line to the protected line at the crossing point. The potential should be measured at the point of maximum exposure if it is different from the crossing point. The bond resistance is set so that the foreign line potential is the same with the affecting rectifier either on or off. This restores the foreign pipeline potential to the original value. *Figure 10* shows a typical foreign line test setup with a bond in place, and the instrument connections used for measuring the structure-to-soil potentials on the affected pipeline.

Step 10 Document readings as required by the operator's procedures.

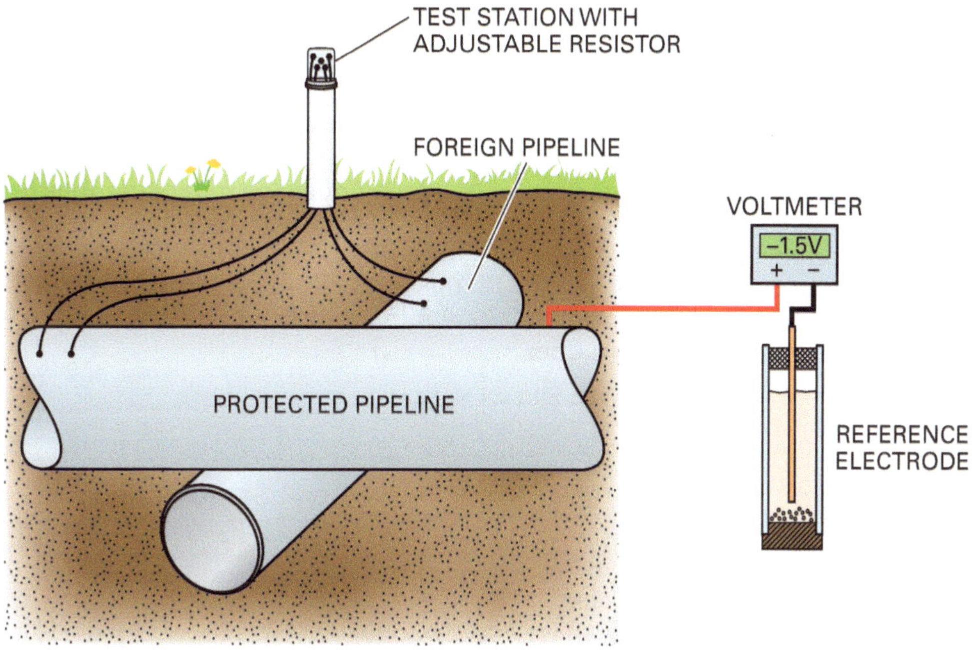

Figure 10 Foreign line test setup with a bond in place.

Additional Resources

It is assumed that you will learn to perform these procedures under the direct supervision of your crew leader, foreman, or instructor. For independent study, refer to the recommended Additional Resources at the back of this module.

2.0.0 Section Review

1. Before installing bonds to eliminate stray current interference, individuals must have knowledge of cathodic protection systems and components comparable to _____.
   a. ANSI Accreditation Phase 2
   b. *NFPA 70* apprenticeship standards
   c. Journeyman electrician classification
   d. NACE Certification Level CP 2

2. One acceptable means of attaching the test leads and bond cables/wires to the structure when installing a resistance bond is _____.
   a. plastic fusion
   b. pit brazing
   c. exothermic welding
   d. epoxy bonding

Summary

Stray currents, or interference, from foreign sources can cause pipeline corrosion. To control this problem, pipeline technicians need to know how to identify when stray currents are present, how to identify the source of the stray currents, and how to use appropriate techniques to reduce the interference.

Techniques for mitigating interference include the use of bonds, coatings, and electrical shields, and galvanic anodes. This module focused on the installation of resistance bonds to reduce or eliminate stray current interference.

Review Questions

1. Natural occurrences of stray current in the earth that are transferred to the protected pipeline are known as _____.
   a. potential gradients
   b. burial pressures
   c. ground water changes
   d. climate variations

2. DC-powered transit systems can cause _____.
   a. potential shifts
   b. static DC interference
   c. dynamic stray currents
   d. impressed currents

3. If a high current density is flowing from a well-coated line through the soil and onto a bare foreign line, it could result in _____.
   a. damaged test leads
   b. CP system shorts
   c. resistance bonding
   d. localized corrosion

4. Electrical shielding can be provided by diverting the current or by _____.
   a. resistance bonding
   b. nonmetallic barriers
   c. current interrupters
   d. galvanic anodes

5. If a pipeline technician discovers a broken or damaged test lead or bond cable when installing a bond, the technician should _____.
   a. notify operator personnel for repair, as required
   b. immediately eliminate potential ignition sources in the area
   c. evacuate all non-essential personnel from the work site
   d. limit access to the work area as much as possible

6. To determine when an installed bond has been properly adjusted, measure the potential using a(n) _____.
   a. ohmmeter
   b. continuity tester
   c. voltmeter
   d. megger

Trade Terms Quiz

Fill in the blank with the correct term that you learned from your study of this module.

1. Connections to the structure being tested, usually wires in a supporting stand with easy connection points for the meter, are called _________.
2. Devices used to allow current to flow in only one direction are _________.
3. Changes in potential with respect to distance, expressed in millivolts per unit of distance are _________.
4. A conductor that joins two points in an electrical circuit to form a parallel or alternative path through which a portion of the current may pass is a(n) _________.
5. Sets of two or more anodes installed in a specific configuration where the anodes are connected together by a header cable/wire are referred to as _________.
6. A portion of a current that flows over a path other than the intended path in an on/off or intermittent fashion and may cause electrochemical corrosion of metals in contact with electrolytes is _________.
7. Electrical connections of two structures that usually have two different cathodic protection levels are _________.
8. The effect on the cathodic protection level of a structure due to the interaction between its cathodic protection system and the cathodic protection system of another structure is called _________.
9. A device used to convert AC current to DC current and which supplies current to the structure in the soil in an impressed cathodic protection system, is a(n) _________.
10. Conditions within a pipeline system, identified by the operator, that may indicate (1) a malfunction of a component or (2) a deviation from normal operations that may indicate a condition exceeding design limits or result in a hazard to persons, property, or the environment are called _________.
11. A substance that ionizes (releases free electrons) in an electric field is a(n) _________.
12. A portion of a current that flows over a path other than the intended path at a constant, unbroken rate and that may cause electrochemical corrosion of metals in contact with electrolytes is _________.
13. A measurement of the voltage or potential difference between a buried or submerged metal structure and an electrolyte, such as soil, is referred to as _________.
14. Current flows through a path other than the intended circuit are called _________.
15. The ability of a material to resist the flow of electric current is called _________.

Trade Terms

Abnormal operating conditions (AOCs)
Anode banks
Bonds
Diodes
Dynamic stray current
Electrolyte
Interference
Potential gradients
Rectifier
Resistivity
Shunt
Static stray current
Stray currents
Structure-to-soil potential
Test leads

Trade Terms Introduced in This Module

Abnormal operating conditions (AOCs): Conditions within a pipeline system, identified by the operator, that may indicate (1) a malfunction of a component or (2) a deviation from normal operations that may indicate a condition exceeding design limits or result in a hazard to persons, property, or the environment.

Anode banks: Sets of two or more anodes installed in a specific configuration where the anodes are connected together by a header cable/wire; also referred to as groundbeds.

Bonds: Electrical connections of two structures that usually have two different cathodic protection levels.

Diodes: Devices used to allow current to flow in only one direction.

Dynamic stray current: A portion of a current that flows over a path other than the intended path in an on/off or intermittent fashion and may cause electrochemical corrosion of metals in contact with electrolytes.

Electrolyte: A substance that ionizes (releases free electrons) in an electric field.

Interference: The effect on the cathodic protection level of a structure due to the interaction between its cathodic protection system and the cathodic protection system of another structure.

Potential gradients: Changes in potential with respect to distance, expressed in millivolts per unit of distance.

Rectifier: A device used to convert AC current to DC current. In an impressed cathodic protection system, it supplies current to the structure in the soil.

Resistivity: The ability of a material to resist the flow of electric current.

Shunt: A conductor that joins two points in an electrical circuit to form a parallel or alternative path through which a portion of the current may pass.

Static stray current: A portion of a current that flows over a path other than the intended path at a constant, unbroken rate and that may cause electrochemical corrosion of metals in contact with electrolytes.

Stray currents: Current flows through a path other than the intended circuit.

Structure-to-soil potential: A measurement of the voltage or potential difference between a buried or submerged metal structure and an electrolyte, such as soil; also referred to as structure-to-electrolyte potential.

Test leads: Connections to the structure being tested, usually wires in a supporting stand with easy connection points for the meter.

Additional Resources

The following recommended resources can provide additional helpful information related to the requirements for this covered task:

API Recommended Practice 1161, Third Edition, © 2014 American Petroleum Institute. API Publishing Services, 1220 L Street NW, Washington, DC 20005. Distributed by Thompson Reuters (Scientific) LLC, **www.techstreet.com**.

49 *CFR* 192, *Subpart I*, **www.ecfr.gov**.

49 *CFR* 195, *Subpart H*, **www.ecfr.gov**.

NCCER Module AOCCC-17, *Abnormal Operating Conditions Control Center*.

NCCER Module AOCFG-17, *Abnormal Operating Conditions Field and Gas*.

NCCER Module 62105-14, *Introduction to Pipeline Documents*.

NCCER Module 64202-02, *Use of Meters and Test Equipment*.

NCCER Module 66101-02, *Introduction to the Pipeline Industry*.

NCCER Module CT1_1-17, *Measure Structure-to-Soil Potential*.

NCCER Module CT1_2-17, *Conduct Close Interval Survey*.

NCCER Module CT1_3-17, *Test and Detect Interference*.

NCCER Module CT1_4-17, *Inspect and Perform Electrical Test of Bonds*.

NCCER Module CT2_2-17, *Repair Damaged Test Leads*.

NCCER Module CT2_3-17, *Install Test Leads by Non-Exothermic Welding Methods*.

NCCER Module CT2_4-17, *Install Test Leads by Exothermic Welding Methods*.

Figure Credits

Tinker & Rasor, Module Opener

Section Review Answer Key

Answer	Section Reference	Objective
Section One		
1. b	1.1.1	1a
2. b	1.2.4	1b
Section Two		
1. d	2.1.1	2a
2. c	2.2.0	2b

NCCER CURRICULA — USER UPDATE

NCCER makes every effort to keep its textbooks up-to-date and free of technical errors. We appreciate your help in this process. If you find an error, a typographical mistake, or an inaccuracy in NCCER's curricula, please fill out this form (or a photocopy), or complete the online form at **www.nccer.org/olf**. Be sure to include the exact module ID number, page number, a detailed description, and your recommended correction. Your input will be brought to the attention of the Authoring Team. Thank you for your assistance.

Instructors – If you have an idea for improving this textbook, or have found that additional materials were necessary to teach this module effectively, please let us know so that we may present your suggestions to the Authoring Team.

NCCER Product Development and Revision

13614 Progress Blvd., Alachua, FL 32615

Email: curriculum@nccer.org
Online: www.nccer.org/olf

❑ Trainee Guide ❑ Lesson Plans ❑ Exam ❑ PowerPoints Other ______________________

Craft / Level: Copyright Date:

Module ID Number / Title:

Section Number(s):

Description:

Recommended Correction:

Your Name:

Address:

Email: Phone:

CT9_2-17

Install Galvanic Anodes

Overview

Various cathodic protection systems—which include the installation of bonds, galvanic anodes, rectifiers, and impressed current groundbeds—are used to mitigate stray current interference and prevent corrosion in pipelines. Galvanic anodes are installed to provide cathodic protection for buried or submerged metallic structures, and may be used for stray current mitigation. This module describes the general procedures for the installation of galvanic anodes.

Trainees with successful module completions may be eligible for credentialing through the NCCER Registry. To learn more, go to **www.nccer.org** or contact us at 1.888.622.3720. Our website has information on the latest product releases and training, as well as online versions of our *Cornerstone* magazine and Pearson's product catalog.

Your feedback is welcome. You may email your comments to **curriculum@nccer.org**, send general comments and inquiries to **info@nccer.org**, or fill in the User Update form at the back of this module.

This information is general in nature and intended for training purposes only. Actual performance of activities described in this manual requires compliance with all applicable operating, service, maintenance, and safety procedures under the direction of qualified personnel. References in this manual to patented or proprietary devices do not constitute a recommendation of their use.

CT9_2-17
Install Galvanic Anodes

Objectives

When you have completed this module, you will be able to do the following:

1. Describe stray current interference and methods for reducing or eliminating it.
   a. Identify types of stray current and describe how they affect cathodic protection.
   b. Identify and describe methods used to reduce or eliminate stray current interference.
2. Describe how to install galvanic anodes (CT9_2-17).
   a. Identify the prerequisites, competencies, and abnormal operating conditions associated with installing galvanic anodes.
   b. Describe the procedure for installing galvanic anodes.

Performance Tasks

Under the supervision of the instructor, you should be able to do the following:

1. Install galvanic anodes (CT9_2-17).
   - Identify potential abnormal operating conditions that may occur during performance of this CT, and know the appropriate actions to take in response to them.
   - Utilize the appropriate personal protective equipment according to relevant company procedures.
   - Verify the location for anode installation. Ensure that the electrolytic environment is moist. Horizontal anodes may require backhoe excavation.
   - Install the anode into an augered hole or horizontal excavation. Wet down the anode prior to backfilling or before placing it in the ground.
   - Uncoil the anode lead and extend it fully. Do not damage or kink the wire.
   - If the design is for direct connection, connect the test lead directly to the pipe. Install shunts for measurement of current flow and resistance inside of the test station.
   - Backfill carefully with approved soil backfill. Use rock-free backfill to pad the anode and the anode lead wire.
   - Complete appropriate documentation as required by operator's procedures.

Trade Terms

Abnormal operating conditions (AOCs)
Anode
Anode banks
Bonds
Diodes
Dynamic stray current
Electrolyte
Exothermic welds
Header cable/wire
Interference
Potential gradients
Rectifier
Resistivity
Shunt
Splice connection
Static stray current
Stray currents
Structure-to-soil potential
Test leads
Ultrasonic tester

Industry Recognized Credentials

If you are training through an NCCER-accredited sponsor, you may be eligible for credentials from NCCER's Registry. The ID number for this module is CT9_2-17. Note that this module may have been used in other NCCER curricula and may apply to other level completions. Contact NCCER's Registry at 888.622.3720 or go to **www.nccer.org** for more information.

Contents

Figures and Tables

SECTION ONE

1.0.0 STRAY CURRENT INTERFERENCE

Objective

Describe stray current interference and methods for reducing or eliminating it.

a. Identify types of stray current and describe how they affect cathodic protection.
b. Identify and describe methods used to reduce or eliminate stray current interference.

Trade Terms

Anode: The electrode of an electrochemical cell at which oxidation occurs. Electrons flow away from the anode in the external circuit. Corrosion usually occurs and metal ions enter the solution at the anode.

Anode banks: Sets of two or more anodes installed in a specific configuration where the anodes are connected together by a header cable/wire; also referred to as groundbeds.

Bonds: Electrical connections of two structures that usually have two different cathodic protection levels.

Diodes: Devices used to allow current to flow in only one direction.

Dynamic stray current: A portion of a current that flows over a path other than the intended path in an on/off or intermittent fashion and may cause electrochemical corrosion of metals in contact with electrolytes.

Electrolyte: A substance that ionizes (releases free electrons) in an electric field.

Interference: The effect on the cathodic protection level of a structure due to the interaction between its cathodic protection system and the cathodic protection system of another structure.

Potential gradients: Changes in potential with respect to distance, expressed in millivolts per unit of distance.

Rectifier: A device used to convert AC current to DC current. In an impressed cathodic protection system, it supplies current to the structure in the soil.

Resistivity: The ability of a material to resist the flow of electric current.

Shunt: A conductor that joins two points in an electrical circuit to form a parallel or alternative path through which a portion of the current may pass.

Static stray current: A portion of a current that flows over a path other than the intended path at a constant, unbroken rate and that may cause electrochemical corrosion of metals in contact with electrolytes.

Stray currents: Current flows through a path other than the intended circuit.

Structure-to-soil potential: A measurement of the voltage or potential difference between a buried or submerged metal structure and an electrolyte, such as soil; also referred to as structure-to-electrolyte potential.

Test leads: Connections to the structure being tested, usually wires in a supporting stand with easy connection points for the meter.

Corrosion on a pipeline can be caused by **stray currents**, in which electrical currents flow through a path other than the intended circuit. The current flow results in the problem called **interference**, or *stray current interference*. This section describes the types and sources of stray currents, explains how to test for interference current, and reviews some of the methods that can be used to eliminate stray current interference.

1.1.0 Stray Current

Stray currents are direct currents flowing in the earth that are from a source other than the corrosion protection source. **Potential gradients** cause natural occurrences of stray current in the earth that are transferred to the protected pipeline. *Figure 1* shows potential gradients around a protected pipeline.

Normally, this outside current enters the pipeline at one point, exits the pipeline at another point, and returns to the earth. The area where the current returns to earth, completing the circuit, is the point at which the corrosion may occur if the material is in contact with an **electrolyte**. The measurement of the voltage or potential difference between a buried or submerged metal structure and an electrolyte, such as soil, is the **structure-to-soil potential**, also referred to as the *structure-to-electrolyte potential.*

1.1.1 Types of Stray Current

Stray currents are either static (constant) or dynamic (intermittent). Sources of stray current include cathodic protection (CP) systems on other pipelines, DC transit systems, DC welding operations, DC mining operations, high-voltage

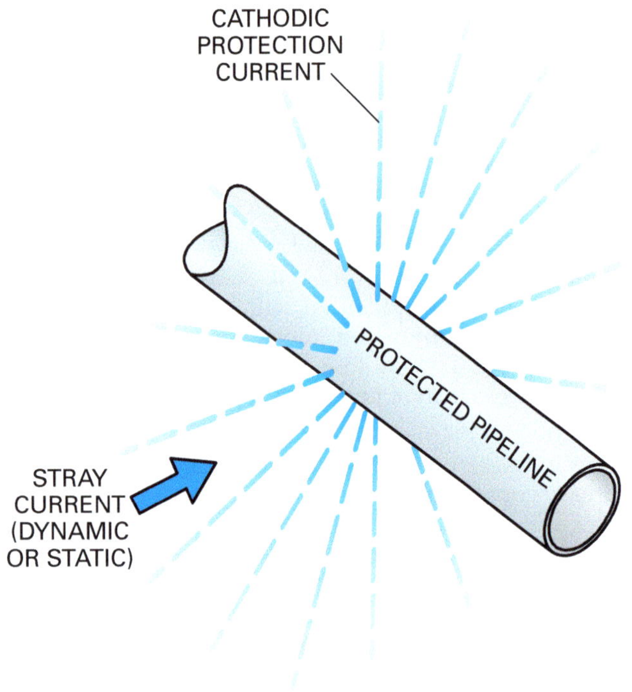

Figure 1 Potential gradients around a protected pipeline.

DC transmission systems, or disturbances in the earth's magnetic field. If environmental conditions allow rectifying junctions such as copper oxide films, AC current flow can be rectified. This results in a direct current that can be mitigated. There are many different sources of interference current in a cathodic protection system. For example, stray current interference on adjacent pipelines can be caused by impressed current cathodic protection systems. This depends on the locations of the **anode banks** and the operating characteristics of the cathodic protection system. Corrosion usually occurs and metal ions enter the solution at the **anode**.

Figure 2 shows stray current to a crossover foreign pipeline. In this example, the foreign pipeline picks up current from the anode bank and transfers the current to the protected pipeline when it crosses under it. The potential for corrosion is greatly increased at the crossing point of the pipelines.

Figure 3 shows stray current to a parallel foreign pipeline. In this example, the parallel foreign pipeline is affected by the anode bank of the protected pipe. This is because the foreign pipeline travels through the anode bank and picks up current. When this occurs, the effectiveness of the CP system will be reduced in this area of the pipeline. In the affected area of the foreign pipeline, it becomes positive to the soil and discharges current to the earth. This increases the potential for corrosion on the foreign pipeline in the affected area.

1.1.2 Testing for Interference Current

Static stray current interference is easy to test for as long as there is no presence of **dynamic stray current**. Dynamic stray current can be caused by sources such as DC-powered transit systems, underground mining, welding, and pipeline interference. In *Figure 4*, a section of coated pipeline has an impressed current CP system applied to it and has two foreign pipelines crossing it. In this example, "B" is a strong indicator of interference because it has the largest difference in voltage. Always look for large changes in voltage to identify interference. Follow company guidelines for determining the need to mitigate interference.

To distinguish the current in the foreign pipeline from the current in the protected line, an automatic current interrupter is installed in the output of the **rectifier**. The current interrupter is a device that automatically opens and closes an electrical circuit at a preset time. With the current interrupter operating, measure the potential of the lines under both on and off conditions. *Figure 5* shows how to measure this potential at the test station. As shown, a **shunt** is installed for measurement of current flow and resistance inside of the test station.

These tests are performed using a copper sulfate electrode placed over the crossing point of the lines. Pipeline locators can be used to determine the exact location of the crossing.

WARNING: Always be alert when testing for interference current. A release could be found while testing.

The data obtained at each pipeline crossing must be properly recorded. *Figure 6* shows typical interference testing potentials. These records are also called field data sheets. Field data sheets should also include any information about the protected line, the date, and the current output of the rectifier.

Telluric Currents

Telluric currents are caused by sunspots and solar flares and can result in inaccurate readings during intense periods of solar activity.

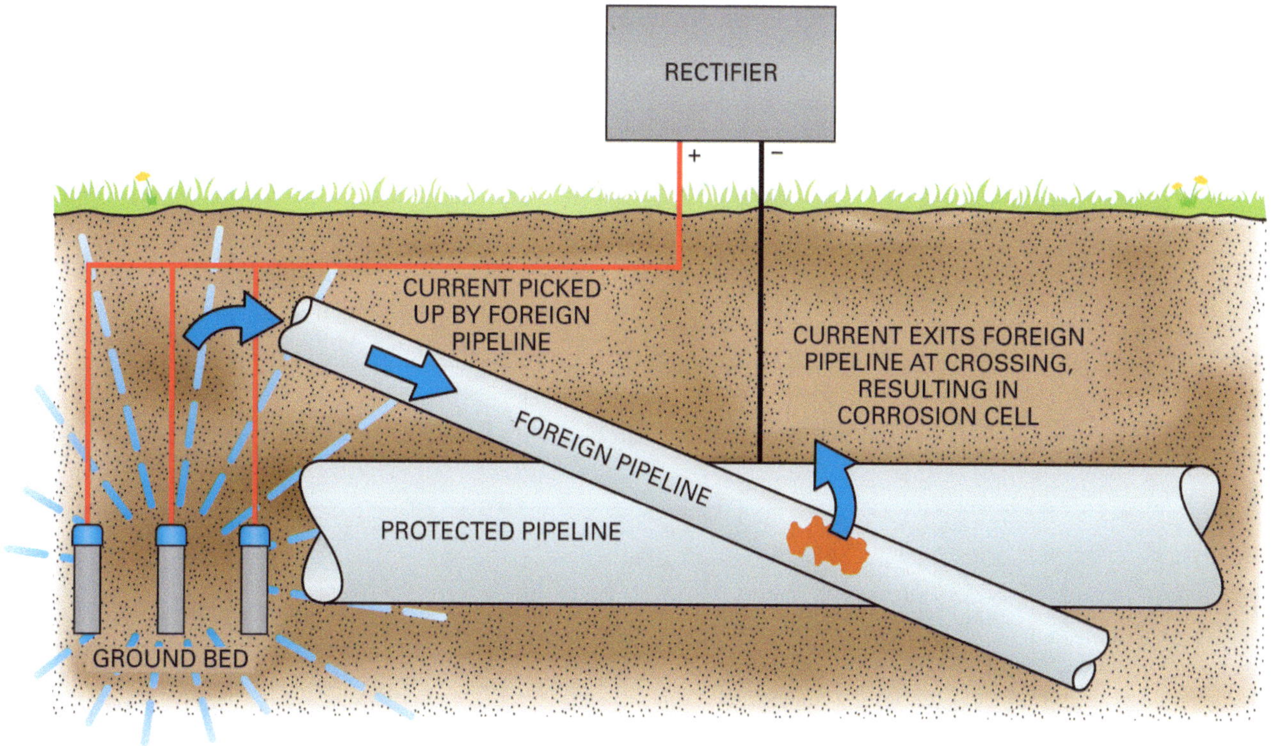

Figure 2 Stray current to a crossover foreign pipeline.

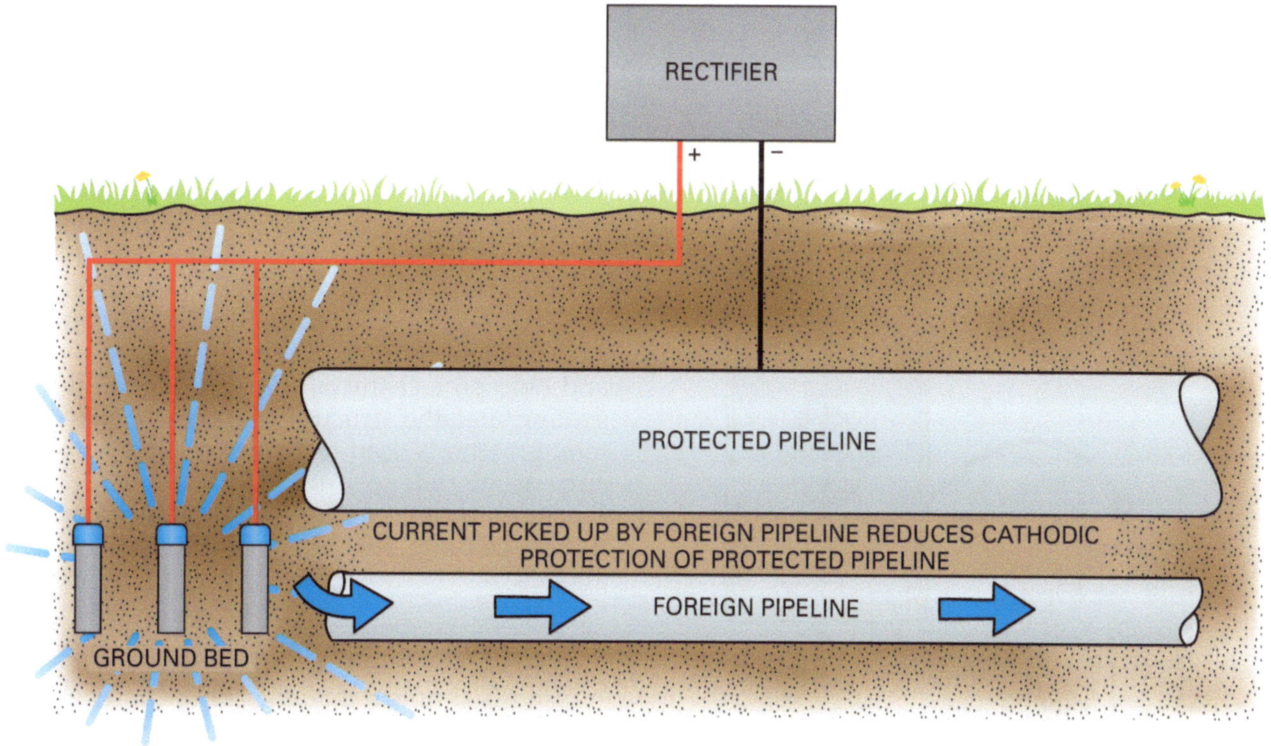

Figure 3 Stray current to a parallel foreign pipeline.

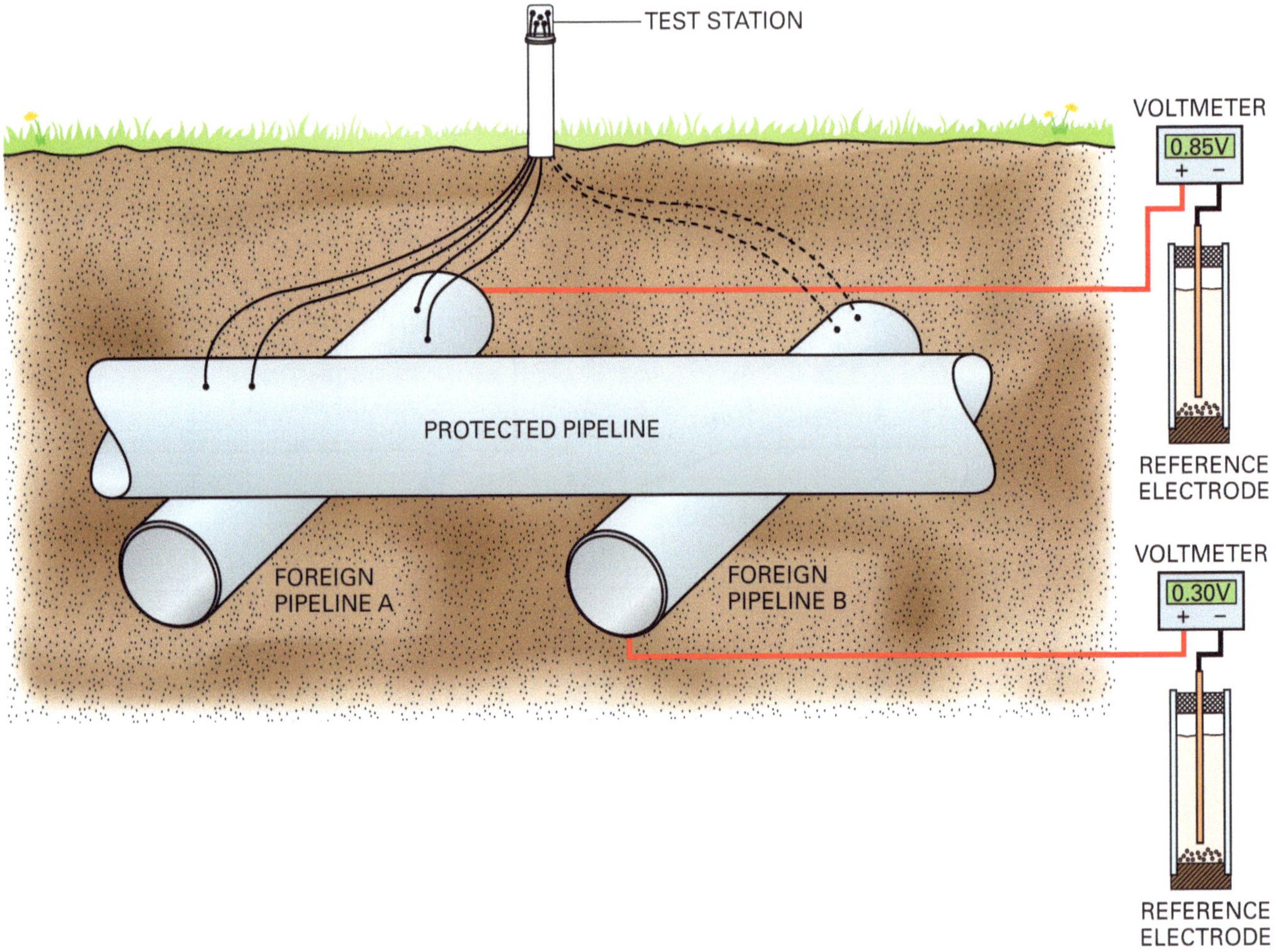

Figure 4 Testing foreign pipeline crossings for stray current interference from CP installations.

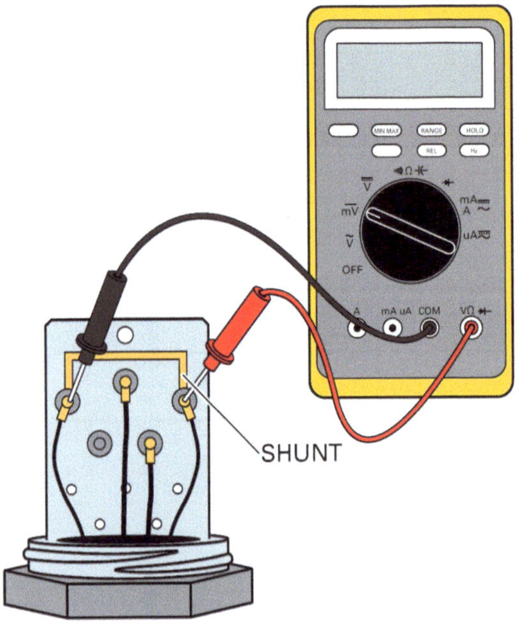

Figure 5 Measuring potential at the test station.

1.2.0 Methods for Reducing or Eliminating Stray Current Interference

According to the *Code of Federal Regulations* 49 *CFR* 192.473 (for gas pipelines) and 49 *CFR* 195.577 (for liquid pipelines), any pipeline system subjected to stray currents must have an ongoing program to minimize the effects of these currents and help protect surrounding or adjacent underground metallic structures. There are many different methods used to reduce or eliminate stray current interference from a CP system. Some of the most common methods include the use of bonds, coatings, electrical shields, and galvanic anodes.

1.2.1 Resistance Bonds

A bond is simply a metallic connection between structures. There are two types of bonds: critical and non-critical. Critical bonds are identified by the operator's procedures and include those bonds that may cause damage to the structure if disconnected. Like test leads, bond leads are terminated in a test station.

Foreign pipelines can be subject to stray current damage from the impressed current cathodic

FOREIGN LINE		PROTECTED PIPELINE (volts)			INTERFERING PIPELINE (volts)		
NAME	PIPELINE STATION	VOLTAGE ON	VOLTAGE OFF	Δ volts	VOLTAGE ON	VOLTAGE OFF	Δ volts
A	140 + 00	-0.91	-0.90	-0.01	-0.85	-0.88	+0.03
B	1210 + 00	-1.94	-1.00	-0.94	-0.30	-0.68	+0.38

Figure 6 Sample of recording survey notes.

protection system on another pipeline. To correct this problem, a resistance bond is connected between the two pipelines. The amount of resistance in the bond is adjusted to drain enough current from the affected line to eliminate the corrosion potential.

CAUTION

For liability reasons, the representative of the foreign pipeline company should be on site before procedures to correct the problem on the affected line begin.

For a resistance bond, **diodes** are installed as basic components of a rectifier circuit. Single-phase rectifiers are probably the most common units used for cathodic protection. The diode is installed to allow current to flow in one direction only. This characteristic of the diode causes all the current waves to be more aligned when entering into the cables. Along with other components, the diodes reduce ripple effects in the power supply and filter out interference.

1.2.2 Coatings

If the testing of a foreign line indicates that the interference is caused by a foreign pipeline that is bare or poorly coated, the bond base may not be able to correct the condition. If a high current density is flowing from a well-coated line through the soil and onto the bare foreign line, localized corrosion could occur. To correct this problem, reduce the density of the current flow to the crossing area of the foreign line. The best way to accomplish this is to apply a coating to the line. *Figure 7* shows the use of a coating to correct interference at foreign pipeline crossings.

With proper coating, the voltage drops in the earth around the foreign line are reduced or eliminated. Over-the-line potential measurements are used to identify the portion of the foreign line subject to interference. This information is then used to determine the area on the foreign line that must be coated to mitigate the interference.

CAUTION

Take special precautions to avoid bare spots in the coating (holidays) to prevent localized corrosion.

1.2.3 Electrical Shields

Electrical shields are defined as barriers to stray current flow. Electrical shielding can be provided by nonmetallic insulating barriers or by diverting the current from the affected pipeline to another structure. Shields are particularly useful for protecting pipeline in congested areas such as pumping stations and tank farms. *Figure 8* shows bare pipe used as electrical shields around a pipeline.

1.2.4 Galvanic Anodes

Galvanic anodes can provide cathodic protection for buried or submerged structures. They can also be used to correct interference, as spot protection, and for alternating current mitigation. When used to correct interference, the anodes are attached to the affected line (*Figure 9*). They are located in the area that is subject to interference from the foreign line. Anodes are buried in a special backfill that must be wet in order for the anodes to operate.

This method uses the anodic potential fields that are around the galvanic anode to offset the potential gradient field around the protected pipeline. Usually an anode bank consisting of a single line of anodes between the pipeline and the foreign pipeline causing the interference will mitigate the effects of the interference. When properly designed, the remaining current will discharge from the anodes instead of the

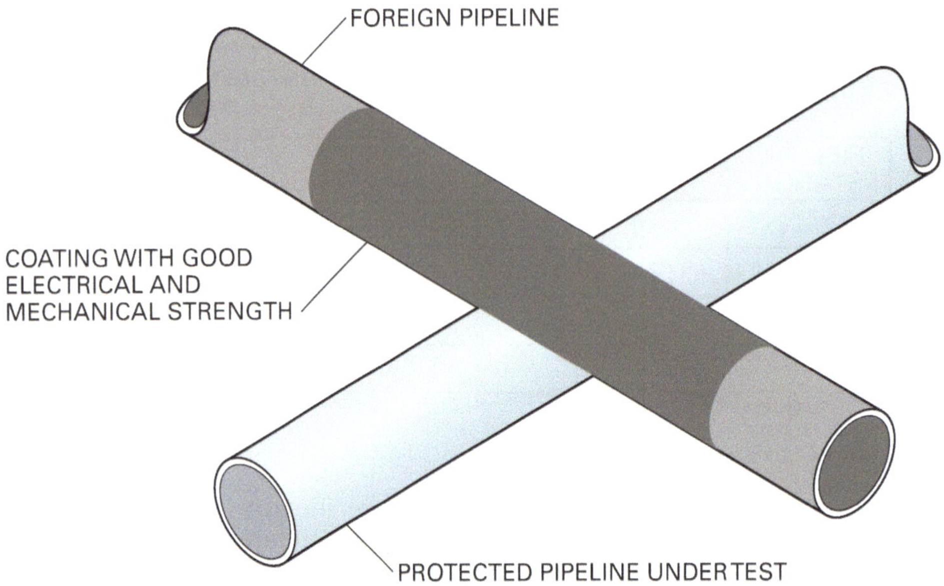

Figure 7 Using a coating to correct interference at foreign pipeline crossings.

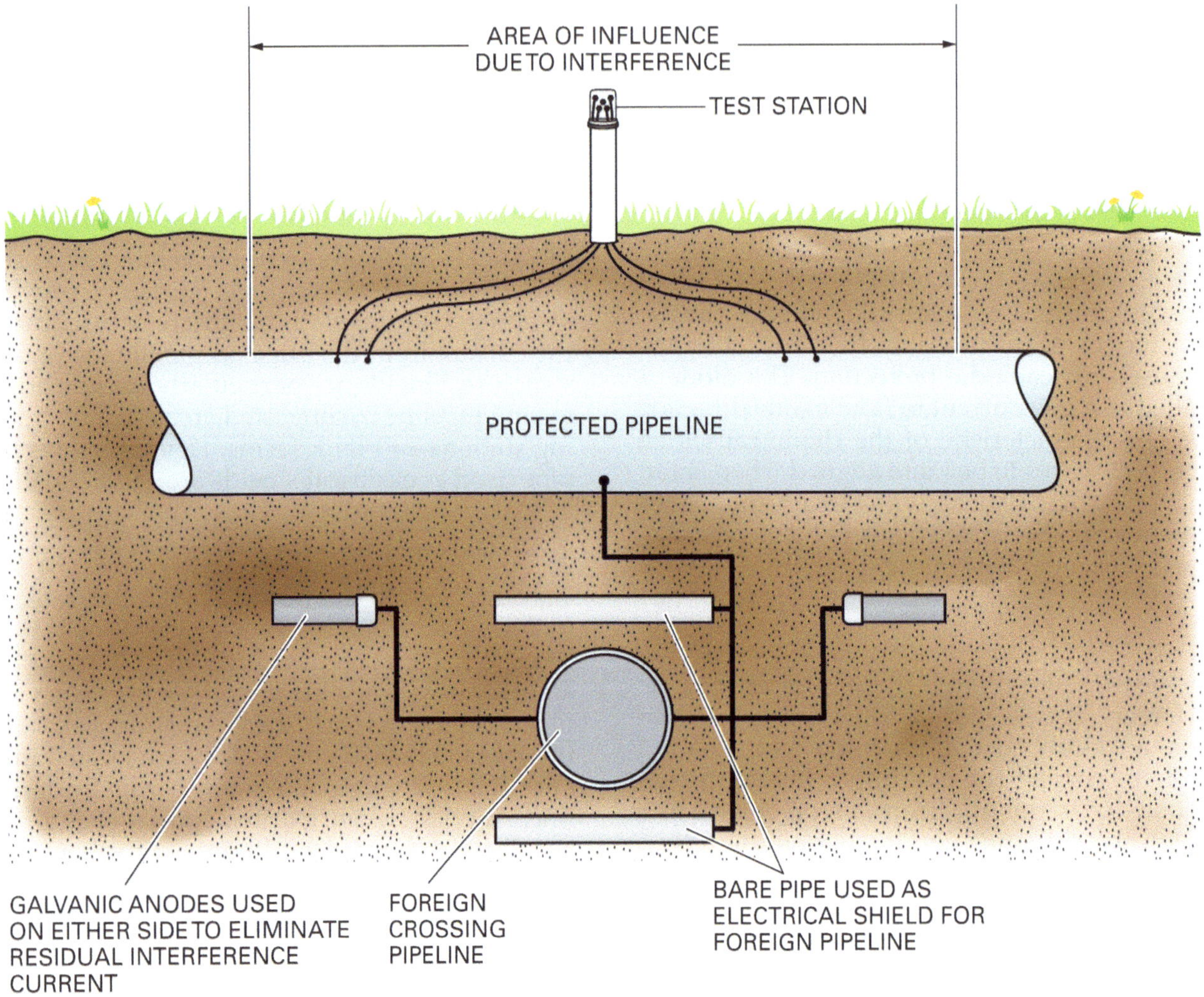

Figure 8 Use of electrical shields to reduce stray current interference.

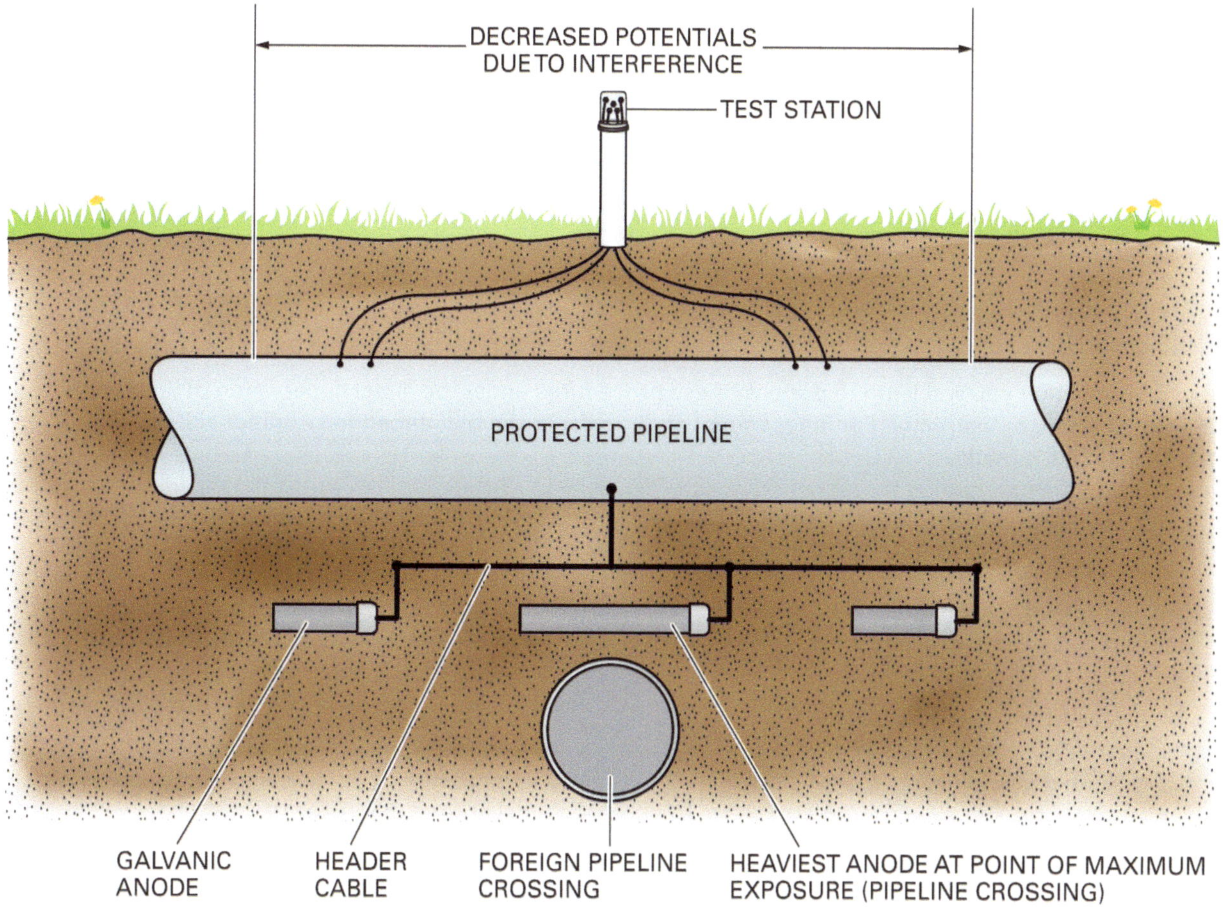

Figure 9 Galvanic anodes attached to a line under test.

pipeline. Several strings of anodes may be required when potential differences are very high. The anode life using this arrangement is short unless the surrounding soil has a high resistivity. Magnesium anodes are preferred to zinc anodes because magnesium anodes have a higher anodic potential gradient field.

1.2.5 Galvanic Anode Banks

Galvanic anode banks work to ensure that the electrical potential of the pipeline is negative with respect to the surrounding earth. Remote anodes, also referred to as *remote groundbeds*, can be used to protect sections of pipeline located at a distance, while distributed anodes are more suited for protecting smaller, nearby areas of pipeline.

A buried pipeline extending across an area influenced by an anode bank may be subject to influence from stray electrical current surrounding the bed. Specific conditions in the area where this takes place determine how serious the effect will be on the foreign pipeline. In cases where the electrical potential of the stray current is high enough, protective measures must be put in place to prevent corrosion damage from current flowing away from anodic areas on the pipe.

One approach is to employ bare pipe shields on either side of the foreign pipeline where it lies within the potential gradient of the anode bank and picks up stray current. If the bare pipe shields are electrically connected to the rectifier for the anode bank, they will emit a negative cathodic field that opposes the positive field of the anodes.

High installation and operating costs must be weighed against other alternatives. It may be more cost effective to relocate the anode bank than to consider the use of a shield.

1.2.6 Notifications

Pipeline operators must notify owners of any nearby structures when planning any changes to existing CP systems or any new installations. The courtesy should be reciprocated. Stray currents may weaken or completely eliminate existing CP on structures within the same field of influence. Open communications between all owners and operators in an area should be encouraged. Owners should provide strip maps to each other, with each owner responsible for marking applicable structures. The exact location, size, and scope of any new projects can have serious effects on others. Test points in the same area may be shared. To avoid future problems, put all communications in writing.

Additional Resources

It is assumed that you will learn to perform these procedures under the direct supervision of your crew leader, foreman, or instructor. For independent study, refer to the recommended Additional Resources at the back of this module.

1.0.0 Section Review

1. An impressed current cathodic protection system can subject an adjacent pipeline to _____.
   a. endwise current
   b. stray current interference
   c. current discharge
   d. potential gradients

2. Galvanic anode banks work to ensure that the electrical potential of the pipeline with respect to the surrounding earth is _____.
   a. negative
   b. positive
   c. neutral
   d. at the same polarity

SECTION TWO

2.0.0 INSTALLING GALVANIC ANODES

Objective

Describe how to install galvanic anodes (CT9_2-17).

a. Identify the prerequisites, competencies, and abnormal operating conditions associated with installing galvanic anodes.
b. Describe the procedure for installing galvanic anodes.

Performance Task

Install galvanic anodes (CT9_2-17).

- Identify potential abnormal operating conditions that may occur, and know the appropriate actions to take in response to them.
- Utilize the appropriate personal protective equipment according to relevant company procedures.
- Verify the location for anode installation. Ensure that the electrolytic environment is moist. Horizontal anodes may require backhoe excavation.
- Install the anode into an augered hole or horizontal excavation. Wet down the anode prior to backfilling or before placing it in the ground.
- Uncoil the anode lead and extend it fully. Do not damage or kink the wire.
- If the design is for direct connection, connect the test lead directly to the pipe. Install shunts for measurement of current flow and resistance inside of the test station.
- Backfill carefully with approved soil backfill. Use rock-free backfill to pad the anode and the anode lead wire.
- Complete appropriate documentation as required by operator's procedures.

Trade Terms

Abnormal operating conditions (AOCs): Conditions within a pipeline system, identified by the operator, that may indicate (1) a malfunction of a component or (2) a deviation from normal operations that may indicate a condition exceeding design limits or result in a hazard to persons, property, or the environment.

Exothermic welds: Connections made to a structure using exothermic (thermite) welding equipment and techniques, usually through the semi-controlled firing of a powered material.

Header cable/wire: A cable or wire to which the anode lead wire is connected.

Splice connection: The electrical connection between the anode lead wire and the header cable/wire.

Ultrasonic tester: An instrument that uses high-frequency sound waves to determine the thickness of a structure.

Galvanic anodes provide cathodic protection for buried or submerged metallic structures and may be used for stray current mitigation. The task of installing galvanic anodes begins with determining the location and method of installation specified in the design requirements and ends when the anode is installed and connected.

This section describes some of the basic skills required to install galvanic anodes and identifies **abnormal operating conditions (AOCs)** that personnel may need to respond to when performing this task. It also describes the general procedure for installing galvanic anodes.

2.1.0 Prerequisites, Competencies, and Abnormal Operating Conditions

Personnel must possess a variety of skills in order to install galvanic anodes safely and efficiently. The various operations require a series of steps, which personnel must be competent to perform. Personnel must also be aware of the particular abnormal operating conditions that may occur during the installation procedure and know the proper responses to them.

2.1.1 Prerequisites and Competencies

Personnel assigned to install galvanic anodes must possess the following basic knowledge and skills prior to performing this task:

- Experience in test station and test lead installation
- Experience in making **exothermic welds**
- Experience in taking pipe-to-soil potentials
- Experience in measuring structure wall thickness and using an **ultrasonic tester**

Personnel must have knowledge of cathodic protection systems and components comparable to

NACE Certification Level CP2. This includes, but is not limited to, the following:

- Knowledge of connection methods, such as using a splice connection to connect the anode lead wire to the header cable/wire, and knowing that:
  - Connections are made in a test station with a lead connected to the structure that is being protected and across a shunt for measurement and testing.
  - Isolation of galvanic anodes may be necessary for additional testing of the structure.
- Knowledge of the different types of galvanic anodes and their applications, including knowing that:
  - Galvanic anodes may be used for direct cathodic protection, shielding of electrical interference, spot protection, or AC mitigation.
  - Galvanic anode applications may be in various soil conditions, underwater or offshore, or where power for impressed current CP systems is unavailable.
- Awareness that galvanic anodes are typically supplied in special backfill (hydrated gypsum, bentonite clay, and sodium sulfate) that must be wet for the anodes to start discharging current.

2.1.2 Abnormal Operating Conditions

Before installing galvanic anodes, a pipeline technician must be able to identify and properly respond to potential AOCs, should they occur. The following are some potential AOCs, and the appropriate action personnel should take in response to them:

- *CP levels are low* – Make proper company notifications.
- *Test lead or bond cable is broken or damaged* – Notify operator personnel for repair or replacement of the damaged lead or cable, as required.
- *Component failure occurs* – Evaluate the extent of the failure/damage and make company notifications as applicable.
- *Fire or explosion occurs from burn-through during exothermic welding* – Limit access to the area as much as possible and make company notifications.
- *Unexpected leak of hazardous liquid or carbon dioxide occurs within or outside casings* – Make immediate notifications per company procedure. Try to identify and isolate the source of the leak if possible and if properly trained to do so.

2.2.0 Procedure

As shown in *Figure 10*, when galvanic anodes are used to correct stray current interference on a pipeline, the anodes are attached to the affected line. They are located in the area that is subject to interference from a foreign line and are buried in a special backfill that must be wet in order for the anodes to operate. The heaviest anode is placed at the point of crossing. This is where the exposure is the most intense. The lighter anodes are used elsewhere on the line.

The length of the affected area of the pipe determines the length of the anode string. A test point and a permanent reference electrode should be installed to periodically measure potentials in order to monitor the interference on the underside of the line at the maximum exposure point. The following steps outline the general procedure for installing galvanic anodes:

NOTE

Although it is possible to fit the interfered pipeline with anode systems alone, it is not possible to achieve a correction by fitting the interfering pipeline only. Strapping the anodes together will reduce the consumption rate.

WARNING!

Always obtain and use the appropriate personal protective equipment for each task. Only qualified individuals may make electrical connections and operate equipment. Always verify the absence of voltage and current before making connections. Follow the manufacturer's instructions for the equipment being installed.

Step 1 Identify potential abnormal operating conditions that may occur, and know the appropriate actions to take in response to them.

Step 2 Utilize the appropriate personal protective equipment according to relevant company procedures.

Step 3 Identify a location near the source of interference with a high sub-surface moisture content.

NOTE

Moisture in the electrolyte is essential for proper operation of the anode.

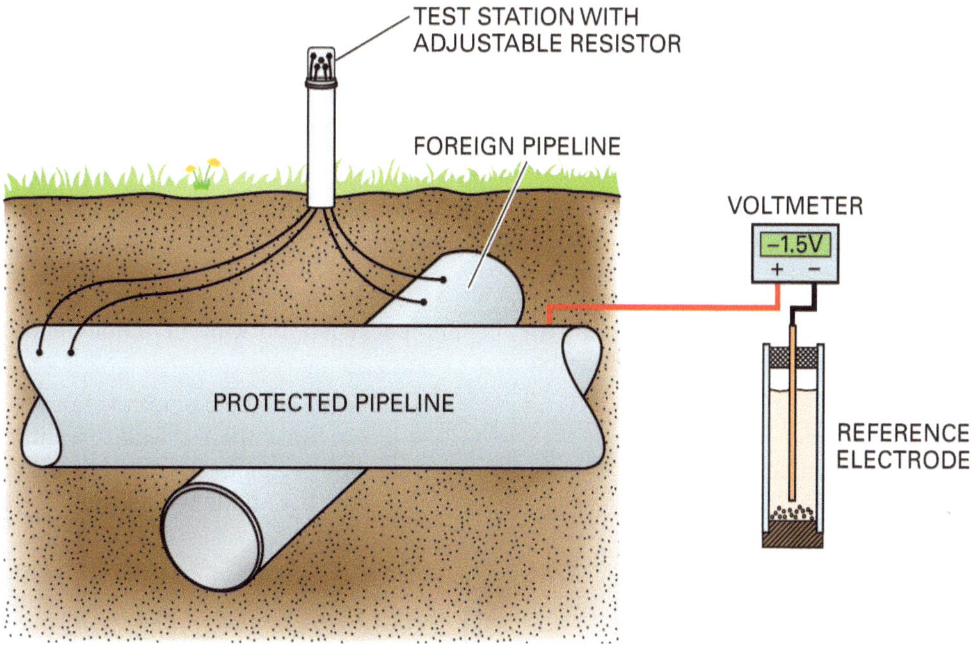

Figure 10 Galvanic anodes attached to a line under test.

Step 4 Ensure that the hole/excavation provides a moist electrolytic environment. Anodes must be buried at the depth of the pipeline and deeper if necessary to maintain a constant moisture level. Vertical anodes may require the use of an auger, while horizontal anodes may require excavation using a backhoe.

NOTE

Excavations should be sufficiently deep so that ground water levels will not dry out. Anode holes should be at least as deep as the pipeline.

Step 5 Remove the anode from its protective packaging.

Step 6 Galvanic anodes are supplied in a special backfill that is designed to facilitate the operation of the device. Wet the anode and prepackaged backfill, and then install it in the hole or excavation.

CAUTION

Take care not to damage the anode or its prepackaged backfill when preparing the anode for installation.

Step 7 Carefully uncoil the anode lead wire (pigtail) to avoid kinking and damage.

Step 8 If the design calls for a direct connection, connect the test lead directly to the pipe. Record the exact location so that each anode can be identified and tested during system checkout and for later maintenance/repair. Where possible, install a shunt inside the test station to monitor the current flow and resistance.

Step 9 Backfill carefully to avoid damaging the anode and lead wires. Use rock-free backfill to pad the anode and the anode lead wire.

Step 10 Document the installation as required by the operator's procedures.

Additional Resources

It is assumed that you will learn to perform these procedures under the direct supervision of your crew leader, foreman, or instructor. For independent study, refer to the recommended Additional Resources at the back of this module.

2.0.0 Section Review

3. Individuals who are assigned to install galvanic anodes must have knowledge of cathodic protection systems and components comparable to _____.
   e. Journeyman electrician classification
   f. NACE Certification Level CP2
   g. ANSI Accreditation Phase 2
   h. NFPA 70® apprenticeship standards

4. Before installing a galvanic anode and its prepackaged backfill in the hole or excavation, _____.
   i. purge the hole or excavation
   j. pin braze the test lead to a shunt
   k. wet the anode and the backfill
   l. coat the anode with a rust inhibitor

SUMMARY

Stray currents, or interference, from foreign sources can cause pipeline corrosion. To control this problem, pipeline technicians need to know how to identify when stray currents are present, how to identify the source of the stray currents, and how to use appropriate techniques to reduce the interference.

Techniques for mitigating interference include the use of bonds, coatings, galvanic anodes, and electrical shields. This module described the procedures for the installation of galvanic anodes for stray current mitigation.

Review Questions

1. Natural occurrences of stray current in the earth that are transferred to the protected pipeline are known as _____.
   a. potential gradients
   b. burial pressures
   c. ground water changes
   d. climate variations

2. DC-powered transit systems can cause _____.
   a. potential shifts
   b. static DC interference
   c. dynamic stray currents
   d. impressed currents

3. Electrical shielding can be provided by diverting the current or by _____.
   a. resistance bonding
   b. nonmetallic barriers
   c. current interrupters
   d. galvanic anodes

4. The type of galvanic anode normally preferred for its higher anodic potential gradient field is a(n) _____.
   a. aluminum anode
   b. magnesium anode
   c. zinc anode
   d. iron anode

5. If a pipeline technician discovers a broken or damaged test lead or bond cable when installing galvanic anodes, the technician should _____.
   a. notify the supervisor and limit access to the work area as much as possible
   b. immediately eliminate potential ignition sources in the area and assess the damage
   c. evacuate all non-essential personnel from the work site and call 911
   d. notify operator personnel for repair, as required

6. The length of the anode string is determined by the _____.
   a. current of the affected pipeline area
   b. length of the affected pipeline area
   c. content of the affected pipeline area
   d. environment of the affected pipeline area

Trade Terms Quiz

Fill in the blank with the correct term that you learned from your study of this module.

1. Connections to the structure being tested, usually wires in a supporting stand with easy connection points for the meter, are called ________.
2. Devices used to allow current to flow in only one direction are ________.
3. Changes in potential with respect to distance, expressed in millivolts per unit of distance are ________.
4. A conductor that joins two points in an electrical circuit to form a parallel or alternative path through which a portion of the current may pass is a(n) ________.
5. Conditions within a pipeline system, identified by the operator, that may indicate (1) a malfunction of a component or (2) a deviation from normal operations that may indicate a condition exceeding design limits or result in a hazard to persons, property, or the environment are called ________.
6. Sets of two or more anodes installed in a specific configuration where the anodes are connected together by a header cable/wire are referred to as ________.
7. A portion of a current that flows over a path other than the intended path in an on/off or intermittent fashion and may cause electrochemical corrosion of metals in contact with electrolytes is ________.
8. Connections made to a structure using exothermic (thermite) welding equipment and techniques are called ________.
9. Electrical connections of two structures that usually have two different cathodic protection levels are ________.
10. The effect on the cathodic protection level of a structure due to the interaction between its cathodic protection system and the cathodic protection system of another structure is called ________.
11. A device used to convert AC current to DC current and which supplies current to the structure in the soil in an impressed cathodic protection system, is a(n) ________.
12. A substance that ionizes (releases free electrons) in an electric field is a(n) ________.
13. A portion of a current that flows over a path other than the intended path at a constant, unbroken rate and that may cause electrochemical corrosion of metals in contact with electrolytes is ________.
14. The electrical connection between the anode lead wire and the header cable/wire is a(n) ________.
15. A measurement of the voltage or potential difference between a buried or submerged metal structure and an electrolyte, such as soil, is referred to as ________.
16. Current flows through a path other than the intended circuit are called ________.
17. A cable or wire to which the anode lead wire is connected is a(n) ________.
18. The ability of a material to resist the flow of electric current is called ________.
19. An instrument that uses high-frequency sound waves to determine the thickness of a structure is a(n) ________.
20. The electrode of an electrochemical cell at which oxidation occurs is called the ________.

Trade Terms

Abnormal operating conditions (AOCs)
Anode
Anode banks
Bonds
Diodes
Dynamic stray current
Electrolyte
Exothermic welds
Header cable/wire
Interference
Potential gradients
Rectifier
Resistivity
Shunt
Splice connection
Static stray current
Stray currents
Structure-to-soil potential
Test leads
Ultrasonic tester

Trade Terms Introduced in This Module

Abnormal operating conditions (AOCs): Conditions within a pipeline system, identified by the operator, that may indicate (1) a malfunction of a component or (2) a deviation from normal operations that may indicate a condition exceeding design limits or result in a hazard to persons, property, or the environment.

Anode: The electrode of an electrochemical cell at which oxidation occurs. Electrons flow away from the anode in the external circuit. Corrosion usually occurs and metal ions enter the solution at the anode.

Anode banks: Sets of two or more anodes installed in a specific configuration where the anodes are connected together by a header cable/wire; also referred to as groundbeds.

Bonds: Electrical connections of two structures that usually have two different cathodic protection levels.

Diodes: Devices used to allow current to flow in only one direction.

Dynamic stray current: A portion of a current that flows over a path other than the intended path in an on/off or intermittent fashion and may cause electrochemical corrosion of metals in contact with electrolytes.

Electrolyte: A substance that ionizes (releases free electrons) in an electric field.

Exothermic welds: Connections made to a structure using exothermic (thermite) welding equipment and techniques, usually through the semi-controlled firing of a powered material.

Header cable/wire: A cable or wire to which the anode lead wire is connected.

Interference: The effect on the cathodic protection level of a structure due to the interaction between its cathodic protection system and the cathodic protection system of another structure.

Potential gradients: Changes in potential with respect to distance, expressed in millivolts per unit of distance.

Rectifier: A device used to convert AC current to DC current. In an impressed cathodic protection system, it supplies current to the structure in the soil.

Resistivity: The ability of a material to resist the flow of electric current.

Shunt: A conductor that joins two points in an electrical circuit to form a parallel or alternative path through which a portion of the current may pass.

Splice connection: The electrical connection between the anode lead wire and the header cable/wire.

Static stray current: A portion of a current that flows over a path other than the intended path at a constant, unbroken rate and that may cause electrochemical corrosion of metals in contact with electrolytes.

Stray currents: Current flows through a path other than the intended circuit.

Structure-to-soil potential: A measurement of the voltage or potential difference between a buried or submerged metal structure and an electrolyte, such as soil; also referred to as structure-to-electrolyte potential.

Test leads: Connections to the structure being tested, usually wires in a supporting stand with easy connection points for the meter.

Ultrasonic tester: An instrument that uses high-frequency sound waves to determine the thickness of a structure.

Additional Resources

The following recommended resources can provide additional helpful information related to the requirements for this covered task:

API Recommended Practice 1161, Third Edition, © 2014 American Petroleum Institute. API Publishing Services, 1220 L Street NW, Washington, DC 20005. Distributed by Thompson Reuters (Scientific) LLC, **www.techstreet.com**.

49 *CFR* 192, *Subpart I*, **www.ecfr.gov**.

49 *CFR* 195,*Subpart H*, **www.ecfr.gov**.

NCCER Module AOCCC-17, *Abnormal Operating Conditions Control Center*.

NCCER Module AOCFG-17, *Abnormal Operating Conditions Field and Gas*.

NCCER Module 62105-14, *Pipeline Documentation*.

NCCER Module 64202-02, *Use of Meters and Test Equipment*.

NCCER Module 66101-14, *Introduction to the Pipeline Industry*.

NCCER Module CT1_1-17, *Measure Structure-to-Soil Potential*.

NCCER Module CT1_2-17, *Conduct Close Interval Survey*.

NCCER Module CT1_3-17, *Test and Detect Interference*.

NCCER Module CT1_4-17, *Inspect and Perform Electrical Test of Bonds*.

NCCER Module CT2_2-17, *Repair Damaged Test Leads*.

NCCER Module CT2_3-17, *Install Test Leads by Non-Exothermic Welding Methods*.

NCCER Module CT2_4-17, *Install Test Leads by Exothermic Welding Methods*.

Figure Credits

Tinker & Rasor, Module Opener

Section Review Answer Key

Answer	Section Reference	Objective
Section One		
1. b	1.1.1	1a
2. a	1.2.5	1b
Section Two		
1. b	2.1.1	2a
2. c	2.2.0	2b

NCCER CURRICULA — USER UPDATE

NCCER makes every effort to keep its textbooks up-to-date and free of technical errors. We appreciate your help in this process. If you find an error, a typographical mistake, or an inaccuracy in NCCER's curricula, please fill out this form (or a photocopy), or complete the online form at **www.nccer.org/olf**. Be sure to include the exact module ID number, page number, a detailed description, and your recommended correction. Your input will be brought to the attention of the Authoring Team. Thank you for your assistance.

Instructors – If you have an idea for improving this textbook, or have found that additional materials were necessary to teach this module effectively, please let us know so that we may present your suggestions to the Authoring Team.

NCCER Product Development and Revision

13614 Progress Blvd., Alachua, FL 32615

Email: curriculum@nccer.org
Online: www.nccer.org/olf

❑ Trainee Guide ❑ Lesson Plans ❑ Exam ❑ PowerPoints Other ____________

Craft / Level: Copyright Date:

Module ID Number / Title:

Section Number(s):

Description:

Recommended Correction:

Your Name:

Address:

Email: Phone:

CT9_3-17

Install Rectifiers

Overview

Various cathodic protection systems—which include the installation of bonds, galvanic anodes, rectifiers, and impressed current groundbeds—are used to mitigate stray current interference and prevent corrosion in pipelines. In an impressed cathodic protection system, the rectifier supplies current to the structure in the soil. This module describes the general procedures for installing rectifiers in cathodic protection systems.

Trainees with successful module completions may be eligible for credentialing through the NCCER Registry. To learn more, go to **www.nccer.org** or contact us at 1.888.622.3720. Our website has information on the latest product releases and training, as well as online versions of our *Cornerstone* magazine and Pearson's product catalog.

Your feedback is welcome. You may email your comments to **curriculum@nccer.org**, send general comments and inquiries to **info@nccer.org**, or fill in the User Update form at the back of this module.

This information is general in nature and intended for training purposes only. Actual performance of activities described in this manual requires compliance with all applicable operating, service, maintenance, and safety procedures under the direction of qualified personnel. References in this manual to patented or proprietary devices do not constitute a recommendation of their use.

CT9_3-17
Install Rectifiers

Objectives

When you have completed this module, you will be able to do the following:

1. Describe stray current interference and methods for mitigating it.
   a. Describe types of stray current and how they affect cathodic protection.
   b. Describe methods used to mitigate stray current interference.
2. Describe how to install rectifiers (CT9_3-17).
   a. Identify the prerequisites, competencies, and abnormal operating conditions associated with installing rectifiers.
   b. Describe the procedure for installing a rectifier.

Performance Tasks

Under the supervision of the instructor, you should be able to do the following:

1. Install rectifiers (CT9_3-17).
   - Identify potential abnormal operating conditions that may occur during performance of this CT, and know the appropriate actions to take in response to them.
   - Utilize the appropriate personal protective equipment according to relevant company procedures.
   - Verify the rectifier required for the location and service.
   - Mount the rectifier at the designated location (pole, post, wall, etc.), making sure that it is securely fastened.
   - Connect the AC power feed wires through an appropriately sized conduit according to the *National Electric Code* (*NEC*®), the *National Electric Safety Code*, and local electric and building codes.
   - Terminate the AC feed wires at terminals on circuit breaker or AC input connection wires.
   - Connect the DC conduits to the rectifier.
   - Install the DC cables from the anode groundbed and the structure in their respective conduits, and terminate on their respective terminals.
   - Test and verify that cables are correctly installed.
   - Complete appropriate documentation as required by operator's procedures.

Trade Terms

Abnormal operating conditions (AOCs)
Anode
Anode banks
Bonds
Cathode
Diodes
Dynamic stray current
Electrolyte
Interference
Potential gradients
Rectifier
Resistivity
Shunt
Static stray current
Stray currents
Structure-to-soil potential
Test leads

Industry Recognized Credentials

If you are training through an NCCER-accredited sponsor, you may be eligible for credentials from NCCER's Registry. The ID number for this module is CT9_3-17. Note that this module may have been used in other NCCER curricula and may apply to other level completions. Contact NCCER's Registry at 888.622.3720 or go to **www.nccer.org** for more information.

Contents

Figures and Tables

SECTION ONE

1.0.0 STRAY CURRENT INTERFERENCE

Objective

Describe stray current interference and methods for mitigating it.

a. Describe types of stray current and how they affect cathodic protection.
b. Describe methods used to mitigate stray current interference.

Trade Terms

Anode: The electrode of an electrochemical cell at which oxidation occurs. Electrons flow away from the anode in the external circuit. Corrosion usually occurs and metal ions enter the solution at the anode.

Anode banks: Sets of two or more anodes installed in a specific configuration where the anodes are connected together by a header cable/wire; also referred to as groundbeds.

Bonds: Electrical connections of two structures that usually have two different cathodic protection levels.

Diodes: Devices used to allow current to flow in only one direction.

Dynamic stray current: A portion of a current that flows over a path other than the intended path in an on/off or intermittent fashion and may cause electrochemical corrosion of metals in contact with electrolytes.

Electrolyte: A substance that ionizes (releases free electrons) in an electric field.

Interference: The effect on the cathodic protection level of a structure due to the interaction between its cathodic protection system and the cathodic protection system of another structure.

Potential gradients: Changes in potential with respect to distance, expressed in millivolts per unit of distance.

Rectifier: A device used to convert AC current to DC current. In an impressed cathodic protection system, the rectifier supplies current to the structure in the soil.

Resistivity: The ability of a material to resist the flow of electric current.

Shunt: A conductor that joins two points in an electrical circuit to form a parallel or alternative path through which a portion of the current may pass.

Static stray current: A portion of a current that flows over a path other than the intended path at a constant, unbroken rate and that may cause electrochemical corrosion of metals in contact with electrolytes.

Stray currents: Current flows through a path other than the intended circuit.

Structure-to-soil potential: A measurement of the voltage or potential difference between a buried or submerged metal structure and an electrolyte, such as soil; also referred to as structure-to-electrolyte potential.

Test leads: Connections to the structure being tested, usually wires in a supporting stand with easy connection points for the meter.

Corrosion on a pipeline can be caused by **stray currents**, in which electrical currents flow through a path other than the intended circuit. The stray current flow results in the problem called **interference**, or *stray current interference*. This section describes the types and sources of stray currents, explains how to test for interference current, and reviews some of the methods that can be used to eliminate stray current interference.

1.1.0 Stray Current

Stray currents are direct currents flowing in the earth that are from a source other than the corrosion protection source. **Potential gradients** cause natural occurrences of stray current in the earth that are transferred to the protected pipeline. *Figure 1* shows potential gradients around a protected pipeline.

Normally, this outside current enters the pipeline at one point, exits the pipeline at another point, and returns to the earth. The area where the current returns to earth, completing the circuit, is the point at which the corrosion may occur if the material is in contact with an **electrolyte**. The measurement of the voltage or potential difference between a buried or submerged metal structure and an electrolyte, such as soil, is the **structure-to-soil potential**, also referred to as the *structure-to-electrolyte potential.*

1.1.1 Types of Stray Current

Stray currents are either static (constant) or dynamic (intermittent). Sources of stray current include cathodic protection (CP) systems on other pipelines, DC transit systems, DC welding operations, DC mining operations, high-voltage

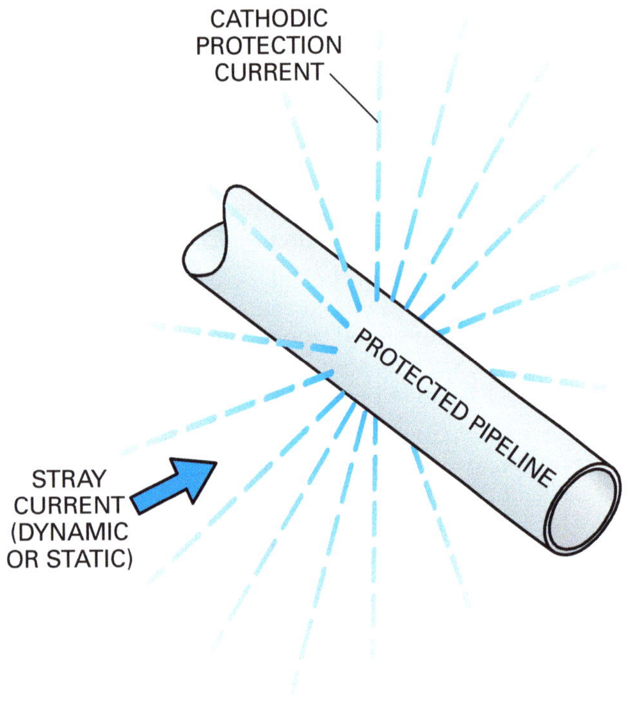

Figure 1 Potential gradients around a protected pipeline.

DC transmission systems, or disturbances in the earth's magnetic field. If environmental conditions allow rectifying junctions such as copper oxide films, AC current flow can be rectified. This results in a direct current that can be mitigated.

There are many different sources of interference current in a cathodic protection system. For example, stray current interference on adjacent pipelines can be caused by impressed current cathodic protection systems. This depends on the locations of the anode banks and the operating characteristics of the cathodic protection system. Corrosion usually occurs and metal ions enter the solution at the anode.

Figure 2 shows stray current to a crossover foreign pipeline. In this example, the foreign pipeline picks up current from the anode bank and transfers the current to the protected pipeline when it crosses under it. The potential for corrosion is greatly increased at the crossing point of the pipelines.

Figure 3 shows stray current to a parallel foreign pipeline. In this example, the parallel foreign pipeline is affected by the anode bank of the protected pipe. This is because the foreign pipeline travels through the anode bank and picks up current. When this occurs, the effectiveness of the CP system will be reduced in this area of the pipeline. In the affected area of the foreign pipeline, it becomes positive to the soil and discharges current to the earth. This increases the potential for corrosion on the foreign pipeline in the affected area.

1.1.2 Testing for Interference Current

Static stray current interference is easy to test for as long as there is no dynamic stray current present. Dynamic stray current can be caused by sources such as DC-powered transit systems, underground mining, welding, and pipeline interference. In *Figure 4*, a section of coated pipeline has an impressed current CP system applied to it and has two foreign pipelines crossing it. In this example, "B" is a strong indicator of interference because it has the largest difference in voltage. Always look for large changes in voltage to identify interference. Follow company guidelines for determining the need to mitigate interference.

To distinguish the current in the foreign pipeline from the current in the protected line, an automatic current interrupter is installed in the output of the rectifier. The current interrupter is a device that automatically opens and closes an electrical circuit at a preset time. With the current interrupter operating, measure the potential of the lines under both on and off conditions. *Figure 5* shows how to measure this potential at the test station. As shown, a shunt is installed for measurement of current flow and resistance inside of the test station.

These tests are performed using a copper sulfate electrode placed over the crossing point of the lines. Pipeline locators can be used to determine the exact location of the crossing.

WARNING!

Always be alert when testing for interference current. A release could be found while testing.

The data obtained at each pipeline crossing must be properly recorded. *Figure 6* shows typical interference testing potentials. These records are also called field data sheets. Field data sheets should also include any information about the protected line, the date, and the current output of the rectifier.

Telluric Currents

Telluric currents are caused by sunspots and solar flares and can result in inaccurate readings during intense periods of solar activity.

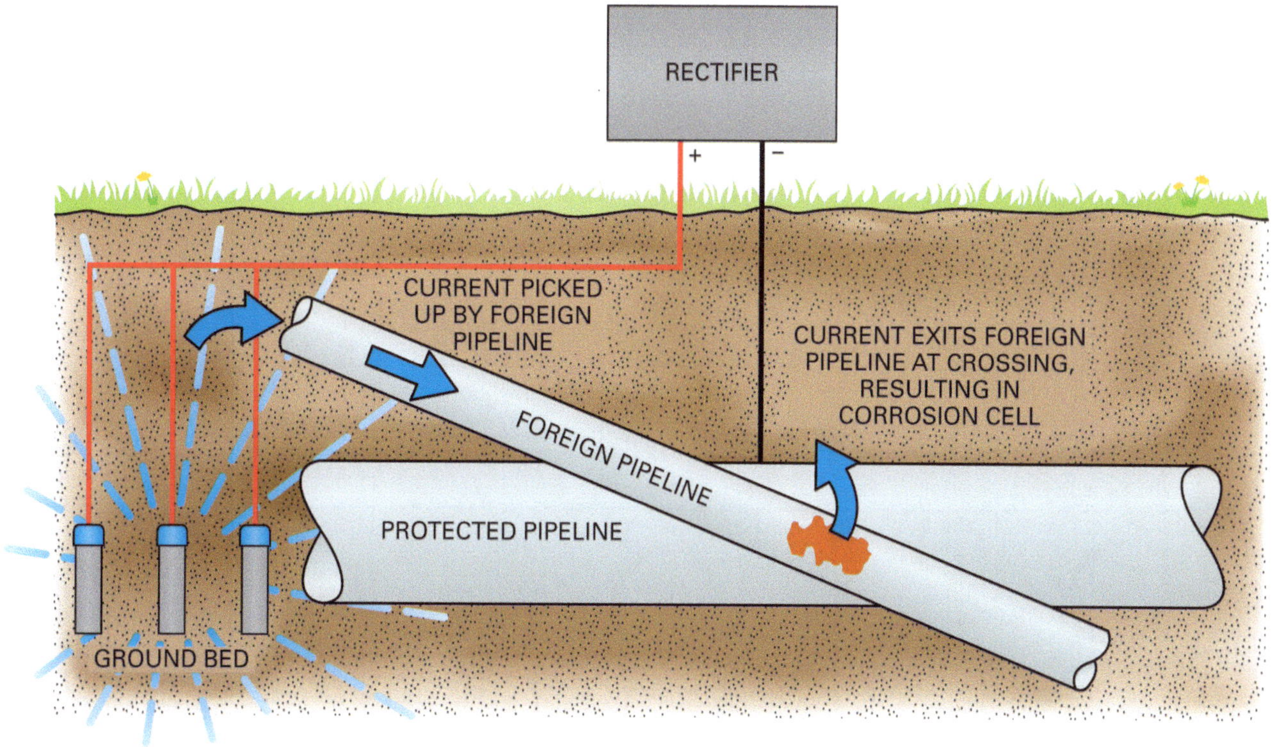

Figure 2 Stray current to a crossover foreign pipeline.

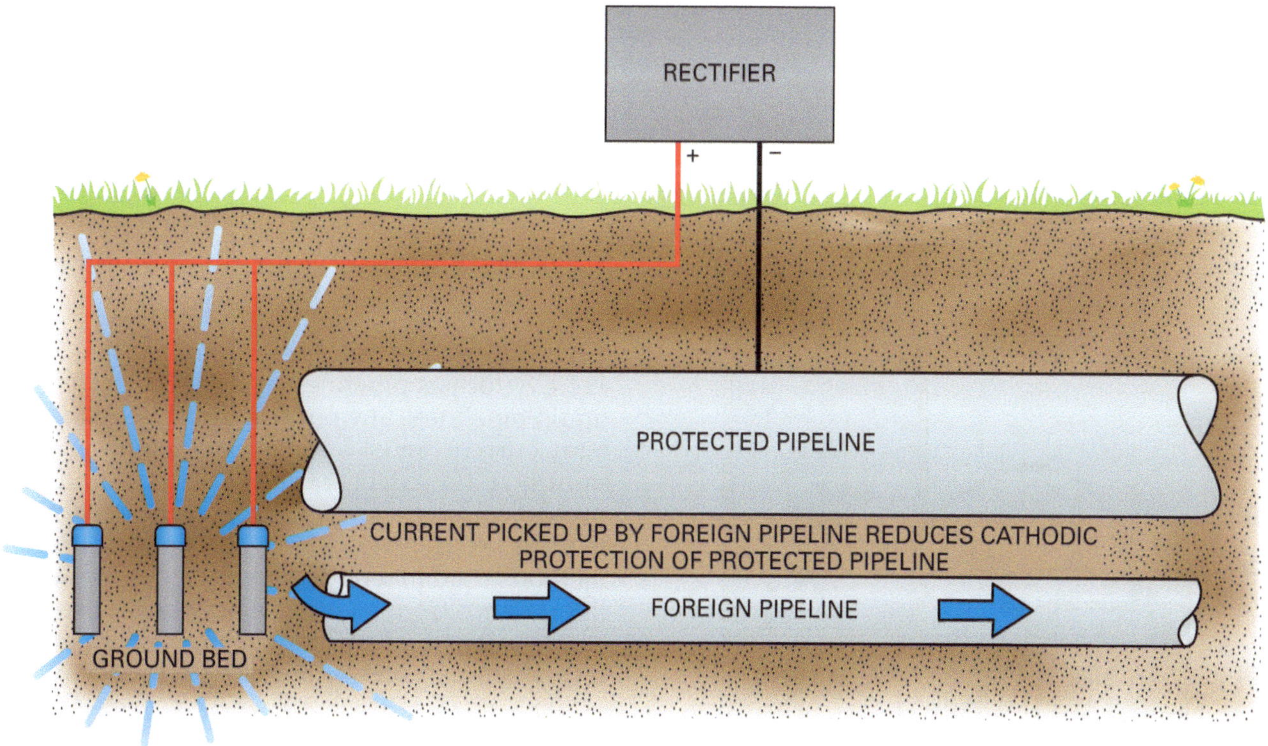

Figure 3 Stray current to a parallel foreign pipeline.

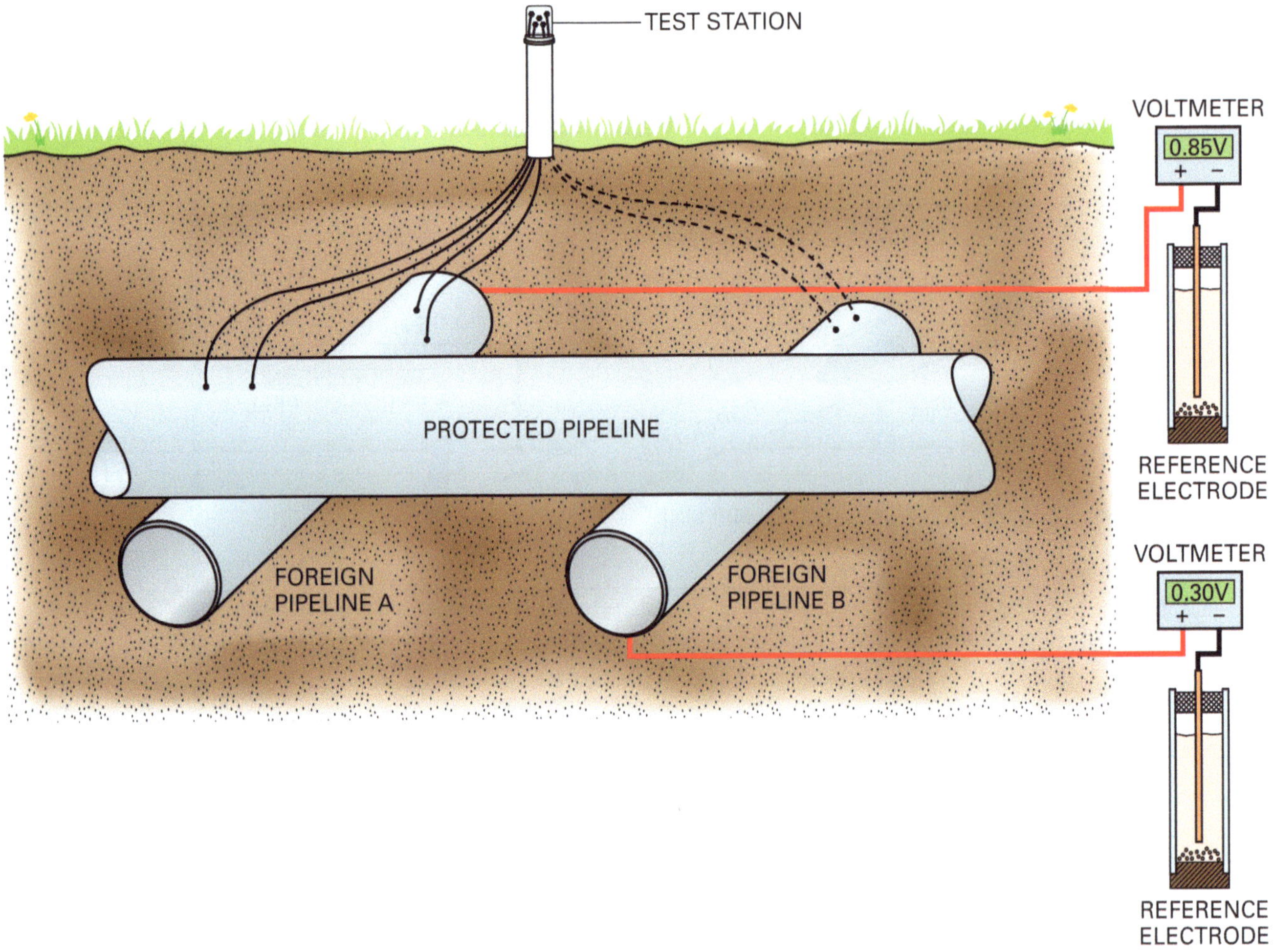

Figure 4 Testing foreign pipeline crossings for stray current interference from CP installations.

Figure 5 Measuring potential at the test station.

1.2.0 Methods for Reducing or Eliminating Stray Current Interference

According to the *Code of Federal Regulations* 49 *CFR* 192.473 (for gas pipelines) and 49*CFR* 195.577 (for liquid pipelines), any pipeline system subjected to stray currents must have an ongoing program to minimize the effects of these currents and help protect surrounding or adjacent underground metallic structures. There are many different methods used to reduce or eliminate stray current interference from a CP system. Some of the most common methods include the use of **bonds**, coatings, electrical shields, and galvanic anodes.

1.2.1 Resistance Bonds

A bond is simply a metallic connection between structures. There are two types of bonds: critical and non-critical. Critical bonds are identified by the operator's procedures and include those

FOREIGN LINE		PROTECTED PIPELINE (volts)			INTERFERING PIPELINE (volts)		
NAME	PIPELINE STATION	VOLTAGE ON	VOLTAGE OFF	Δ volts	VOLTAGE ON	VOLTAGE OFF	Δ volts
A	140 + 00	-0.91	-0.90	-0.01	-0.85	-0.88	+0.03
B	1210 + 00	-1.94	-1.00	-0.94	-0.30	-0.68	+0.38

Figure 6 Sample of recording survey notes.

bonds that may cause damage to the structure if disconnected. Like test leads, bond leads are terminated in a test station.

Foreign pipelines can be subject to stray current damage from the impressed current cathodic protection system on another pipeline. To correct this problem, a resistance bond is connected between the two pipelines. The amount of resistance in the bond is adjusted to drain enough current from the affected line to eliminate the corrosion potential.

CAUTION

For liability reasons, the representative of the foreign pipeline company should be on site before procedures to correct the problem on the affected line begin.

For a resistance bond, diodes are installed as basic components of a rectifier circuit. Single-phase rectifiers are probably the most common units used for cathodic protection. The diode is installed to allow current to flow in one direction only. This characteristic of the diode causes all the current waves to be more aligned when entering into the cables. Along with other components, the diodes reduce ripple effects in the power supply and filter out interference.

1.2.2 Coatings

If the testing of a foreign line indicates that the interference is caused by a foreign pipeline that is bare or poorly coated, the bond base may not be able to correct the condition. If a high current density is flowing from a well-coated line through the soil and onto the bare foreign line, localized corrosion could occur. To correct this problem, reduce the density of the current flow to the crossing area of the foreign line. The best way to accomplish this is to apply a coating to the line. *Figure 7* shows the use of a coating to correct interference at foreign pipeline crossings.

With proper coating, the voltage drops in the earth around the foreign line are reduced or eliminated. Over-the-line potential measurements are used to identify the portion of the foreign line subject to interference. This information is then used to determine the area on the foreign line that must be coated to mitigate the interference.

CAUTION

Take special precautions to avoid bare spots in the coating (holidays) to prevent localized corrosion.

1.2.3 Electrical Shields

Electrical shields are defined as barriers to stray current flow. Electrical shielding can be provided by nonmetallic insulating barriers or by diverting the current from the affected pipeline to another structure. Shields are particularly useful for protecting pipeline in congested areas such as pumping stations and tank farms. *Figure 8* shows bare pipe used as electrical shields around a pipeline.

1.2.4 Galvanic Anodes

Galvanic anodes can provide cathodic protection for buried or submerged structures. They can also be used to correct interference, as spot protection, and for alternating current mitigation. When used to correct interference, the anodes are attached to the affected line (see *Figure 9*). They are located in the area that is subject to interference from the foreign line. Anodes are buried in a special backfill that must be wet in order for the anodes to operate.

This method uses the anodic potential fields that are around the galvanic anode to offset

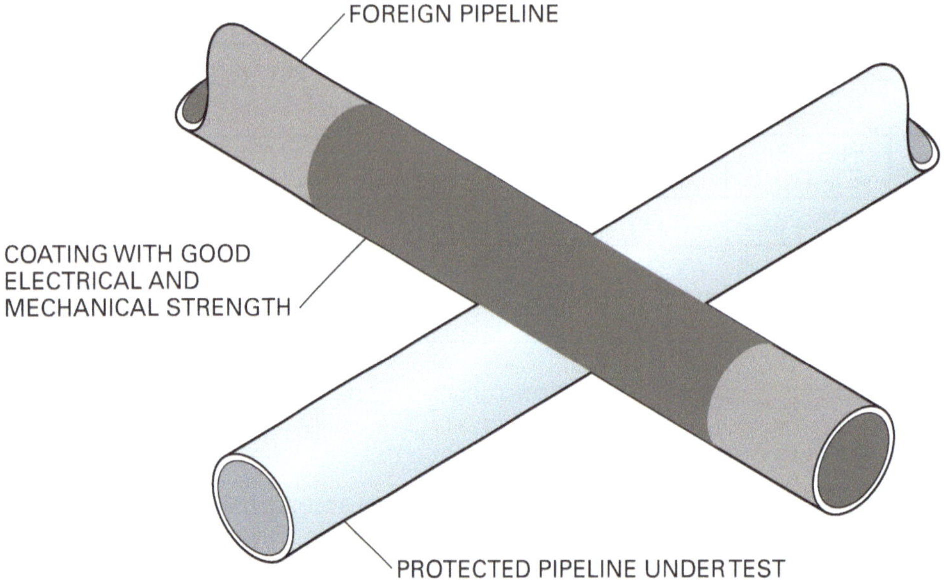

Figure 7 Using a coating to correct interference at foreign pipeline crossings.

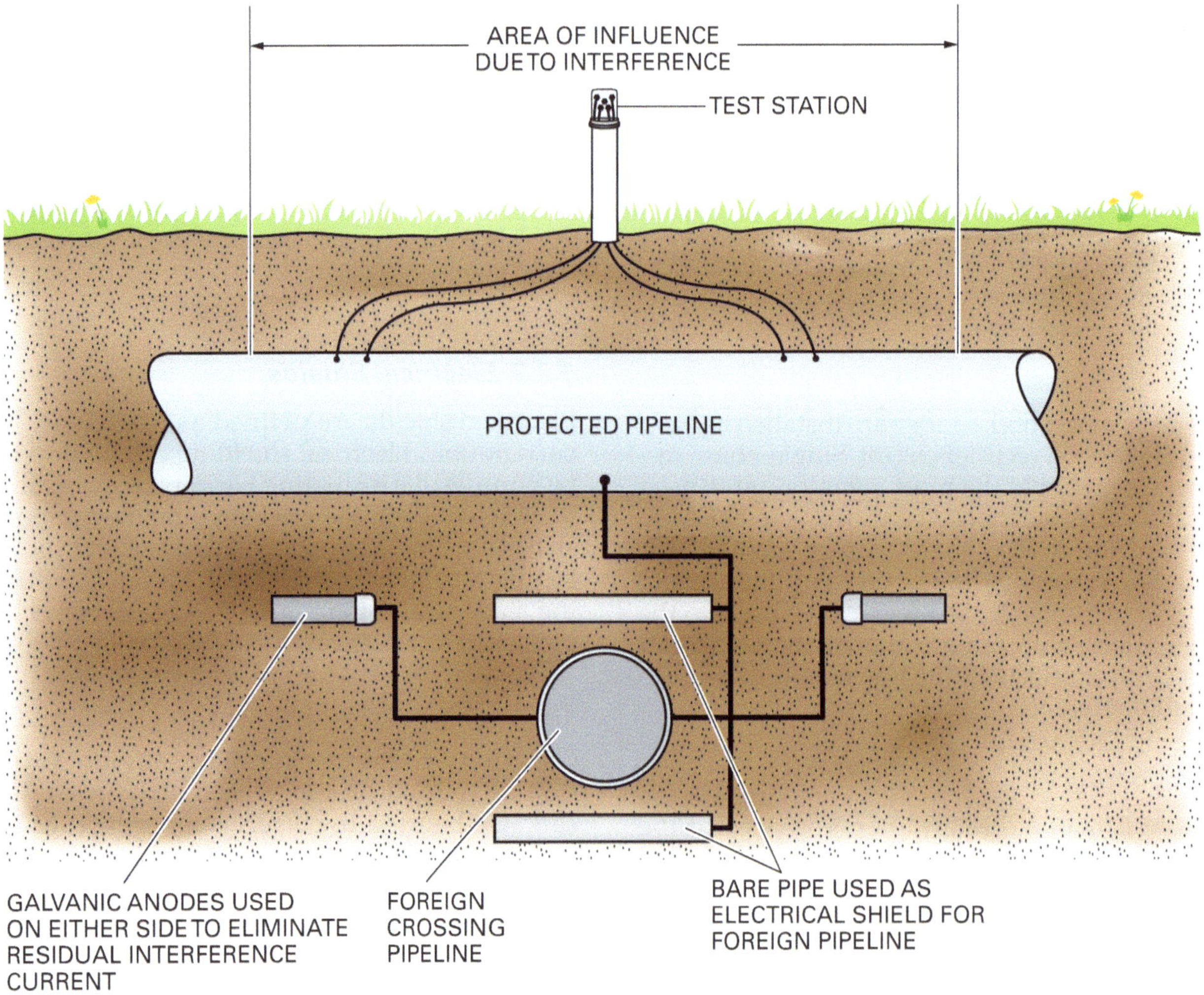

Figure 8 Use of electrical shields to reduce stray current interference.

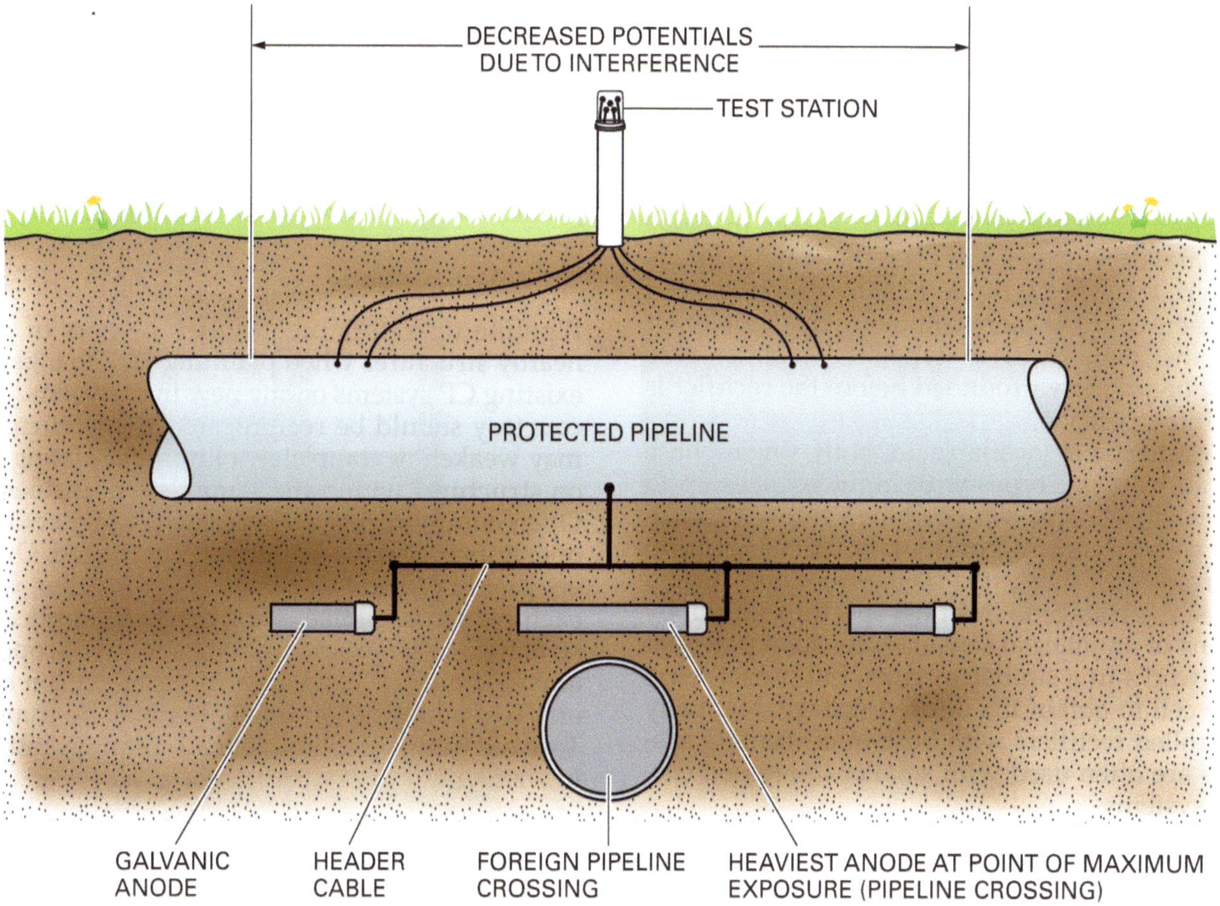

Figure 9 Galvanic anodes attached to a line under test.

the potential gradient field around the protected pipeline. Usually an anode bank consisting of a single line of anodes between the pipeline and the foreign pipeline causing the interference will mitigate the effects of the interference. When properly designed, the remaining current will discharge from the anodes instead of the pipeline. Several strings of anodes may be required when potential differences are very high. The anode life using this arrangement is short unless the surrounding soil has a high **resistivity**. Magnesium anodes are preferred to zinc anodes because magnesium anodes have a higher anodic potential gradient field.

1.2.5 Galvanic Anode Banks

Galvanic anode banks work to ensure that the electrical potential of the pipeline is negative with respect to the surrounding earth. Remote anodes, also referred to as *remote groundbeds*, can be used to protect sections of pipeline located at a distance, while distributed anodes are more suited for protecting smaller, nearby areas of pipeline.

A buried pipeline extending across an area influenced by an anode bank may be subject to influence from stray electrical current surrounding the bed. Specific conditions in the area where this takes place determine how serious the effect will be on the foreign pipeline. In cases where the electrical potential of the stray current is high enough, protective measures must be put in place to prevent corrosion damage from current flowing away from anodic areas on the pipe.

One approach is to employ bare pipe shields on either side of the foreign pipeline where it lies within the potential gradient of the anode bank and picks up stray current. If the bare pipe shields are electrically connected to the rectifier for the anode bank, they will emit a negative cathodic field that opposes the positive field of the anodes.

High installation and operating costs must be weighed against other alternatives. It may be more cost effective to relocate the anode bank than to consider the use of a shield.

SECTION TWO

2.0.0 INSTALLING RECTIFIERS

Objective

Describe how to install a rectifier (CT9_3-17).

a. Identify the prerequisites, competencies, and abnormal operating conditions associated with installing rectifiers.
b. Describe the procedure for installing a rectifier.

Performance Task

Install rectifiers (CT9_3-17).

- Identify potential abnormal operating conditions that may occur, and know the appropriate actions to take in response to them.
- Utilize the appropriate personal protective equipment according to relevant company procedures.
- Verify the rectifier required for the location and service.
- Mount the rectifier at the designated location (pole, post, wall, etc.), making sure that it is securely fastened.
- Connect the AC power feed wires through an appropriately sized conduit according to the *National Electric Code* (*NEC*®), the *National Electric Safety Code*, and local electric and building codes.
- Terminate the AC feed wires at terminals on circuit breaker or AC input connection wires.
- Connect the DC conduits to the rectifier.
- Install the DC cables from the anode groundbed and the structure in their respective conduits, and terminate on their respective terminals.
- Test and verify that cables are correctly installed.
- Complete appropriate documentation as required by operator's procedures.

Trade Terms

Abnormal operating conditions (AOCs): Conditions within a pipeline system, identified by the operator, that may indicate (1) a malfunction of a component or (2) a deviation from normal operations that may indicate a condition exceeding design limits or result in a hazard to persons, property, or the environment.

Cathode: The electrode of an electrochemical cell at which reduction is the principal reaction. Electrons flow toward the cathode in the external circuit.

The installation of impressed current cathodic protection rectifiers is essential in protecting facilities against external corrosion. The task of installing a rectifier begins with selecting the location and making the initial connections between the structure that is being protected and the groundbed. The task does not end until the rectifier has been installed, connected to the structure/anodes, and the installation has been documented.

This section describes some of the basic skills required to install rectifiers and identifies abnormal operating conditions (AOCs) that personnel may need to respond to when performing this task. It also describes the general procedure for installing rectifiers.

2.1.0 Prerequisites, Competencies, and Abnormal Operating Conditions

Personnel must possess a variety of skills in order to install rectifiers safely and efficiently. The various operations require a series of steps, which personnel must be competent to perform. Personnel must also be aware of the particular abnormal operating conditions that may occur during the installation procedure and know the proper responses to them.

2.1.1 Prerequisites and Competencies

Personnel assigned to install rectifiers must have knowledge of cathodic protection systems and components comparable to NACE Certification Level CP2. This includes, but is not limited to, the following:

- Understanding of basic electricity and electrical circuits and the associated terminology (anode, cathode, etc.)
- Knowledge of the type of rectifier being installed (air-cooled, oil-cooled, explosion-proof, solar-powered, etc.)
- Knowledge of mounting requirements (pole mount, rack mount, concrete pad, etc.)
- Awareness that terminating each cable at the proper terminal is essential for safe and effective cathodic protection

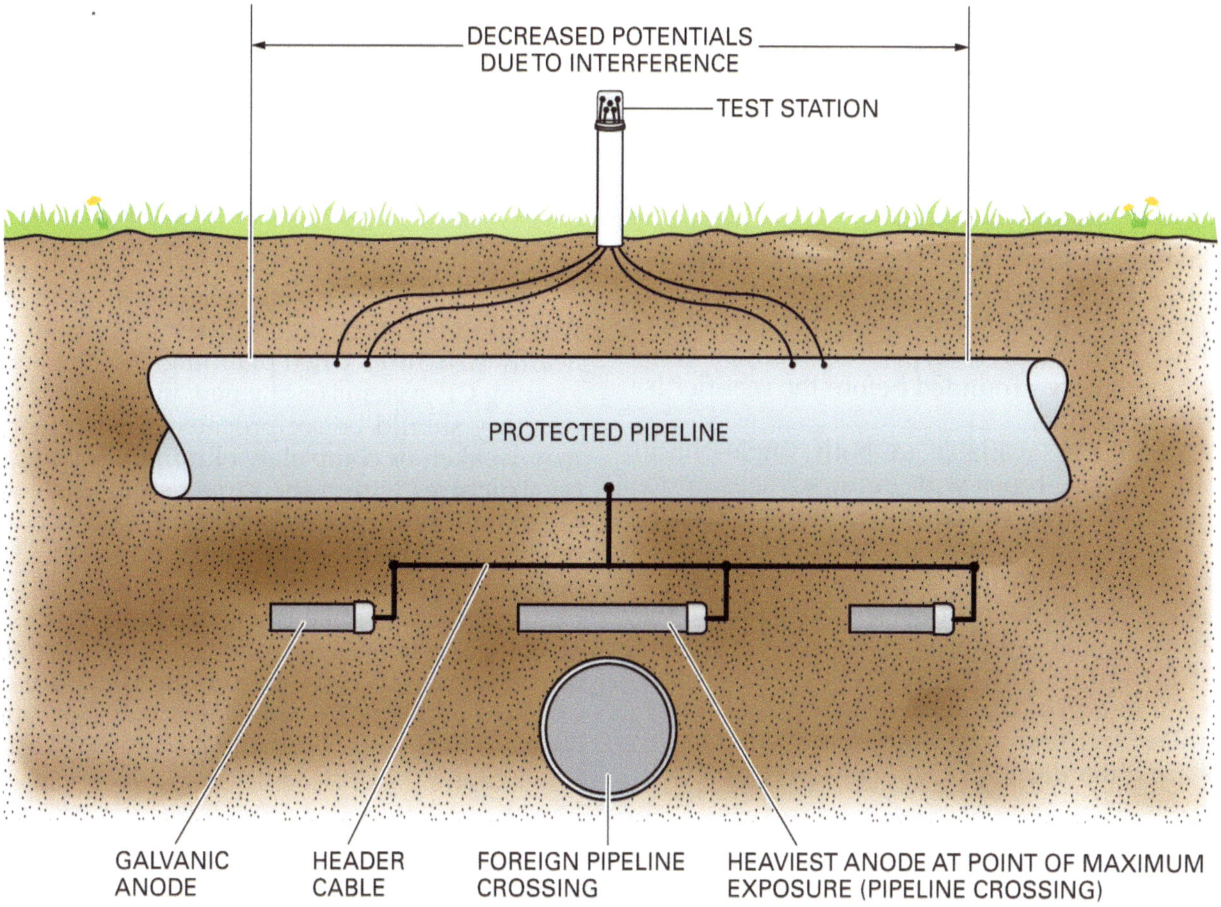

Figure 9 Galvanic anodes attached to a line under test.

the potential gradient field around the protected pipeline. Usually an anode bank consisting of a single line of anodes between the pipeline and the foreign pipeline causing the interference will mitigate the effects of the interference. When properly designed, the remaining current will discharge from the anodes instead of the pipeline. Several strings of anodes may be required when potential differences are very high. The anode life using this arrangement is short unless the surrounding soil has a high **resistivity**. Magnesium anodes are preferred to zinc anodes because magnesium anodes have a higher anodic potential gradient field.

1.2.5 Galvanic Anode Banks

Galvanic anode banks work to ensure that the electrical potential of the pipeline is negative with respect to the surrounding earth. Remote anodes, also referred to as *remote groundbeds*, can be used to protect sections of pipeline located at a distance, while distributed anodes are more suited for protecting smaller, nearby areas of pipeline.

A buried pipeline extending across an area influenced by an anode bank may be subject to influence from stray electrical current surrounding the bed. Specific conditions in the area where this takes place determine how serious the effect will be on the foreign pipeline. In cases where the electrical potential of the stray current is high enough, protective measures must be put in place to prevent corrosion damage from current flowing away from anodic areas on the pipe.

One approach is to employ bare pipe shields on either side of the foreign pipeline where it lies within the potential gradient of the anode bank and picks up stray current. If the bare pipe shields are electrically connected to the rectifier for the anode bank, they will emit a negative cathodic field that opposes the positive field of the anodes.

High installation and operating costs must be weighed against other alternatives. It may be more cost effective to relocate the anode bank than to consider the use of a shield.

1.2.6 Types of Rectifiers

Rectifiers are devices used to convert AC current to DC current. In an impressed cathodic protection system, the rectifier supplies current to the structure in the soil. Various types of rectifiers are available for use in cathodic protection systems.

A typical air-cooled rectifier is shown in *Figure 10*. Smaller rectifiers are often pole-mounted, as shown in *Figure 11*. *Figure 12* shows an oil-cooled rectifier, which is more commonly used in larger systems. Regardless of type, the rectifier case must be properly grounded before the rectifier is put into operation.

Rectifiers are available in both single-phase and three-phase types with input voltages up to 480 volts AC, and can produce various DC output voltage/current configurations. Some units, referred to as *constant current*, will automatically adjust the voltage to meet the specified output by means of an internal sensor. A constant potential unit works in a similar manner, but uses a reference cell to obtain the desired voltage. Manually controlled rectifiers with built-in air-cooling systems are often used with smaller impressed current systems.

Rectifiers should be installed in safe locations away from traffic. They can be located inside a protective structure or outside, if rated for the environment. Good air circulation is essential to prevent overheating. For example, air-cooled rectifiers are often designed with a pedestal to promote air circulation around the unit.

1.2.7 Notifications

Pipeline operators must notify owners of any nearby structures when planning any changes to existing CP systems or any new installations. The courtesy should be reciprocated. Stray currents may weaken or completely eliminate existing CP on structures within the same field of influence. Open communications between all owners and operators in an area should be encouraged. Owners should provide strip maps to each other, with each owner responsible for marking applicable structures. The exact location, size, and scope of any new projects can have serious effects on others. Test points in the same area may be shared. To avoid future problems, put all communications in writing.

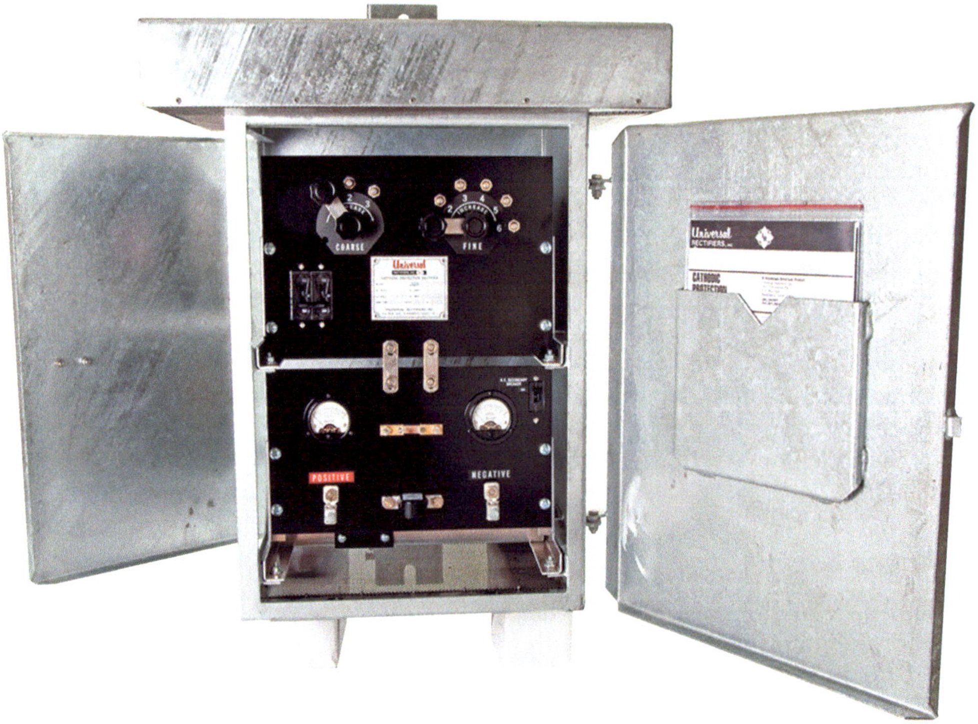

Figure 10 Air-cooled rectifier.

Figure 11 Pole-mounted rectifier.

Figure 12 Oil-cooled rectifier.

Additional Resources

It is assumed that you will learn to perform these procedures under the direct supervision of your crew leader, foreman, or instructor. For independent study, refer to the recommended Additional Resources at the back of this module.

1.0.0 Section Review

1. An impressed current cathodic protection system can subject an adjacent pipeline to _____.
   a. endwise current
   b. stray current interference
   c. current discharge
   d. potential gradients

2. Which of the following is true of rectifiers?
   a. Rectifiers must be installed outdoors.
   b. Rectifiers require good air circulation to prevent overheating.
   c. Constant-current units use a reference cell to obtain the desired voltage.
   d. Rectifiers do not require grounding.

SECTION TWO

2.0.0 INSTALLING RECTIFIERS

Objective

Describe how to install a rectifier (CT9_3-17).

a. Identify the prerequisites, competencies, and abnormal operating conditions associated with installing rectifiers.
b. Describe the procedure for installing a rectifier.

Performance Task

Install rectifiers (CT9_3-17).

- Identify potential abnormal operating conditions that may occur, and know the appropriate actions to take in response to them.
- Utilize the appropriate personal protective equipment according to relevant company procedures.
- Verify the rectifier required for the location and service.
- Mount the rectifier at the designated location (pole, post, wall, etc.), making sure that it is securely fastened.
- Connect the AC power feed wires through an appropriately sized conduit according to the *National Electric Code* (*NEC®*), the *National Electric Safety Code*, and local electric and building codes.
- Terminate the AC feed wires at terminals on circuit breaker or AC input connection wires.
- Connect the DC conduits to the rectifier.
- Install the DC cables from the anode groundbed and the structure in their respective conduits, and terminate on their respective terminals.
- Test and verify that cables are correctly installed.
- Complete appropriate documentation as required by operator's procedures.

Trade Terms

Abnormal operating conditions (AOCs): Conditions within a pipeline system, identified by the operator, that may indicate (1) a malfunction of a component or (2) a deviation from normal operations that may indicate a condition exceeding design limits or result in a hazard to persons, property, or the environment.

Cathode: The electrode of an electrochemical cell at which reduction is the principal reaction. Electrons flow toward the cathode in the external circuit.

The installation of impressed current cathodic protection rectifiers is essential in protecting facilities against external corrosion. The task of installing a rectifier begins with selecting the location and making the initial connections between the structure that is being protected and the groundbed. The task does not end until the rectifier has been installed, connected to the structure/anodes, and the installation has been documented.

This section describes some of the basic skills required to install rectifiers and identifies abnormal operating conditions (AOCs) that personnel may need to respond to when performing this task. It also describes the general procedure for installing rectifiers.

2.1.0 Prerequisites, Competencies, and Abnormal Operating Conditions

Personnel must possess a variety of skills in order to install rectifiers safely and efficiently. The various operations require a series of steps, which personnel must be competent to perform. Personnel must also be aware of the particular abnormal operating conditions that may occur during the installation procedure and know the proper responses to them.

2.1.1 Prerequisites and Competencies

Personnel assigned to install rectifiers must have knowledge of cathodic protection systems and components comparable to NACE Certification Level CP2. This includes, but is not limited to, the following:

- Understanding of basic electricity and electrical circuits and the associated terminology (anode, cathode, etc.)
- Knowledge of the type of rectifier being installed (air-cooled, oil-cooled, explosion-proof, solar-powered, etc.)
- Knowledge of mounting requirements (pole mount, rack mount, concrete pad, etc.)
- Awareness that terminating each cable at the proper terminal is essential for safe and effective cathodic protection

2.1.2 Abnormal Operating Conditions

Before installing rectifiers a pipeline technician must be able to identify and properly respond to potential AOCs, should they occur. The following are some potential AOCs, and the appropriate action personnel should take in response to them:

- *Component failure occurs* – Evaluate the extent of the failure/damage and make company notifications as applicable.
- *Fire or explosion occurs from burn-through during exothermic welding* – Limit access to the area as much as possible and make company notifications.
- *Unexpected leak of hazardous liquid or carbon dioxide occurs within or outside casings* – Make immediate notifications per company procedure. Try to identify and isolate the source of the leak if possible and if properly trained to do so.

2.2.0 Procedure

The rectifier or other power source should be installed so that the possibility of damage or vandalism is minimized. All wiring must meet the applicable codes and specifications, and any other applicable standards. An external disconnect with appropriate overcurrent protection should be installed so that the power can be turned off before work begins on the rectifier. If a rectifier requires replacement, use proper installation procedures and observe all safety rules.

The following steps outline the general procedure for installing rectifiers:

WARNING!

Always obtain and use the appropriate personal protective equipment for each task. Only qualified individuals may make electrical connections and operate equipment. Always verify the absence of voltage and current before making connections. Follow the manufacturer's instructions for the equipment being installed.

Step 1 Identify potential abnormal operating conditions that may occur, and know the appropriate actions to take in response to them.

Step 2 Utilize the appropriate personal protective equipment according to relevant company procedures.

Step 3 Select an appropriate rectifier for the location, application, and class of service. Rectifiers used in classified (hazardous) locations must be explosion-proof and listed for the application.

Step 4 Mount the rectifier securely at the designated location. Rectifiers may be mounted on poles, posts, walls, panels, racks, concrete pads, etc.

CAUTION

The installation must meet all applicable building and electrical codes. Rectifiers must be mounted securely using appropriately sized fasteners. Insecure fastening could lead to injury or equipment damage.

Step 5 Connect the AC power feed wires through appropriately sized conduit in accordance with the appropriate sections of the *National Electrical Code®* (*NEC®*), the *National Electrical Safety Code®*, and local electric and building codes. The AC is typically supplied through a safety switch or circuit breaker panel. Follow all safety requirements when making connections.

CAUTION

Consult the applicable codes and requirements to prevent electrical shock. The AC conduit is usually connected to the "knockout" supplied for that purpose.

Step 6 Terminate the AC feed wires at the terminal board, circuit breaker, or AC input connection. Follow the instructions provided in the manufacturer's manual.

Step 7 Connect the DC conduits to the rectifier. DC conduits are used to carry the DC output cables from the DC output terminals to the anode bank (positive) and structure (negative).

CAUTION

In classified (hazardous) installations, sealing fittings are required and may only be installed and poured by a qualified individual.

Step 8 Install DC cables from the anode bank and the structure in their respective conduits, and terminate on their respective terminals. The positive cable is connected to the anodes and the negative cable is connected to the structure.

WARNING!

Make sure to terminate each cable at the proper terminal. Connecting the anode groundbed to the negative and the structure to the positive can have disastrous consequences.

Step 9 Verify that the cables have been installed correctly and test the operation of the rectifier.

WARNING!

Incorrect cable connections will cause the protected pipeline to act as an anode and result in rapid corrosion of the pipe.

Step 10 Document the installation per the company/operator requirements.

Additional Resources

It is assumed that you will learn to perform these procedures under the direct supervision of your crew leader, foreman, or instructor. For independent study, refer to the recommended Additional Resources at the back of this module.

2.0.0 Section Review

1. Before installing rectifiers, individuals must have knowledge of cathodic protection systems and components comparable to _____.
   a. *NFPA 70*® apprenticeship standards
   b. Journeyman electrician classification
   c. NACE Certification Level CP2
   d. ANSI Accreditation Phase 2

2. Rectifiers used in hazardous locations must be _____.
   a. rated at 240 volts AC
   b. listed as explosion-proof
   c. rated at 120 volts DC
   d. oil-filled

SUMMARY

Stray currents, or interference, from foreign sources can cause pipeline corrosion. To control this problem, pipeline technicians need to know how to identify when stray currents are present, how to identify the source of the stray currents, and how to use appropriate techniques to reduce the interference. This module described methods and devices used to reduce or eliminate stray current interference and described procedures for the installation of rectifiers in cathodic protection systems.

Review Questions

1. Natural occurrences of stray current in the earth that are transferred to the protected pipeline are known as _____.
   a. potential gradients
   b. burial pressures
   c. ground water changes
   d. climate variations

2. DC-powered transit systems can cause _____.
   a. potential shifts
   b. static DC interference
   c. dynamic stray currents
   d. impressed currents

3. Smaller rectifiers are often _____.
   a. oil-cooled
   b. pole-mounted
   c. intrinsically safe
   d. dielectrically neutral

4. To obtain the desired voltage, a constant potential rectifier uses a(n) _____.
   a. internal sensor
   b. interrupter
   c. diode
   d. reference cell

5. If a pipeline technician discovers a broken or damaged component when installing a rectifier, the technician should _____.
   a. evaluate the extent of the damage/failure and make applicable company notifications
   b. immediately eliminate potential ignition sources in the area
   c. evacuate all non-essential personnel from the work site
   d. limit access to the work area as much as possible

6. When installing a rectifier, a pipeline technician must connect the DC conduits to the _____.
   a. interrupter
   b. terminal board
   c. rectifier
   d. circuit breaker

Trade Terms Quiz

Fill in the blank with the correct term that you learned from your study of this module.

1. The electrode of an electrochemical cell at which reduction is the principal reaction is the ________.
2. Changes in potential with respect to distance, expressed in millivolts per unit of distance are ________.
3. A conductor that joins two points in an electrical circuit to form a parallel or alternative path through which a portion of the current may pass is a(n) ________.
4. Sets of two or more anodes installed in a specific configuration where the anodes are connected together by a header cable/wire are referred to as ________.
5. A portion of a current that flows over a path other than the intended path in an on/off or intermittent fashion and may cause electrochemical corrosion of metals in contact with electrolytes is ________.
6. The electrode of an electrochemical cell at which oxidation occurs is the ________.
7. The effect on the cathodic protection level of a structure due to the interaction between its cathodic protection system and the cathodic protection system of another structure is called ________.
8. A device used to convert AC current to DC current and which supplies current to the structure in the soil in an impressed cathodic protection system is a(n) ________.
9. A substance that ionizes (releases free electrons) in an electric field is a(n) ________.
10. A portion of a current that flows over a path other than the intended path at a constant, unbroken rate and that may cause electrochemical corrosion of metals in contact with electrolytes is ________.
11. A measurement of the voltage or potential difference between a buried or submerged metal structure and an electrolyte, such as soil, is referred to as ________.
12. Current flows through a path other than the intended circuit are called ________.
13. Conditions within a pipeline system, identified by the operator, that may indicate (1) a malfunction of a component or (2) a deviation from normal operations that may indicate a condition exceeding design limits or result in a hazard to persons, property, or the environment are called ________.
14. Devices used to allow current to flow in only one direction are ________.
15. Connections to the structure being tested, usually wires in a supporting stand with easy connection points for the meter, are called ________.
16. The ability of a material to resist the flow of electric current is called ________.
17. Electrical connections of two structures that usually have two different cathodic protection levels are ________.

Trade Terms

Abnormal operating conditions (AOCs)
Anode
Anode banks
Bonds
Cathode
Diodes
Dynamic stray current
Electrolyte
Interference
Potential gradients
Rectifier
Resistivity
Shunt
Static stray current
Stray currents
Structure-to-soil potential
Test leads

Trade Terms Introduced in This Module

Abnormal operating conditions (AOCs): Conditions within a pipeline system, identified by the operator, that may indicate (1) a malfunction of a component or (2) a deviation from normal operations that may indicate a condition exceeding design limits or result in a hazard to persons, property, or the environment.

Anode: The electrode of an electrochemical cell at which oxidation occurs. Electrons flow away from the anode in the external circuit. Corrosion usually occurs and metal ions enter the solution at the anode.

Anode banks: Sets of two or more anodes installed in a specific configuration where the anodes are connected together by a header cable/wire; also referred to as groundbeds.

Bonds: Electrical connections of two structures that usually have two different cathodic protection levels.

Cathode: The electrode of an electrochemical cell at which reduction is the principal reaction. Electrons flow toward the cathode in the external circuit.

Diodes: Devices used to allow current to flow in only one direction.

Dynamic stray current: A portion of a current that flows over a path other than the intended path in an on/off or intermittent fashion and may cause electrochemical corrosion of metals in contact with electrolytes.

Electrolyte: A substance that ionizes (releases free electrons) in an electric field.

Interference: The effect on the cathodic protection level of a structure due to the interaction between its cathodic protection system and the cathodic protection system of another structure.

Potential gradients: Changes in potential with respect to distance, expressed in millivolts per unit of distance.

Rectifier: A device used to convert AC current to DC current. In an impressed cathodic protection system, it supplies current to the structure in the soil.

Resistivity: The ability of a material to resist the flow of electric current.

Shunt: A conductor that joins two points in an electrical circuit to form a parallel or alternative path through which a portion of the current may pass.

Static stray current: A portion of a current that flows over a path other than the intended path at a constant, unbroken rate and that may cause electrochemical corrosion of metals in contact with electrolytes.

Stray currents: Current flows through a path other than the intended circuit.

Structure-to-soil potential: A measurement of the voltage or potential difference between a buried or submerged metal structure and an electrolyte, such as soil; also referred to as structure-to-electrolyte potential.

Test leads: Connections to the structure being tested, usually wires in a supporting stand with easy connection points for the meter.

Additional Resources

The following recommended resources can provide additional helpful information related to the requirements for this covered task:

API Recommended Practice 1161, Third Edition, © 2014 American Petroleum Institute. API Publishing Services, 1220 L Street NW, Washington, DC 20005. Distributed by Thompson Reuters (Scientific) LLC, **www.techstreet.com**.

National Electrical Code® (NFPA 70®), **www.nfpa.org**.

National Electrical Safety Code® (*NESC®*), **www.ieee.org**.

49 *CFR* 192, **www.ecfr.gov**.

49 *CFR* 195, **www.ecfr.gov**.

NCCER Module AOCCC-17, *Abnormal Operating Conditions Control Center*.

NCCER Module AOCFG-17, *Abnormal Operating Conditions Field and Gas*.

NCCER Module 62105-14, *Pipeline Documentation*.

NCCER Module 64202-02, *Use of Meters and Test Equipment*.

NCCER Module 66101-14, *Introduction to the Pipeline Industry*.

NCCER Module CT1_1-17, *Measure Structure-to-Soil Potential*.

NCCER Module CT1_2-17, *Conduct Close Interval Survey*.

NCCER Module CT1_3-17, *Test and Detect Interference*.

NCCER Module CT1_4-17, *Inspect and Perform Electrical Test of Bonds*.

NCCER Module CT2_2-17, *Repair Damaged Test Leads*.

NCCER Module CT2_3-17, *Install Test Leads by Non-Exothermic Welding Methods*.

NCCER Module CT4_3-17, *Adjustment of Rectifier*.

NCCER Module CT9_1-17, *Install Bonds*.

NCCER Module CT9_2-17, *Install Galvanic Anodes*.

NCCER Module CT9_4-17, *Install Impressed Current Groundbeds*.

Figure Credits

Tinker & Rasor, Module Opener

Universal Rectifiers, Inc., Figures 10 and 11

IRT Integrated Rectifier Technologies, Inc., Figure 12

Section Review Answer Key

Answer	Section Reference	Objective
Section One		
1. b	1.1.1	1a
2. b	1.2.6	1b
Section Two		
1. c	2.1.1	2a
2.	2.2.0	2b

NCCER CURRICULA — USER UPDATE

NCCER makes every effort to keep its textbooks up-to-date and free of technical errors. We appreciate your help in this process. If you find an error, a typographical mistake, or an inaccuracy in NCCER's curricula, please fill out this form (or a photocopy), or complete the online form at **www.nccer.org/olf**. Be sure to include the exact module ID number, page number, a detailed description, and your recommended correction. Your input will be brought to the attention of the Authoring Team. Thank you for your assistance.

Instructors – If you have an idea for improving this textbook, or have found that additional materials were necessary to teach this module effectively, please let us know so that we may present your suggestions to the Authoring Team.

NCCER Product Development and Revision

13614 Progress Blvd., Alachua, FL 32615

Email: curriculum@nccer.org
Online: www.nccer.org/olf

❑ Trainee Guide ❑ Lesson Plans ❑ Exam ❑ PowerPoints Other ______________________

Craft / Level: Copyright Date:

Module ID Number / Title:

Section Number(s):

Description:

Recommended Correction:

Your Name:

Address:

Email: Phone:

CT9_4-17

Install Impressed Current Groundbeds

Overview

Various cathodic protection systems—which include the installation of bonds, galvanic anodes, rectifiers, and impressed current groundbeds—are used to mitigate stray current interference and prevent corrosion in pipelines. Impressed current groundbeds are installed to provide cathodic protection for buried or submerged metallic structures. This module describes the general procedures for installing impressed current groundbeds in cathodic protection systems.

Trainees with successful module completions may be eligible for credentialing through the NCCER Registry. To learn more, go to **www.nccer.org** or contact us at 1.888.622.3720. Our website has information on the latest product releases and training, as well as online versions of our *Cornerstone* magazine and Pearson's product catalog.

Your feedback is welcome. You may email your comments to **curriculum@nccer.org**, send general comments and inquiries to **info@nccer.org**, or fill in the User Update form at the back of this module.

This information is general in nature and intended for training purposes only. Actual performance of activities described in this manual requires compliance with all applicable operating, service, maintenance, and safety procedures under the direction of qualified personnel. References in this manual to patented or proprietary devices do not constitute a recommendation of their use.

CT9_4-17
INSTALL IMPRESSED CURRENT GROUNDBEDS

Objectives

When you have completed this module, you will be able to do the following:

1. Describe stray current interference and methods for mitigating it.
   a. Describe types of stray current and how they affect cathodic protection.
   b. Describe methods used to mitigate stray current interference.
2. Describe how to install impressed current groundbeds (CT9_4-17).
   a. Identify the prerequisites, competencies, and abnormal operating conditions associated with installing impressed current groundbeds.
   b. Describe procedures for installing impressed current groundbeds.

Performance Task

Under the supervision of the instructor, you should be able to do the following:

1. Install impressed current groundbeds (CT9_4-17).
   - Identify potential abnormal operating conditions that may occur during performance of this CT, and know the appropriate actions to take in response to them.
   - Utilize the appropriate personal protective equipment according to relevant company procedures.
   - Confirm that the location and materials meet the requirements of the design criteria.
   - Lay out the number, spacing, and configuration of the anodes according to the design criteria.
   - Excavate a vertical hole or horizontal ditch.
   - Install an anode in the excavated hole, being careful to avoid damage.
   - Install the anode header cable between the groundbed and the rectifier.
   - Backfill the hole or ditch with appropriate backfill material.
   - Complete appropriate documentation as required by operator's procedures.

Trade Terms

Abnormal operating conditions (AOCs)
Anodes
Anode banks
Bonds
Carbonaceous
Casing
Diodes
Dynamic stray current
Electrolyte
Header cable/wire
Interference
Potential gradients
Rectifier
Resistivity
Shunt
Splice connection
Static stray current
Stray currents
Structure-to-soil potential
Test leads

Industry Recognized Credentials

If you are training through an NCCER-accredited sponsor, you may be eligible for credentials from NCCER's Registry. The ID number for this module is CT9_4-17. Note that this module may have been used in other NCCER curricula and may apply to other level completions. Contact NCCER's Registry at 888.622.3720 or go to **www.nccer.org** for more information.

Contents

Figures

SECTION ONE

1.0.0 STRAY CURRENT INTERFERENCE

Objective

Describe stray current interference and methods for mitigating it.

a. Describe types of stray current and how they affect cathodic protection.
b. Describe methods used to mitigate stray current interference.

Trade Terms

Anodes: The electrodes of electrochemical cells at which oxidation occurs. Electrons flow away from the anodes in the external circuits. Corrosion usually occurs and metal ions enter the solution at the anodes.

Anode banks: Sets of two or more anodes installed in a specific configuration where the anodes are connected together by a header cable/wire; also referred to as groundbeds or anode groundbeds.

Bonds: Electrical connections of two structures that usually have two different cathodic protection levels.

Carbonaceous: Relating to or composed of carbon.

Diodes: Devices used to convert AC to DC by allowing current to flow in only one direction.

Dynamic stray current: A portion of a current that flows over a path other than the intended path in an on/off or intermittent fashion and may cause electrochemical corrosion of metals in contact with electrolytes.

Electrolyte: A substance that ionizes (releases free electrons) in an electric field.

Interference: The effect on the cathodic protection level of a structure due to the interaction between its cathodic protection system and the cathodic protection system of another structure.

Potential gradients: Changes in potential with respect to distance, expressed in millivolts per unit of distance.

Rectifier: A device used to convert AC current to DC current. In an impressed cathodic protection system, it supplies current to the structure in the soil.

Resistivity: The ability of a material to impede the flow of electric current.

Shunt: A conductor that joins two points in an electrical circuit to form a parallel or alternate path through which a portion of the current may pass.

Static stray current: A portion of a current that flows over a path other than the intended path at a constant, unbroken rate and that may cause electrochemical corrosion of metals in contact with electrolytes.

Stray currents: Current that flows through a path other than the intended circuit.

Structure-to-soil potential: A measurement of the voltage or potential difference between a buried or submerged metal structure and an electrolyte, such as soil; also referred to as structure-to-electrolyte potential.

Test leads: Connections to the structure being tested, usually wires in a supporting stand with easy connection points for the meter.

Corrosion on a pipeline can be caused by stray currents, in which electrical currents flow through a path other than the intended circuit. The stray current flow results in the problem called interference, or *stray current interference*. Impressed current groundbeds are installed to provide cathodic protection for buried or submerged metallic structures, such as gas and oil pipelines. This section describes the vertical and horizontal placement of anodes in groundbeds and describes three common anode system configurations—distributed anode banks, surface groundbeds, and deep groundbeds.

1.1.0 Stray Current

Stray currents are direct currents flowing in the earth that are from a source other than the corrosion protection source. Potential gradients cause natural occurrences of stray current in the earth that are transferred to the protected pipeline. *Figure 1* shows potential gradients around a protected pipeline.

Normally, this outside current enters the pipeline at one point, exits the pipeline at another point, and returns to the earth. The area where the current returns to earth, completing the circuit, is the point at which the corrosion may occur if the material is in contact with an electrolyte. The measurement of the voltage or potential difference between a buried or submerged metal structure and an electrolyte, such as soil, is the structure-to-soil potential, also referred to as the *structure-to-electrolyte potential.*

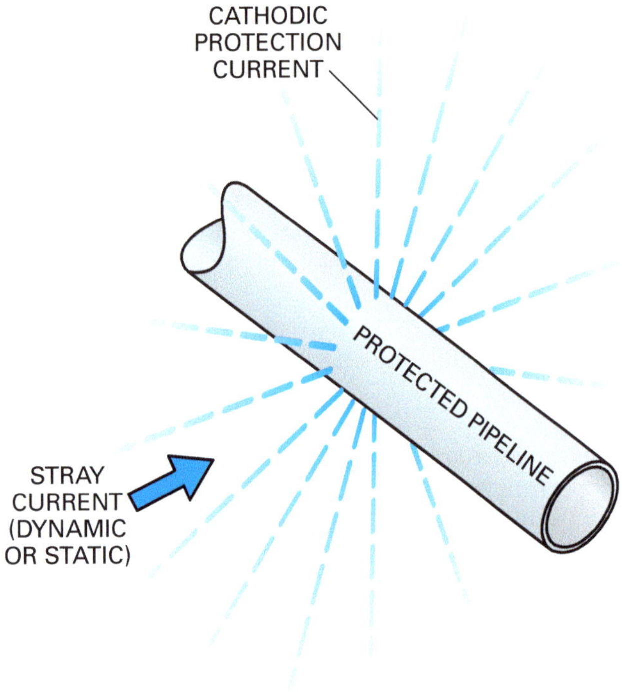

Figure 1 Potential gradients around a protected pipeline.

1.1.1 Types of Stray Current

Stray currents are either static (constant) or dynamic (intermittent). Sources of stray current include cathodic protection (CP) systems on other pipelines, DC transit systems, DC welding operations, DC mining operations, high-voltage DC transmission systems, or disturbances in the earth's magnetic field. If environmental conditions allow rectifying junctions such as copper oxide films, AC current flow can be rectified. This results in a direct current that can be mitigated. There are many different sources of interference current in a cathodic protection system. For example, stray current interference on adjacent pipelines can be caused by impressed current cathodic protection systems. This depends on the locations of the anode banks and the operating characteristics of the cathodic protection system.

Figure 2 shows stray current to a crossover foreign pipeline. In this example, the foreign pipeline picks up current from the anode bank and transfers the current to the protected pipeline when it crosses under it. The potential for corrosion is greatly increased at the crossing point of the pipelines.

Figure 3 shows stray current to a parallel foreign pipeline. In this example, the parallel foreign pipeline is affected by the anode bank of the protected pipe. This is because the foreign pipeline travels through the anode bank and picks up current. When this occurs, the effectiveness of the CP system will be reduced in this area of the pipeline. In the affected area of the foreign pipeline, it becomes positive to the soil and discharges current to the earth. This increases the potential for corrosion on the foreign pipeline in the affected area.

1.1.2 Testing for Interference Current

Static stray current interference is easy to test for as long as there is no **dynamic stray current** present. Dynamic stray current can be caused by sources such as DC-powered transit systems, underground mining, welding, and pipeline interference. In *Figure 4*, a section of coated pipeline has an impressed current CP system applied to it and has two foreign pipelines crossing it. In this example, "B" is a strong indicator of interference because it has the largest difference in voltage. Always look for large changes in voltage to identify interference. Follow company guidelines for determining the need to mitigate interference.

To distinguish the current in the foreign pipeline from the current in the protected line, an automatic current interrupter is installed in the output of the **rectifier**. The current interrupter is a device that automatically opens and closes an electrical circuit at a preset time. With the current interrupter operating, measure the potential of the lines under both on and off conditions. *Figure 5* shows how to measure this potential at the test station. As shown, a **shunt** is installed for measurement of current flow and resistance inside of the test station.

These tests are performed using a copper sulfate electrode placed over the crossing point of the lines. Pipeline locators can be used to determine the exact location of the crossing.

> **WARNING!**
>
> Always be alert when testing for interference current. A release could be found while testing.

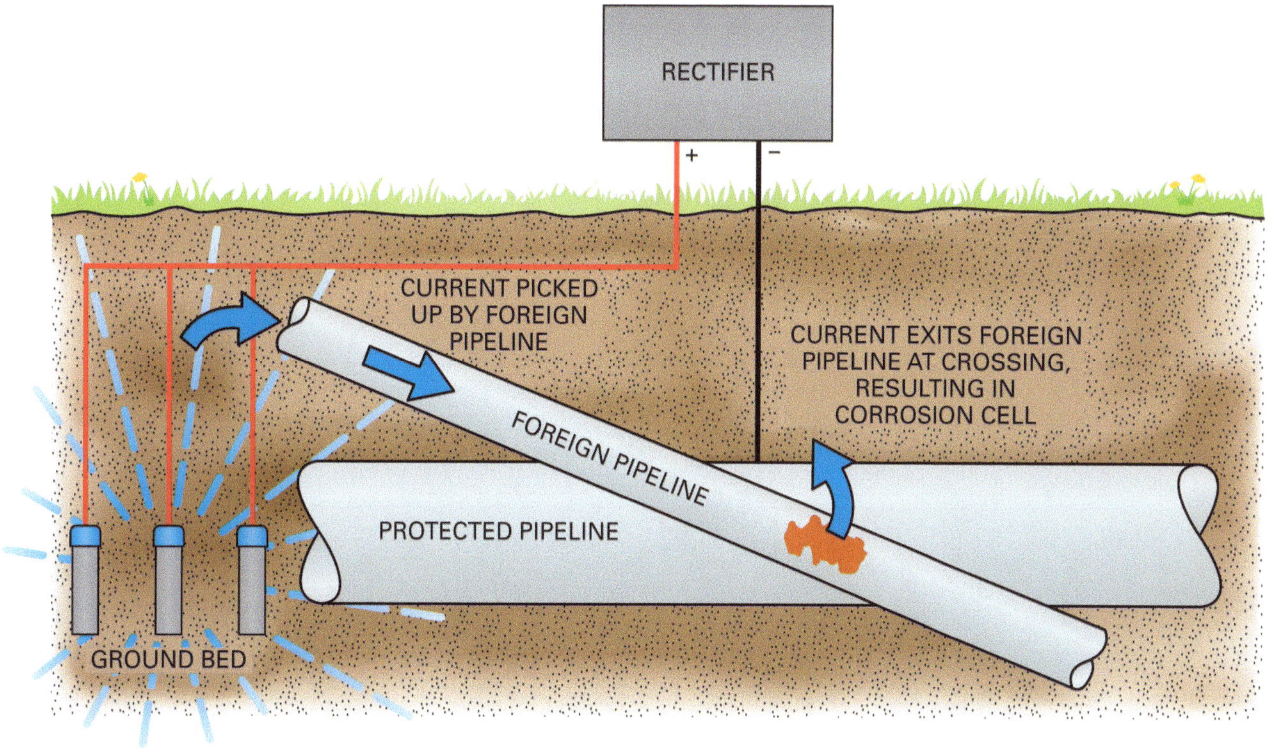

Figure 2 Stray current to a crossover foreign pipeline.

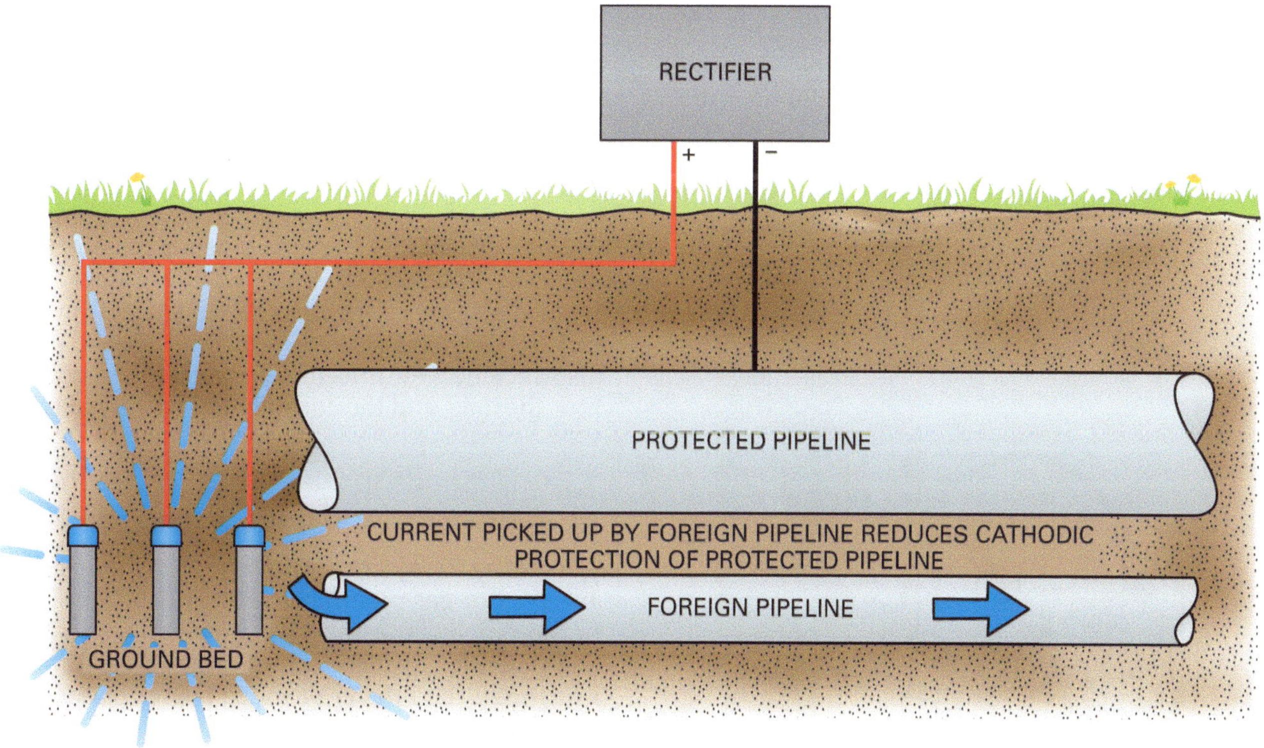

Figure 3 Stray current to a parallel foreign pipeline.

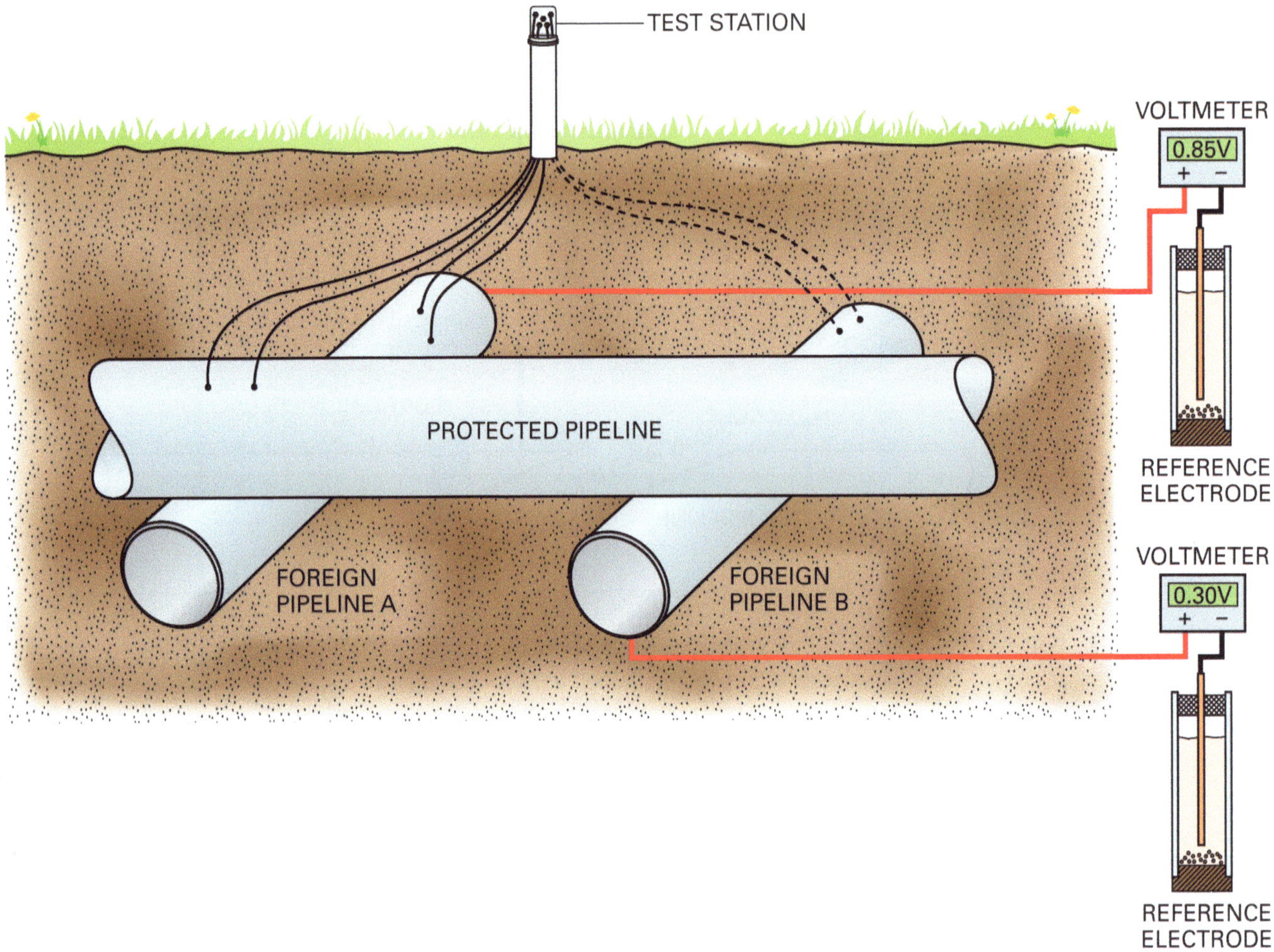

Figure 4 Testing foreign pipeline crossings for stray current interference from CP installations.

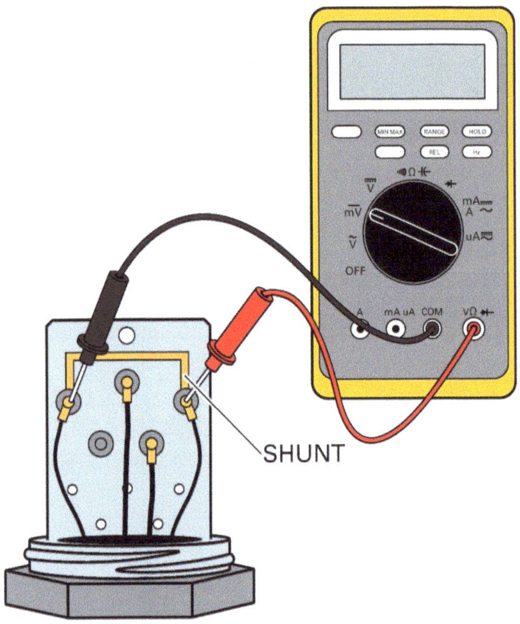

Figure 5 Measuring potential at the test station.

The data obtained at each pipeline crossing must be properly recorded. *Figure 6* shows typical interference testing potentials. These records are also called field data sheets. Field data sheets should also include any information about the protected line, the date, and the current output of the rectifier.

1.2.0 Methods for Reducing or Eliminating Stray Current Interference

According to the *Code of Federal Regulations* 49 *CFR* 192.473 (for gas pipelines) and 49 *CFR* 195.577 (for liquid pipelines), any pipeline system subjected to stray currents must have an ongoing program to minimize the effects of these currents and help protect surrounding or adjacent underground metallic structures. There are many different methods used to reduce or eliminate stray current interference from a CP system. Some of the most common methods include the use of bonds, coatings, electrical shields, galvanic anodes, and impressed current groundbeds.

FOREIGN LINE		PROTECTED PIPELINE (volts)			INTERFERING PIPELINE (volts)		
NAME	PIPELINE STATION	VOLTAGE ON	VOLTAGE OFF	Δ volts	VOLTAGE ON	VOLTAGE OFF	Δ volts
A	140+00	-0.91	-0.90	-0.01	-0.85	-0.88	+0.03
B	1210+00	-1.94	-1.00	-0.94	-0.30	-0.68	+0.38

Figure 6 Sample of recording survey notes.

1.2.1 Resistance Bonds

A bond is simply a metallic connection between structures. There are two types of bonds: critical and non-critical. Critical bonds are identified by the operator's procedures and include those bonds that may cause damage to the structure if disconnected. Like test leads, bond leads are terminated in a test station.

Foreign pipelines can be subject to stray current damage from the impressed current cathodic protection system on another pipeline. To correct this problem, a resistance bond is connected between the two pipelines. The amount of resistance in the bond is adjusted to drain enough current from the affected line to eliminate the corrosion potential.

CAUTION

For liability reasons, the representative of the foreign pipeline company should be on site before procedures to correct the problem on the affected line begin.

For a resistance bond, diodes are installed as basic components of a rectifier circuit. Single-phase rectifiers are probably the most common units used for cathodic protection. The diode is installed to allow current to flow in one direction only. This characteristic of the diode causes all the current waves to be more aligned when entering into the cables. Along with other components, the diodes reduce ripple effects in the power supply and filter out interference.

1.2.2 Coatings

If the testing of a foreign line indicates that the interference is caused by a foreign pipeline that is bare or poorly coated, the bond base may not be able to correct the condition. If a high current density is flowing from a well-coated line through the soil and onto the bare foreign line, localized corrosion could occur. To correct this problem, reduce the density of the current flow to the crossing area of the foreign line. The best way to accomplish this is to apply a coating to the line. *Figure 7* shows the use of a coating to correct interference at foreign pipeline crossings.

With proper coating, the voltage drops in the earth around the foreign line are reduced or eliminated. Over-the-line potential measurements are used to identify the portion of the foreign line subject to interference. This information is then used to determine the area on the foreign line that must be coated to mitigate the interference.

Telluric Currents

Telluric currents are caused by sunspots and solar flares and can result in inaccurate readings during intense periods of solar activity.

CAUTION

Take special precautions to avoid bare spots in the coating (holidays) to prevent localized corrosion.

1.2.3 Electrical Shields

Electrical shields are defined as barriers to stray current flow. Electrical shielding can be provided by nonmetallic insulating barriers or by diverting the current from the affected pipeline to another structure. Shields are particularly useful for protecting pipeline in congested areas such as pumping stations and tank farms. *Figure 8* shows bare pipe used as electrical shields around a pipeline.

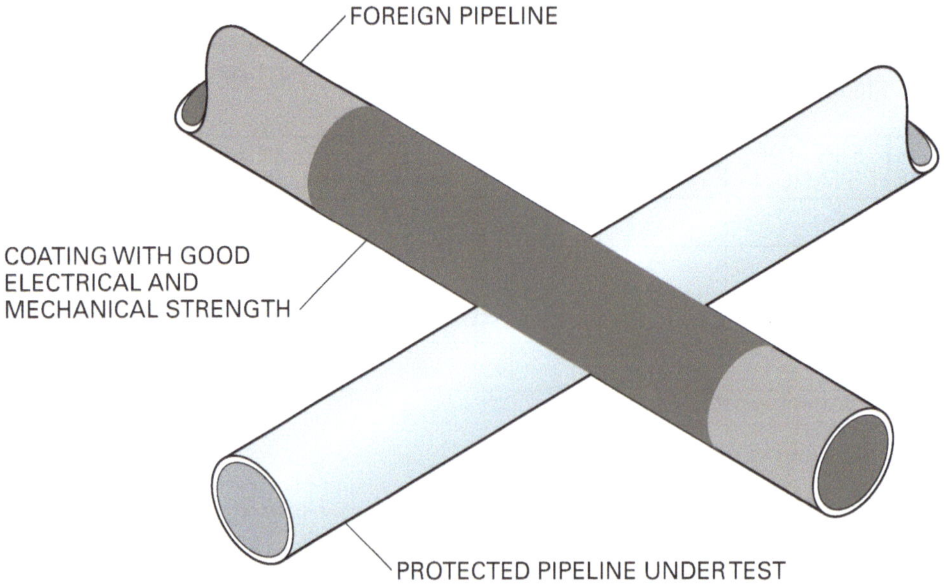

Figure 7 Using a coating to correct interference at foreign pipeline crossings.

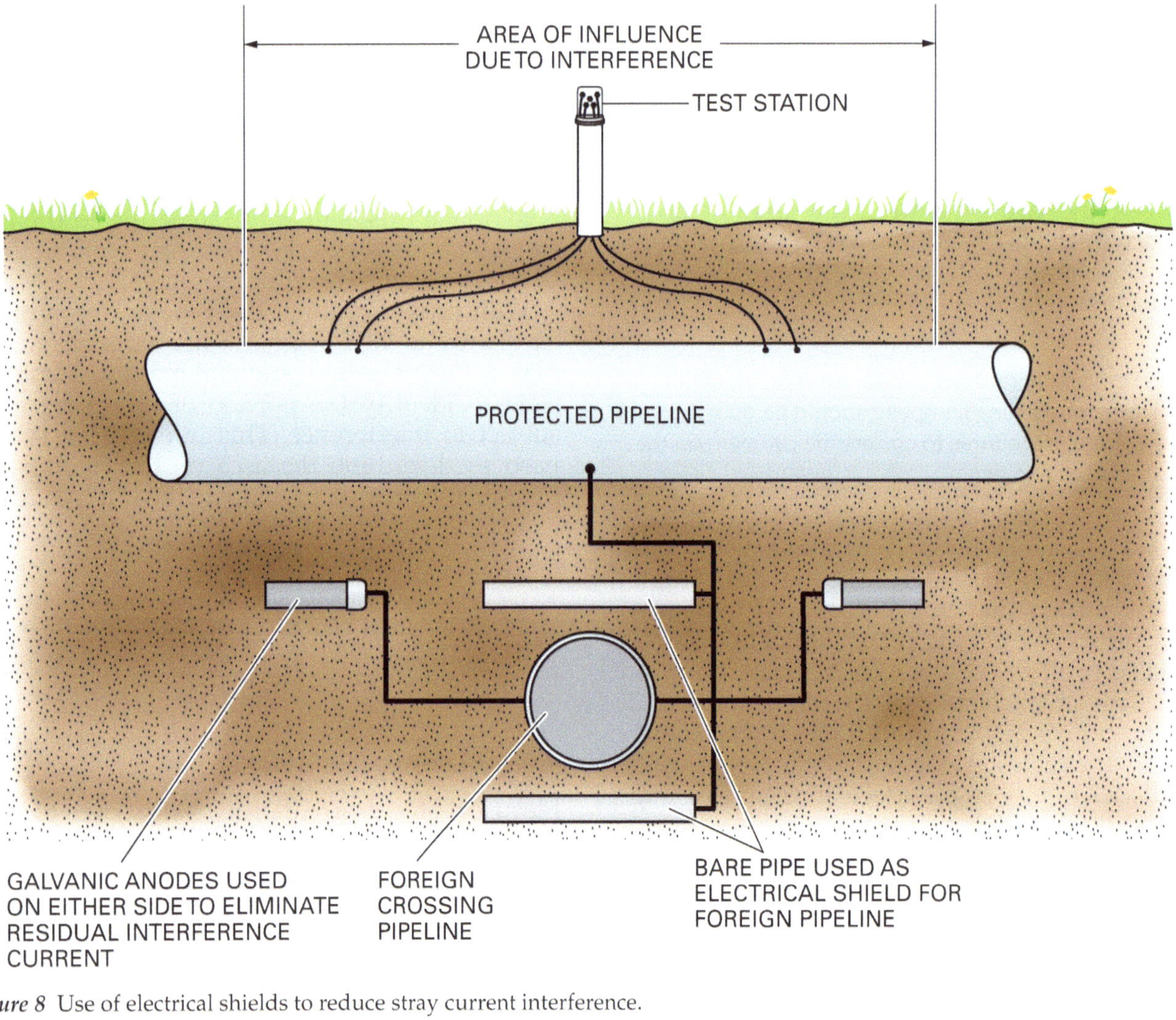

Figure 8 Use of electrical shields to reduce stray current interference.

1.2.4 Galvanic Anodes

Galvanic anodes can provide cathodic protection for buried or submerged structures. They can also be used to correct interference, as spot protection, and for alternating current mitigation. When used to correct interference, the anodes are attached to the affected line (see *Figure 9*). They are located in the area that is subject to interference from the foreign line. Anodes are buried in a special backfill that must be wet in order for the anodes to operate.

This method uses the anodic potential fields that are around the galvanic anode to offset the potential gradient field around the protected pipeline. Usually, an anode bank consisting of a single line of anodes between the pipeline and the foreign pipeline causing the interference will mitigate the effects of the interference. When properly designed, the remaining current will discharge from the anodes instead of the pipeline. Several strings of anodes may be required when potential differences are very high. The anode life using this arrangement is short unless the surrounding soil has a high **resistivity**. Magnesium anodes are preferred to zinc anodes because magnesium anodes have a higher anodic potential gradient field.

1.2.5 Galvanic Anode Banks

Galvanic anode banks work to ensure that the electrical potential of the pipeline is negative with respect to the surrounding earth. Remote anodes, also referred to as *remote groundbeds*, can be used to protect sections of pipeline located at a distance. Distributed anodes are more suited for protecting smaller, nearby areas of pipeline.

A buried pipeline extending across an area influenced by an anode bank may be subject to influence from stray electrical current surrounding the bed. Specific conditions in the area where this takes place determine how serious the effect will be on the foreign pipeline. In cases where the electrical potential of the stray current is high enough, protective measures must be put in place to prevent corrosion damage from current flowing away from anodic areas on the pipe.

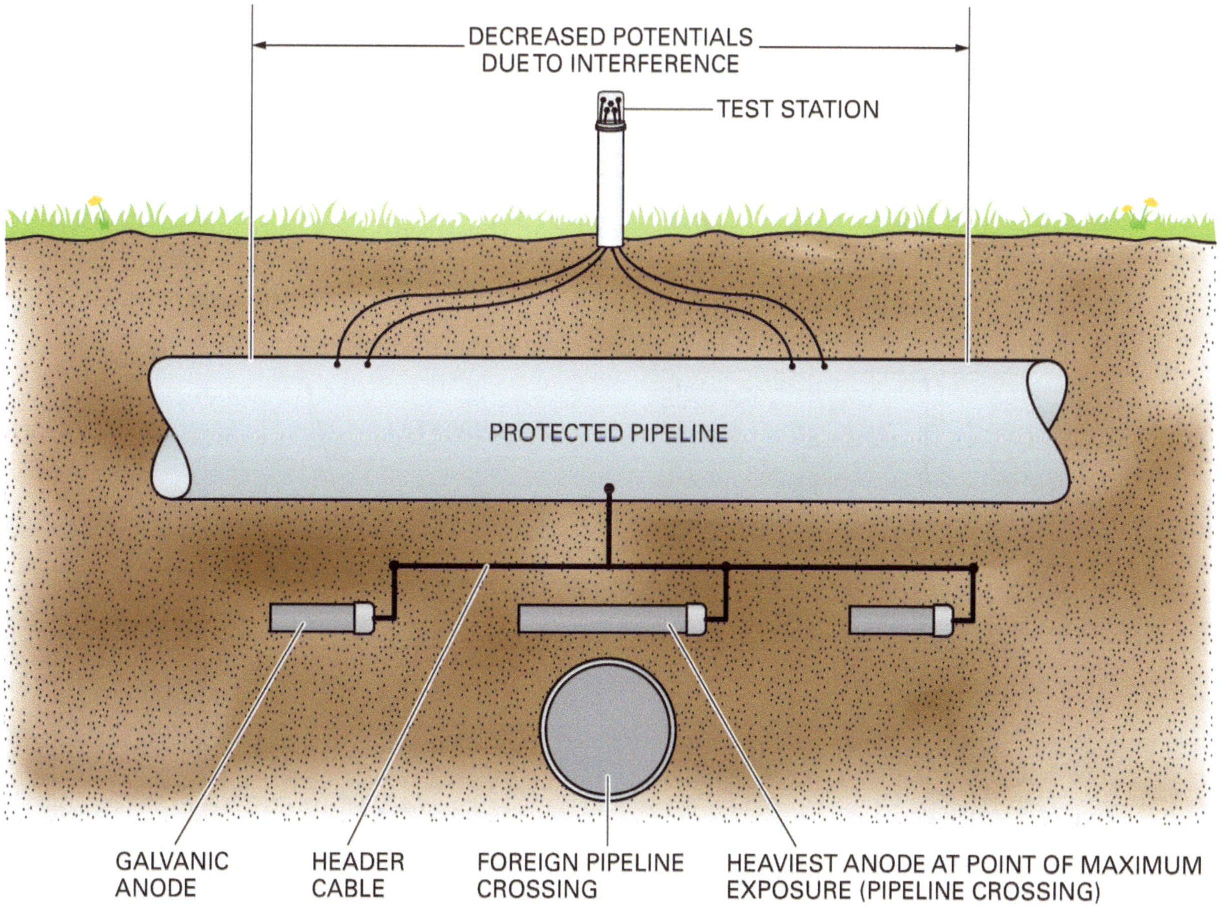

Figure 9 Galvanic anodes attached to a line under test.

One approach is to employ bare pipe shields on either side of the foreign pipeline where it lies within the potential gradient of the anode bank and picks up stray current. If the bare pipe shields are electrically connected to the rectifier for the anode bank, they will emit a negative cathodic field that opposes the positive field of the anodes.

High installation and operating costs must be weighed against other alternatives. It may be more cost effective to relocate the anode bank than to consider the use of a shield.

1.2.6 Impressed Current Groundbeds

Impressed current CP systems also use groundbeds, or anode banks, in which the anodes are installed either vertically or horizontally and are connected to form a bank or groundbed. Depending on the application, the anodes can be installed just below the surface or at much deeper locations.

Figure 10 shows a typical vertical impressed current anode installation. The anode is placed vertically in the center of a drilled hole, then backfill is added around the anode to a specified height and the installation is topped with soil. The dimensions shown are typical of a vertical surface groundbed. Always use the dimensions specified on the installation plan.

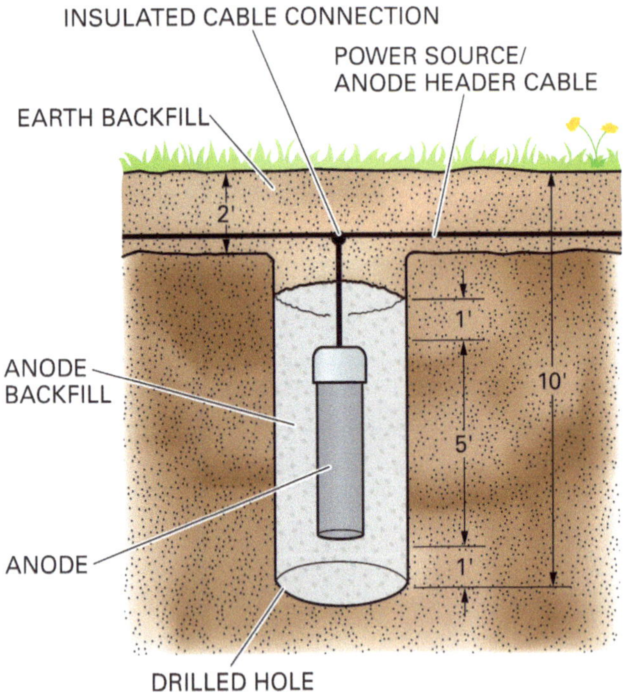

Figure 10 Vertical impressed current anode installation.

A typical horizontal anode installation is shown in *Figure 11*. The anode is placed horizontally in the center of a drilled hole. Then, similar to the installation of a vertical anode, backfill is added around the horizontal anode to a specified height and the installation is topped with soil.

The basic anode configurations for impressed current CP systems are as follows:

- Distributed anode banks
- Surface groundbeds
- Deep anode groundbeds

Distributed Anode Banks – Distributed anodes are used with both galvanic and impressed current systems. This type of anode system provides good current distribution and is often installed to minimize the effects of stray currents in congested areas with a number of nearby structures. *Figure 12* shows a typical distributed anode installation.

Surface Groundbed – When anodes are installed from about 5 to 15 feet (1.5 to 4.6 meters) below ground level, they are referred to as surface groundbeds. The anodes may be installed either vertically or horizontally. When there are multiple anodes in the bed, the lead wires are usually connected to a header cable of sufficient size. The header cable is connected to the positive terminal of the rectifier.

On transmission pipelines and places where few structures exist to be affected by stray current, remote surface groundbeds are often used. In these groundbeds, the anodes are grouped together in a single location, and the current flows from the groundbed directly to the structure. *Figure 13* shows a typical remote surface groundbed installation.

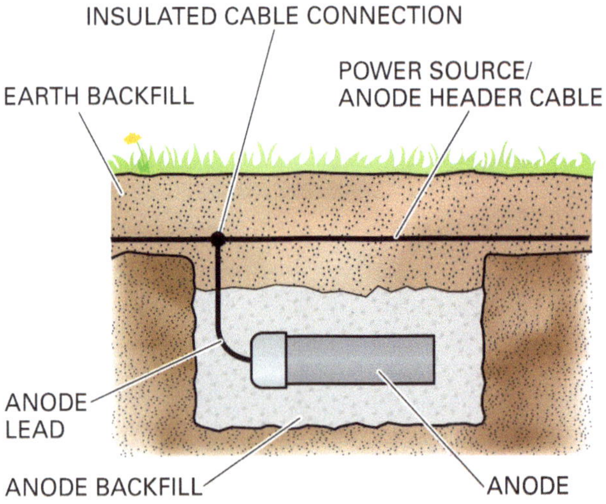

Figure 11 Horizontal impressed current anode installation.

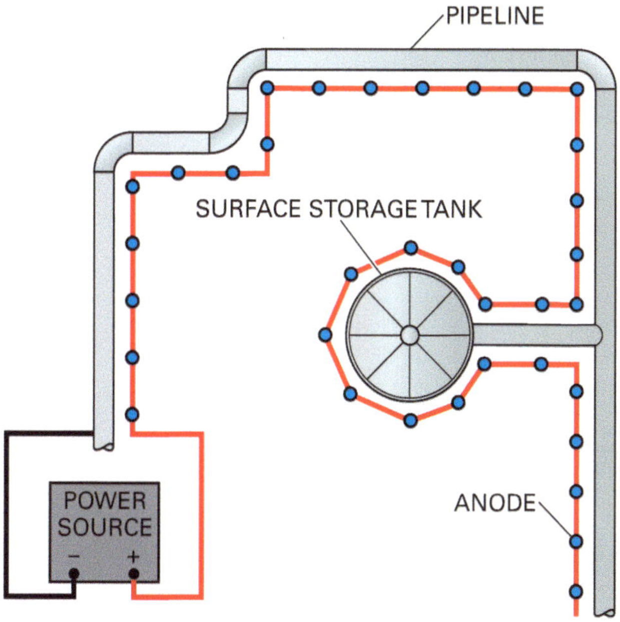

Figure 12 Distributed anode bank installation.

Deep Anode Groundbed – Some anode beds may be hundreds of feet/meters in the ground. In these deep anode installations, a machine auger, usually truck-mounted, drills the hole. Then the anode is placed at the bottom.

Figure 14 shows a typical deep anode installation. **Carbonaceous** backfill is added to the level specified and dropped on the top of the lead wire or the cable header. Above the backfill, the hole may be cased or it may be filled with soil, bentonite, or gravel. As shown, the terminals may sometimes be connected through a shunt to a positive bus bar in order to allow the monitoring of the current output to each of the individual anodes. The bus bar is then connected by cable to the positive terminal in the rectifier.

The equipment used for deep anode installations may include a compressed air circulating system, portable rotary drill rig, mud circulation system, water truck, mud pit to contain the drill mud and cuttings, and various support systems and equipment. *Figure 15* shows a typical portable rotary mud drill rig.

1.2.7 Notifications

Pipeline operators must notify owners of any nearby structures when planning any changes to existing CP systems or any new installations. The courtesy should be reciprocated. Stray currents may weaken or completely eliminate existing CP on structures within the same field of influence. Open communication between all owners and operators in an area should be encouraged. Owners should provide strip maps to each other, with each owner responsible for marking applicable structures. The exact location, size, and scope of any new projects can have serious effects on others. Test points in the same area may be shared. To avoid future problems, put all communications in writing.

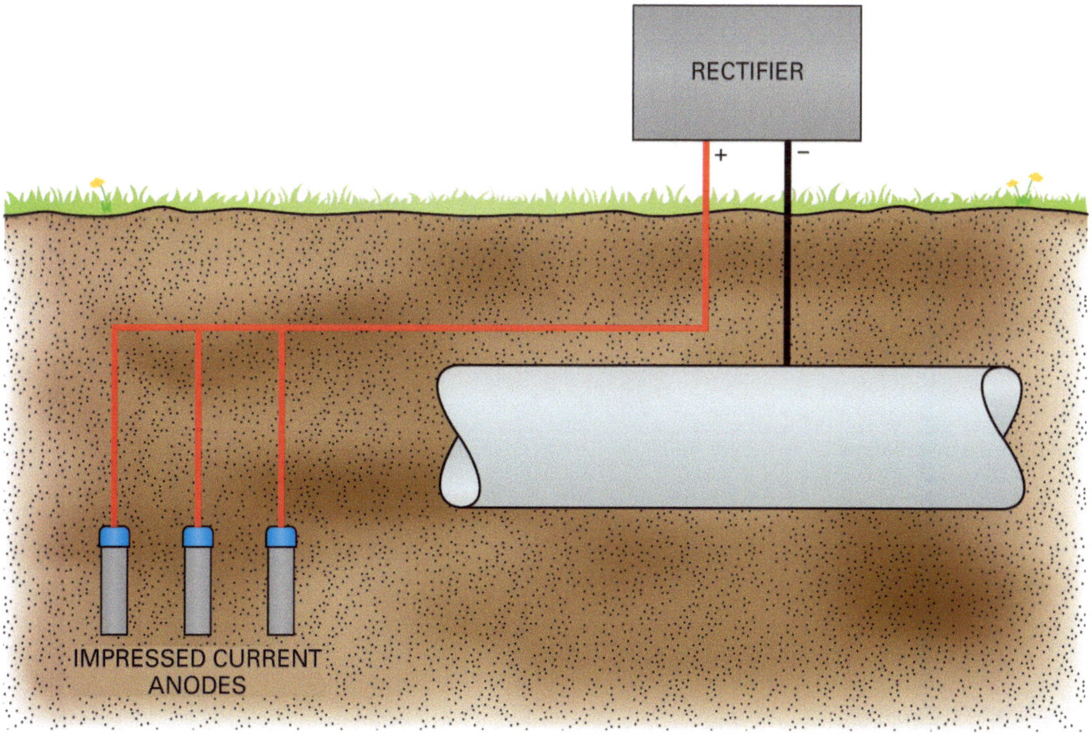

Figure 13 Remote surface groundbed installation.

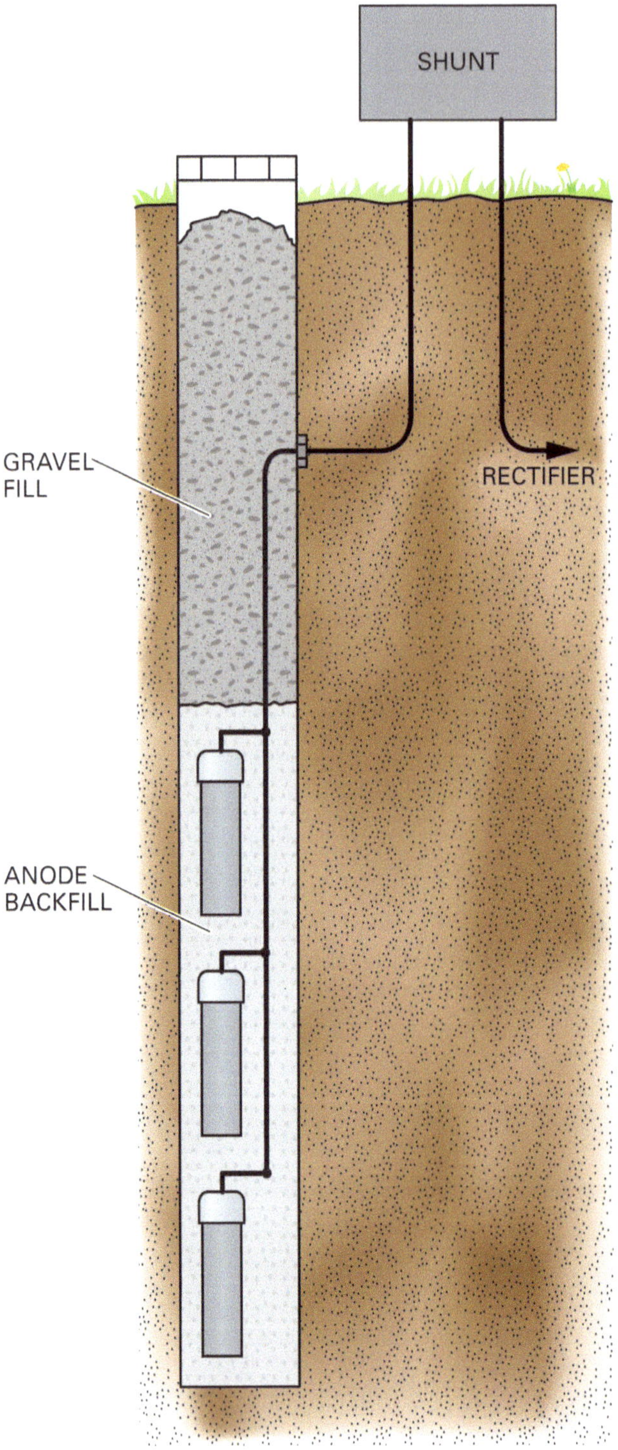

Figure 14 Deep anode installation.

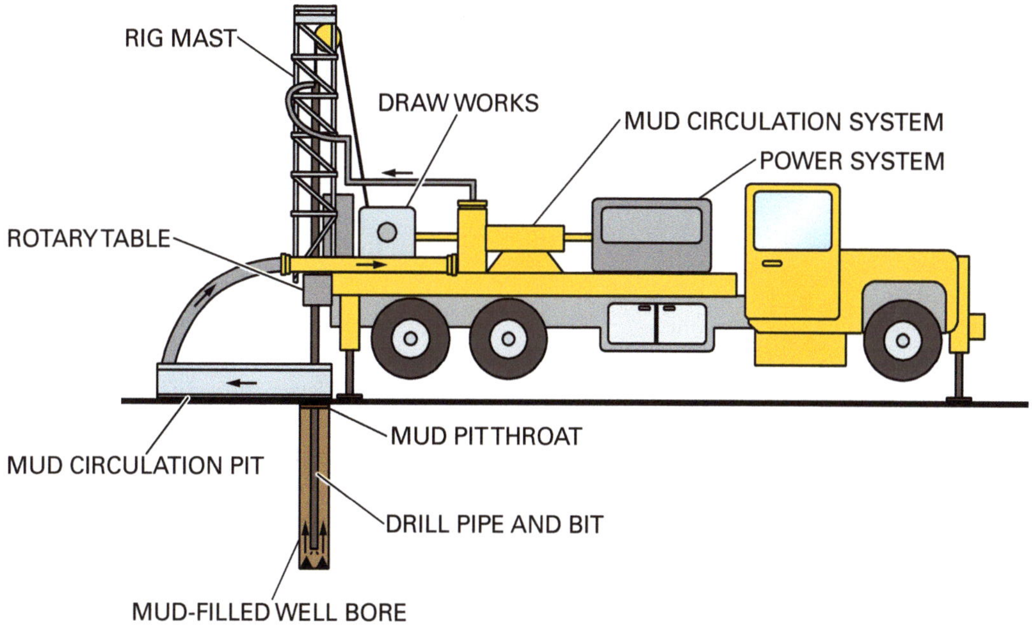

Figure 15 Portable rotary mud drill rig.

Additional Resources

It is assumed that you will learn to perform these procedures under the direct supervision of your crew leader, foreman, or instructor. For independent study, refer to the recommended Additional Resources at the back of this module.

1.0.0 Section Review

1. An impressed-current cathodic protection system can subject an adjacent pipeline to _____.
   a. endwise current
   b. stray current interference
   c. current discharge
   d. potential gradients

2. The depth at which anodes are installed vertically in impressed current groundbeds depends on the _____.
   a. application
   b. resistivity
   c. voltage
   d. polarity

Section Two

2.0.0 Installation of Groundbeds

Objective

Describe how to install impressed current groundbeds (CT9_4-17).

a. Identify the prerequisites, competencies, and abnormal operating conditions associated with installing impressed current groundbeds.
b. Describe procedures for installing impressed current groundbeds.

Performance Task

Install impressed current groundbeds (CT9_4-17).

- Identify potential abnormal operating conditions that may occur, and know the appropriate actions to take in response to them.
- Utilize the appropriate personal protective equipment according to relevant company procedures.
- Confirm that the location and materials meet the requirements of the design criteria.
- Lay out the number, spacing, and configuration of the anodes according to the design criteria.
- Excavate a vertical hole or horizontal ditch.
- Install an anode in the excavated hole, being careful to avoid damage.
- Install the anode header cable between the groundbed and the rectifier.
- Backfill the hole or ditch with appropriate backfill material.
- Complete appropriate documentation as required by operator's procedures.

Trade Terms

Abnormal operating conditions (AOCs): Conditions within a pipeline system, identified by the operator, that may indicate (1) a malfunction of a component or (2) a deviation from normal operations that may indicate a condition exceeding design limits or result in a hazard to persons, property, or the environment.

Casing: A sleeve installed under a roadway, railroad tracks, etc., by boring, which will allow another pipe to be slid through it.

Header cable/wire: A cable or wire to which the anode lead wire is connected.

Splice connection: The electrical connection between the anode lead wire and the header cable/wire.

Impressed current groundbeds are installed to provide cathodic protection for buried or submerged metallic structures. The task begins with verifying that the site location, material, and method of installation all comply with design requirements. The task does not end until the impressed current anodes have been installed and the required documentation has been completed. This section describes some of the basic skills required to install impressed current groundbeds and identifies abnormal operating conditions (AOCs) that personnel may need to respond to when performing this task. It also describes general procedures for installing various types of impressed current groundbeds.

2.1.0 Prerequisites, Competencies, and Abnormal Operating Conditions

Personnel must possess a variety of skills in order to install impressed current groundbeds safely and efficiently. The various operations require a series of steps, which personnel must be competent to perform. Personnel must also be aware of the particular abnormal operating conditions that may occur during the installation procedure and know the proper responses to them.

2.1.1 Prerequisites and Competencies

Personnel assigned to install impressed current groundbeds must possess the following basic knowledge and skills prior to performing this task:

- Experience in test station and test lead installation
- Experience in exothermic (thermite) welding
- Experience in taking pipe-to-soil potentials

Individuals who are assigned to install impressed current groundbeds must have knowledge of cathodic protection systems and components comparable to NACE Certification Level

CP2. This includes, but is not limited to, the following:

- Knowledge of connection methods, such as knowing that:
  - Impressed current anodes are connected together to form an anode bed.
  - Connections are made based on configuration and design of the bed, which may include a termination box with a lead connected to the positive lead of the rectifier.
  - Shunts may be used for measurement and testing of individual anodes. Isolation of individual anodes may be necessary for additional testing of the anodes.
- Awareness that impressed current anodes are installed in special backfill (coke breeze or other filler material).
- Awareness that a **header cable/wire** is a cable or wire to which the anode lead wires are connected.
- Awareness that a **splice connection** is the electrical connection between the anode lead wire and the header cable/wire or between anodes. These splice connections must be carefully insulated to prevent the connection from oxidation.
- Awareness that adjustment of the rectifier is performed after the anode system is energized in order to set cathodic protection levels.

2.1.2 Abnormal Operating Conditions

Before installing impressed current groundbeds, a pipeline technician must be able to identify and properly respond to potential AOCs, should they occur. The following are some potential AOCs, and the appropriate action personnel should take in response to them:

- *Pipeline system is damaged due to exothermic welding burn-through, excessive corrosion, dents and/or gouges* – Anticipate the proper response for unexpected hazardous liquid or carbon dioxide encounter or fire/explosion, if appropriate. Determine the extent of the damage and make proper company notifications.
- *Fire or explosion occurs* – Limit access to the area as much as possible and make company notifications.
- *Unexpected leak of hazardous liquid or carbon dioxide occurs* – Make immediate notifications per company procedure. Try to identify and isolate the source of the leak if possible and if properly trained to do so.

2.2.0 Installing Anodes in Groundbeds

Certain general concerns apply to installing anodes in any type of impressed current groundbed. All connections in the groundbed must be in good condition, and both the connections and the wiring must be sealed against moisture. This includes the connection between the anode and its lead, the anode lead wire and header cable insulation, and the insulation around the connections for the anode lead and header cable.

If the insulation is not maintained, current will be discharged through the holes or holidays, and the wires will corrode and break apart, disabling the cathodic protection. When the connections between the header cable and the lead wires are broken, the system will cease to function.

The anode bank and rectifier must be installed exactly as specified in the installation plans, otherwise, the system may fail in short order. To ensure successful installation, observe the following:

- All the connections must be mechanically secure.
- The connections to the rectifier must be electrically conductive.
- The negative lead wire must be properly connected to the structure.

Verify that the connections between the header cable and the anode leads are secure and conductive. Before energizing power, always be sure the negative conductor is connected to the structure to be protected and the positive conductor is connected to the anodes. Verify proper polarity after energizing the power. Always refer to company specifications to check polarity.

If underground splices on the header cable cannot be avoided, use the fewest number possible. Buried or submerged connections must be completely sealed against moisture to isolate them electrically from the surrounding environment. Leave all wiring with some slack to prevent strain. Avoid damaging the buried cable header; the insulation must remain intact or the system will fail from electrolytic corrosion within a short time.

Before backfilling, sift any reused backfill or native soil to remove any rocks or debris that could cause damage to the wiring or insulation. Follow all guidelines and use extra care when plowing-in a cable.

CAUTION

All work must follow all applicable state and/or local code requirements.

WARNING!

Always obtain and use the appropriate personal protective equipment (PPE) for each task. Only qualified individuals may make electrical connections and operate equipment. Always verify the absence of voltage and current before making connections. Follow the manufacturer's instructions for the equipment being installed.

2.2.1 Installing Anodes in a Surface Groundbed

The following steps provide a general procedure for installing impressed current anodes in a surface groundbed:

Step 1 Identify potential abnormal operating conditions that may occur, and know the appropriate actions to take in response to them.

Step 2 Utilize the appropriate personal protective equipment according to relevant company procedures.

Step 3 Ensure that the selected materials and location meet the system design requirements. Location considerations include soil resistivity, geography/topography, and distance from other structures.

Step 4 Lay out the anodes in accordance with the system design plan. Remote anodes are typically located more than 100 feet (30 meters) from the pipeline, while distributed anodes are located within 10 feet (3 meters) of the structure.

Step 5 Anodes must either be installed in soil or submerged in water that is electrically continuous with the pipeline. If the anode is to be buried, excavate a vertical hole or horizontal ditch for the anode installation.

Step 6 Install the anode in the excavation. Ensure that anodes are placed flat in horizontal installations or centered in a vertical hole installation. Exercise caution to avoid damage to the anode or lead wires.

CAUTION

Do not lift or lower the anode by the lead wire.

Step 7 Install the anode header cable between the groundbed and the rectifier. Exercise caution to avoid damage to the wire insulation.

CAUTION

Insulation damage during installation can result in premature failure of the system. DC current will be discharged at any breaks in the cable insulation.

Step 8 Backfill using rock-free fill to avoid damage to the cables.

Step 9 Document the installation per the company/operator requirements, including the number of anodes and the manner or spacing of the installation.

2.2.2 Installing Vertical Anodes

When installing vertical anodes, drill holes according to the specifications on the plan. If bare anodes are being installed, put some carbonaceous backfill in the bottom of the hole and tamp it down. Carefully lower the anode into the hole. Place the anode in the center of the hole and fill around it with backfill. Tamp down the backfill even with the top of the anode. Then fill in above the anode to the specified height with more backfill, being careful to avoid damaging the lead wire or the insulation. Fill the remaining space to the bottom of the header cable with soil and tamp it down.

CAUTION

Do not hold the anode by its lead wire to avoid breaking or damaging the connection or the wire.

Use the same method for installing packaged anodes. Plans usually call for additional carbonaceous backfill to be placed around the package as well as above and below it. Be sure the installation meets the plan specifications.

2.2.3 Installing Horizontal Anodes

Installing horizontal anodes is similar to installing vertical anodes. Horizontal installations can use either bare or packaged anodes along with carbonaceous backfill. Check the spacing of the anodes in the installation plan. If the spacing of the anodes is less than 10 feet (3 meters), digging a single, long trench that can accommodate both the anodes and the header cable is usually recommended. However, when the spacing exceeds 10 feet (3 meters), a separate hole is prepared for each anode. Check the specifications for the specific installation plans.

2.2.4 Installing Deep Anode Groundbeds

Holes for deep anodes are usually drilled to a specified depth and diameter. Rotary drills can use water-based drilling mud to preserve the integrity of the hole; these drills also remove the contents from the hole. Water-based drilling muds (gels) are used to preserve the integrity of the hole when the soil type would allow the walls to collapse. When the anodes are deeper in the earth (as opposed to surface groundbeds), the potential gradient effects and seasonal output variations from moisture are reduced. There is also less likelihood of damage to the connecting wires from other excavations in the area.

Conduct an electrical log test of the hole to locate the area of lowest resistivity, and measure the circuit resistance. Keep a log of these measurements to find the most advantageous depth for the anode.

Anodes may be lowered by their leads, but this must done very carefully. A centering device keeps the anodes centered in the hole.

Compressed-air circulation systems may be advantageous in limited situations in which downhole formations allow their use. Installing a vent tube in the hole allows any gases to escape to the atmosphere. The tube is usually a length of slotted or perforated PVC or polyethylene with a screened top at ground level. If gases are trapped in the hole, they can increase the resistance of the anode and interfere with its performance.

The integrity of the upper portion of the hole is sometimes maintained by a length of surface casing. After the final backfill has been added, the surface casing is pulled or capped. When all the components are in place, carbonaceous backfill is added to the level specified and dropped on the top of the lead wire or the cable header. Above the backfill, the hole may be filled with soil, gravel, or bentonite, or it may be cased.

Each of the anode lead wires is connected to a separate terminal in a junction box. Each lead should be marked for identification in the junction box. The terminals are sometimes connected through a shunt to a positive bus bar to allow the monitoring of the current output to each of the individual anodes. The bus bar is then connected by cable to the positive terminal in the rectifier.

State and/or local codes govern deep anode installations. Be sure to obtain the proper permits before beginning the work and adhere to all applicable regulations.

Additional Resources

It is assumed that you will learn to perform these procedures under the direct supervision of your crew leader, foreman, or instructor. For independent study, refer to the recommended Additional Resources at the back of this module.

2.0.0 Section Review

1. Before installing impressed current groundbeds, individuals must have knowledge of cathodic protection systems and components comparable to _____.
   a. NFPA 70® apprenticeship standards
   b. Journeyman electrician classification
   c. NACE Certification Level CP2
   d. ANSI Accreditation Phase 2

2. A separate hole is prepared for each horizontal anode when the spacing exceeds _____.
   a. 2 feet (0.6 meters)
   b. 5 feet (1.5 meters)
   c. 8 feet (2.4 meters)
   d. 10 feet (3.0 meters)

SUMMARY

Stray currents, or interference, from foreign sources can cause pipeline corrosion. To control this problem, pipeline technicians need to know how to identify when stray currents are present, how to identify the source of the stray currents, and how to use appropriate techniques to reduce the interference. This module described methods and devices used to reduce or eliminate stray current interference and described procedures for the installation of impressed current groundbeds.

Review Questions

1. Natural occurrences of stray current in the earth that are transferred to the protected pipeline are known as _____.
   a. potential gradients
   b. burial pressures
   c. ground water changes
   d. climate variations

2. DC-powered transit systems can cause _____.
   a. potential shifts
   b. static DC interference
   c. dynamic stray currents
   d. impressed currents

3. For a pipeline in a congested area with multiple nearby structures, a good means of cathodic protection is often a _____.
   a. surface groundbed
   b. distributed anode bank
   c. metallic barrier
   d. current interrupter

4. An anode configuration in which anodes are installed from about 5 to 15 feet (1.5 to 4.6 meters) below ground level is a _____.
   a. deep anode groundbed
   b. backfilled vertical bank
   c. surface groundbed
   d. distributed anode bank

5. If a pipeline technician discovers pipeline system damage, such as dents or gouges, when installing an impressed current ground bed, the technician should _____.
   a. evaluate the damage and make applicable company notifications
   b. immediately initiate a purge of the pipeline system and call 911
   c. evacuate all non-essential personnel from the work site
   d. limit access to the work area as much as possible

6. Documentation for the installation of an impressed current groundbed must include the _____.
   a. type and amount of backfill
   b. location of each underground splice
   c. rectifier type and voltage
   d. the number of anodes and the manner or spacing of the installation

Trade Terms Quiz

Fill in the blank with the correct term that you learned from your study of this module.

1. Connections to the structure being tested, usually wires in a supporting stand with easy connection points for the meter, are called ________.
2. Devices used to allow current to flow in only one direction are ________.
3. Changes in potential with respect to distance, expressed in millivolts per unit of distance are ________.
4. A conductor that joins two points in an electrical circuit to form a parallel or alternative path through which a portion of the current may pass is a(n) ________.
5. Sets of two or more anodes installed in a specific configuration where the anodes are connected together by a header cable/wire are referred to as ________.
6. A portion of a current that flows over a path other than the intended path in an on/off or intermittent fashion and may cause electrochemical corrosion of metals in contact with electrolytes is ________.
7. Electrical connections of two structures that usually have two different cathodic protection levels are ________.
8. The effect on the cathodic protection level of a structure due to the interaction between its cathodic protection system and the cathodic protection system of another structure is called ________.
9. A device used to convert AC current to DC current and which supplies current to the structure in the soil in an impressed cathodic protection system is a(n) ________.
10. A substance that ionizes (releases free electrons) in an electric field is ________.
11. A portion of a current that flows over a path other than the intended path at a constant, unbroken rate and that may cause electrochemical corrosion of metals in contact with electrolytes is ________.
12. A measurement of the voltage or potential difference between a buried or submerged metal structure and an electrolyte, such as soil, is referred to as ________.
13. Current flows through a path other than the intended circuit are called ________.
14. The ability of a material to impede the flow of electric current is called ________.
15. Conditions within a pipeline system, identified by the operator, that may indicate (1) a malfunction of a component or (2) a deviation from normal operations that may indicate a condition exceeding design limits or result in a hazard to persons, property, or the environment are called ________.
16. The electrodes of electrochemical cells at which oxidation occurs are the ________.
17. A product that is composed of carbon is described as ________.
18. A sleeve installed over another pipe is called a(n) ________.
19. A cable or wire to which the anode lead wire is connected is a(n) ________.
20. The electrical connection between the anode lead wire and the header cable/wire is a(n) ________.

Trade Terms

Abnormal operating conditions (AOCs)	Carbonaceous	Electrolyte	Resistivity	Structure-to-soil potential
Anodes	Casing	Header cable/wire	Shunt	Test leads
Anode banks	Diodes	Interference	Splice connection	
Bonds	Dynamic stray current	Potential gradients	Static stray current	
		Rectifier	Stray currents	

Trade Terms Introduced in This Module

Abnormal operating conditions (AOCs): Conditions within a pipeline system, identified by the operator, that may indicate (1) a malfunction of a component or (2) a deviation from normal operations that may indicate a condition exceeding design limits or result in a hazard to persons, property, or the environment.

Anodes: The electrodes of electrochemical cells at which oxidation occurs. Electrons flow away from the anodes in the external circuit. Corrosion usually occurs and metal ions enter the solution at the anodes.

Anode banks: Sets of two or more anodes installed in a specific configuration where the anodes are connected together by a header cable/wire; also referred to as groundbeds or anode groundbeds.

Bonds: Electrical connections of two structures that usually have two different cathodic protection levels.

Carbonaceous: Relating to or composed of carbon.

Casing: A sleeve installed under a roadway, railroad tracks, etc., by boring, which will allow another pipe to be slid through it.

Diodes: Devices used to allow current to flow in only one direction.

Dynamic stray current: A portion of a current that flows over a path other than the intended path in an on/off or intermittent fashion and may cause electrochemical corrosion of metals in contact with electrolytes.

Electrolyte: A substance that ionizes (releases free electrons) in an electric field.

Header cable/wire: A cable or wire to which the anode lead wire is connected.

Interference: The effect on the cathodic protection level of a structure due to the interaction between its cathodic protection system and the cathodic protection system of another structure.

Potential gradients: Changes in potential with respect to distance, expressed in millivolts per unit of distance.

Rectifier: A device used to convert AC current to DC current. In an impressed cathodic protection system, it supplies current to the structure in the soil.

Resistivity: The ability of a material to impede the flow of electric current.

Shunt: A conductor that joins two points in an electrical circuit to form a parallel or alternative path through which a portion of the current may pass.

Splice connection: The electrical connection between the anode lead wire and the header cable/wire.

Static stray current: A portion of a current that flows over a path other than the intended path at a constant, unbroken rate and that may cause electrochemical corrosion of metals in contact with electrolytes.

Stray currents: Current flows through a path other than the intended circuit.

Structure-to-soil potential: A measurement of the voltage or potential difference between a buried or submerged metal structure and an electrolyte, such as soil; also referred to as structure-to-electrolyte potential.

Test leads: Connections to the structure being tested, usually wires in a supporting stand with easy connection points for the meter.

Additional Resources

The following recommended resources can provide additional helpful information related to the requirements for this covered task:

API Recommended Practice 1161, Third Edition, © 2014 American Petroleum Institute. API Publishing Services, 1220 L Street NW, Washington, DC 20005. Distributed by Thompson Reuters (Scientific) LLC, **www.techstreet.com**.

49 *CFR* 192, *Subpart I*,**www.ecfr.gov**.

49 *CFR* 195,*Subpart H*, **www.ecfr.gov**.

NCCER Module AOCCC-17, *Abnormal Operating Conditions Control Center*.

NCCER Module AOCFG-17, *Abnormal Operating Conditions Field and Gas*.

NCCER Module 26112-14, *Electrical Test Equipment*.

NCCER Module 62105-14, *Introduction to Pipeline Documents*.

NCCER Module 64202-02, *Use of Meters and Test Equipment*.

NCCER Module 66101-14, *Introduction to the Pipeline Industry*.

NCCER Module CT1_1-17, *Measure Structure-to-Soil Potential.*

NCCER Module CT1_2-17, *Conduct Close Interval Survey*.

NCCER Module CT1_3-17, *Test and Detect Interference*.

NCCER Module CT1_4-17, *Inspect and Perform Electrical Test of Bonds*.

NCCER Module CT2_2-17, *Repair Damaged Test Leads*.

NCCER Module CT4_3-17, *Adjustment of Rectifier*.

NCCER Module CT9_1-17, *Install Bonds*.

NCCER Module CT9_2-17, *Install Galvanic Anodes*.

NCCER Module CT9_3-17, *Install Rectifiers*.

Figure Credits

Tinker & Rasor, Module Opener

Section Review Answer Key

Answer	Section Reference	Objective
Section One		
1. b	1.1.1	1a
2. a	1.2.6	1b
Section Two		
1. c	2.1.1	2a
2. d	2.2.3	2b

NCCER CURRICULA — USER UPDATE

NCCER makes every effort to keep its textbooks up-to-date and free of technical errors. We appreciate your help in this process. If you find an error, a typographical mistake, or an inaccuracy in NCCER's curricula, please fill out this form (or a photocopy), or complete the online form at **www.nccer.org/olf**. Be sure to include the exact module ID number, page number, a detailed description, and your recommended correction. Your input will be brought to the attention of the Authoring Team. Thank you for your assistance.

Instructors – If you have an idea for improving this textbook, or have found that additional materials were necessary to teach this module effectively, please let us know so that we may present your suggestions to the Authoring Team.

NCCER Product Development and Revision

13614 Progress Blvd., Alachua, FL 32615

Email: curriculum@nccer.org
Online: www.nccer.org/olf

❑ Trainee Guide ❑ Lesson Plans ❑ Exam ❑ PowerPoints Other ______________

Craft / Level: Copyright Date:

Module ID Number / Title:

Section Number(s):

Description:

Recommended Correction:

Your Name:

Address:

Email: Phone:

CT9_5-17

Repair Shorted Casings

Overview

Pipelines are often run through casings, especially at crossings. These pipelines can potentially experience metallic contact between the casing and the pipe that is carrying the product. This contact is referred to as a *shorted casing*, and it compromises the cathodic protection of a pipeline. This module describes how to test for shorted casings and explains how to take corrective action to repair or replace them.

Trainees with successful module completions may be eligible for credentialing through the NCCER Registry. To learn more, go to **www.nccer.org** or contact us at 1.888.622.3720. Our website has information on the latest product releases and training, as well as online versions of our *Cornerstone* magazine and Pearson's product catalog.

Your feedback is welcome. You may email your comments to **curriculum@nccer.org**, send general comments and inquiries to **info@nccer.org**, or fill in the User Update form at the back of this module.

This information is general in nature and intended for training purposes only. Actual performance of activities described in this manual requires compliance with all applicable operating, service, maintenance, and safety procedures under the direction of qualified personnel. References in this manual to patented or proprietary devices do not constitute a recommendation of their use.

CT9_5-17
Repair Shorted Casings

Objectives

When you have completed this module, you will be able to do the following:

1. Describe the use of electrical isolation in CP systems.
   a. Describe the purpose of electrical insulation/isolation in CP systems.
   b. Describe how to test for shorted casings.
2. Describe how to take corrective actions for a shorted casing (CT9_5-17).
   a. Identify the prerequisites, competencies, and abnormal operating conditions associated with repairing shorted casings.
   b. Describe the procedures for repairing shorted casings.

Performance Task

Under the supervision of the instructor, you should be able to do the following:

1. Repair shorted casings (CT9_5-17).
   - Identify potential abnormal operating conditions that may occur during performance of this CT, and know the appropriate actions to take in response to them.
   - Utilize the appropriate personal protective equipment according to relevant company procedures.
   - Clear a workable area and support the pipeline as needed.
   - Remove the end seal.
   - Visually inspect the ends of the carrier pipe and casing for metallic contact.
   - If the pipe has settled, center the carrier pipe within the casing, if possible.
   - If the pipeline is coated, ensure the coating is bonded to the carrier pipe (CT5_3-17).
   - Install casing insulator and centering cradle while providing adequate support.
   - If no metallic contact is found, an electrolyte condition may be the cause of elevated potentials on the casing. This may require removal of the electrolyte material in the casing, if possible, while the end seals are removed.
   - Replace the end seal.
   - Install the test leads as required (CT1_1-17 and CT2_4-17).
   - Conduct a pipe-to-casing potential difference test to determine that the pipe and casing are isolated.
   - Complete appropriate documentation as required by operator's procedures.

Trade Terms

Abnormal operating conditions (AOCs)
Carrier pipe
Casing
Casing insulators
Casing-to-soil potential
Depolarization
Electrolytic short
High dielectric material
Holiday testing
Metallic short
Pipe-to-soil potential
Polarization
Shorted casings

Industry Recognized Credentials

If you are training through an NCCER-accredited sponsor, you may be eligible for credentials from NCCER's Registry. The ID number for this module is CT9_5-17. Note that this module may have been used in other NCCER curricula and may apply to other level completions. Contact NCCER's Registry at 888.622.3720 or go to **www.nccer.org** for more information.

Contents

Figures

SECTION ONE

1.0.0 ELECTRICAL ISOLATION IN CP SYSTEMS

Objective

Describe the use of electrical isolation in CP systems.

a. Describe the purpose of electrical insulation/isolation in CP systems.
b. Describe how to test for shorted casings.

Trade Terms

Carrier pipe: A pipe that carries the product. The carrier pipe is installed through the casing.

Casing: A sleeve installed under a roadway, railroad tracks, etc., by boring, which will allow another pipe to be slid through it.

Casing-to-soil potential: The voltage potential of the casing in the soil or electrolyte as related to a reference electrode.

Depolarization: The removal of factors resisting the current in an electrochemical cell.

Metallic short: Indirect or direct metallic contact between two metallic structures.

Pipe-to-soil potential: A measurement of the voltage or potential difference between a buried or submerged metal structure (pipe) and an electrolyte, such as soil; also referred to as structure-to-electrolyte potential or structure-to-soil potential.

Polarization: A change from the open-circuit potential as a result of current across the electrode/electrolyte interface.

Shorted casings: Casings in which a metallic contact exists between the casing and the carrier pipe that could cause the casing to shield cathodic protection from the carrier pipe.

The pipe and the pipe casing in a cathodic protection (CP) system are designed to be electrically isolated from each other to allow cathodic protection for the pipe. If metallic contact occurs between the casing and the pipe, protection is drawn away from the pipe, which may result in inadequate protection in the cased area. This section identifies common causes of shorted casings and describes methods used to test for shorted casings.

1.1.0 Electrical Insulation/Isolation

The effectiveness of a cathodic protection (CP) system can be compromised if the necessary electrical insulation/isolation fails and the pipeline comes into contact with the casing. The most common causes of shorted casings include the following:

- Settling of the soil around the casing or carrier pipe, usually occurring at each end of the casing where the carrier pipe enters or leaves the casing
- Test station wire with damaged insulation touching the casing
- Carrier pipe movement caused by internal stresses, such as expansion or contraction
- Stresses remaining from jacking the pipe for lineup during new construction
- Improper use of casing insulation spacers
- Misaligned casing pipes
- Mechanical contacts installed during construction
- Foreign objects left inside the casing during installation
- The presence of an electrolyte, such as water or soil

Pipelines inside casings must have spacers to prevent any electrical contact between the carrier pipe and the casing (*Figure 1*). In addition, casing seals should be installed to prevent water from entering the area between the pipe and the casing. Water in this area can cause corrosion damage. On an in-service pipeline, it is important to be able to perform testing to identify the existence of a cased crossing short. Shorts that are located quickly and accurately will save time and expense.

1.2.0 Testing for Shorted Casings

Various tests are used to identify shorted casings. They include the electrical isolation test, potential shift test, internal resistance test, casing depolarization test, and four-wire IR drop test. Each of these tests involves a procedure that consists of several steps.

1.2.1 Electrical Isolation Test

The most common method for identifying possible shorted casings is the electrical isolation test, which is a comparison of casing and carrier potential readings. The annual survey should include electrical isolation testing of all casings using the following procedure:

> **WARNING!**
> Always obtain and use the appropriate personal protective equipment (PPE) for each task. Only qualified individuals may make electrical connections and operate equipment. Always verify the absence of voltage and current before making connections.

Step 1 Measure the carrier **pipe-to-soil potential** (P) with respect to a suitable reference electrode (for example, copper-copper sulfate).

Step 2 With the reference electrode placed at the same location as in Step 1, measure the **casing-to-soil potential** (C).

Step 3 Subtract the casing-to-soil reading from the carrier pipe-to-soil reading (P – C).

Step 4 If the absolute difference (P – C) is equal to or greater than 100 millivolts (0.1 volt), the casing is considered electrically isolated (non-shorted). If the absolute difference (P – C) is less than 100 millivolts (0.1 volt), the casing may be shorted and further testing is required to confirm the possible short.

Step 5 When the absolute difference is less than 100 millivolts, perform one or a combination of the following test(s) to determine if a casing short is present:

- Potential shift test
- Casing depolarization test
- Internal resistance test
- Four-wire IR drop test

Step 6 Document all test results per company requirements.

1.2.2 Potential Shift Test

The potential shift test is easy to perform and requires minimal equipment for pipelines equipped with impressed-current CP systems. A potential shift test is shown in *Figure 2*. To perform the potential shift test, use the following general procedure:

> **WARNING!**
> Always obtain and use the appropriate PPE for each task. Only qualified individuals may make electrical connections and operate equipment. Always verify the absence of voltage and current before making connections.

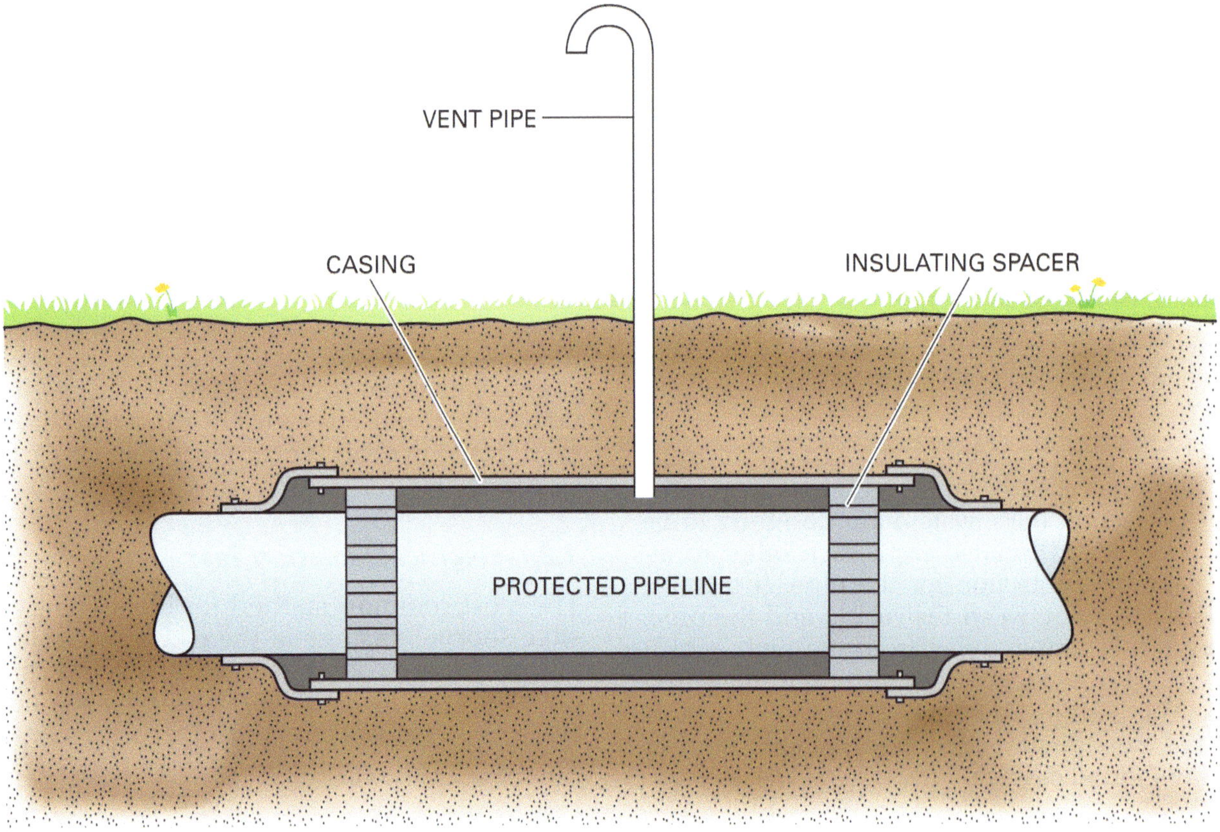

Figure 1 Casing installation requirements.

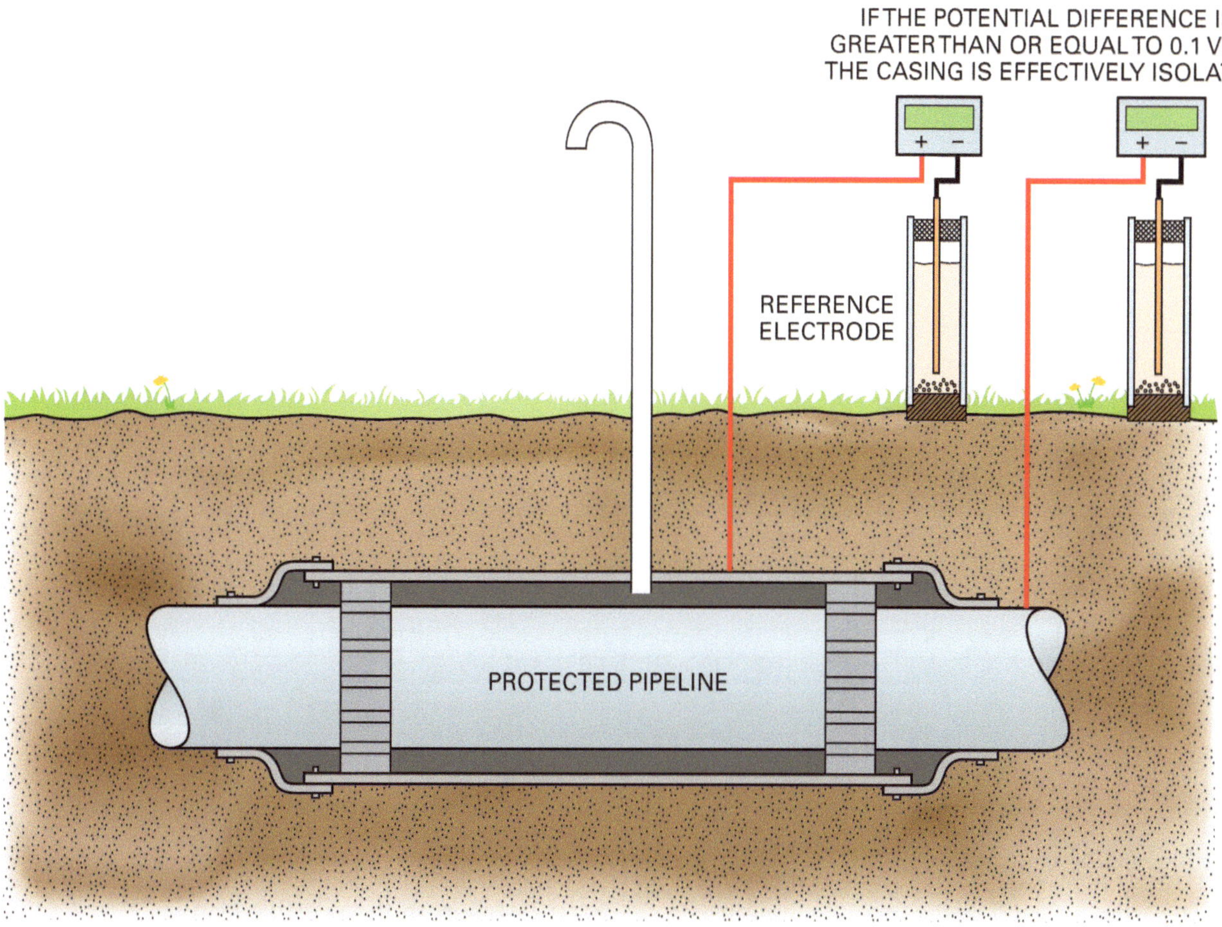

Figure 2 Potential shift test.

Step 1 With the reference electrode placed at the same location, measure the cathodic protection potential for the casing (C1) and the carrier pipe (P1).

Step 2 Increase the cathodic protection current on the system by adjusting the rectifier having the most cathodic protection influence on the carrier pipe.

Step 3 Measure and record the new casing potential (C2) and the pipe potential (P2) resulting from the increase in cathodic protection current.

NOTE

The rectifier selected must be sufficiently remote from the casing under test so that the groundbed voltage gradients do not influence the potential measurements. Distances greater than a mile are preferable.

Step 4 The casing is considered electrically isolated (non-shorted) if one or both of the following conditions exist:

- The pipe potential increases (becomes more negative), while the casing potential remains constant.
- The potential difference between the pipe and casing potential (P2 – C2) becomes more than 100 millivolts (0.1 volt) after the increase in cathodic protection current.

Step 5 If both the pipe and casing increase in the same direction (both become more negative) and the potential difference is less than 100 millivolts, the casing may be shorted and further testing is required to verify the effectiveness of the electrical isolation.

1.2.3 Internal Resistance Test

The internal resistance test is used to determine if metal-to-metal contact exists between the casing and the carrier pipe. This is done by determining the electrical resistance between the casing and the carrier pipe. An internal resistance test is shown in *Figure 3*.

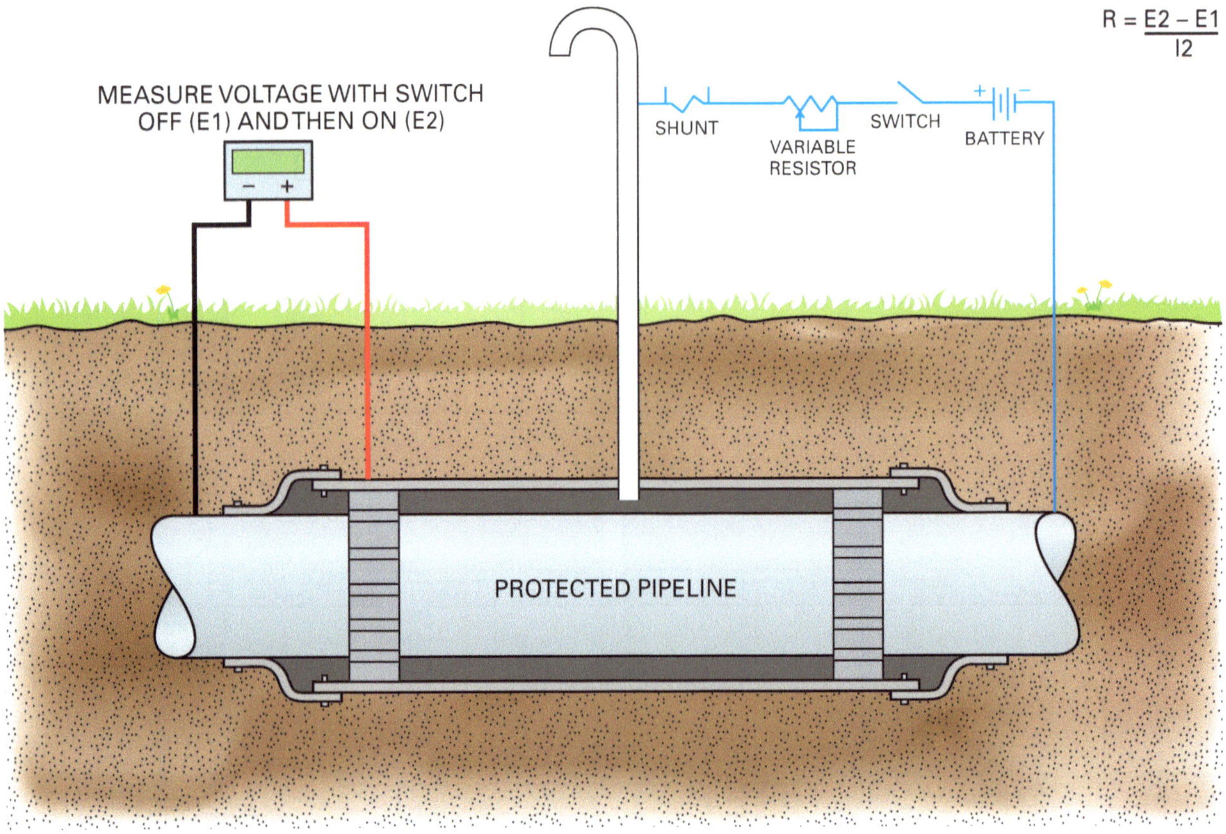

Figure 3 Internal resistance test.

Perform the internal resistance test using the following procedure:

> **WARNING!**
> Always obtain and use the appropriate PPE for each task. Only qualified individuals may make electrical connections and operate equipment. Always verify the absence of voltage and current before making connections.

> **WARNING!**
> When conducting this test, large current flows may be generated by small voltages if a short exists between the carrier and the casing. Use dry cell batteries when conducting this test. A direct short could cause an explosion if wet cell batteries are used.

Step 1 Connect the positive terminal of a DC source (6- or 12-volt battery) to the casing vent, and the negative terminal to the carrier pipe.

> **NOTE**
> A $4\frac{1}{2}$ digit voltmeter is recommended for making the voltage drop measurements.

Step 2 Insert in the positive lead:

- An On/Off switch to interrupt the circuit
- A variable resistor to limit and adjust the test current
- A shunt to measure the circuit current

Step 3 With the circuit switch in the Off position (no current), measure and record the voltage between the casing and the carrier pipe (E1) in volts.

> **NOTE**
> To avoid introducing an error into the measurement for the casing-to-carrier voltage, do not use the current-carrying leads.

Step 4 Use an insulated probe bar to make casing contact for the voltage measurement if there is no separate casing lead available.

Step 5 Use the carrier pipe test lead to connect the negative battery terminal, and use an insulated probe bar to make pipe contact for the voltage measurement.

Step 6 Use the probe bars for voltage measurements, ensuring good metallic contact.

CAUTION

To avoid arcing, do not use the insulated probe bar as a contact on a current-carrying lead.

Step 7 With the circuit switch in the On position, simultaneously measure and record the voltage between the casing and the carrier pipe (E2) in volts, and the circuit current (I2) in amps.

Step 8 Calculate the internal casing-to-pipe resistance (R) by dividing the change in voltage (E2 – E1) by the circuit current applied (I2).

$$R = \frac{E2 - E1}{I2}$$

NOTE

If the internal casing-to-pipe resistance is less than or equal to 0.01 ohm, metal-to-metal contact is indicated and the casing is shorted. If the resistance is greater than 0.01 ohm, the casing may be shorted and further testing is required.

Step 9 Repeat the previous steps to obtain a minimum of three casing-to-pipe resistance calculations. Increase the circuit current at least 0.5 amp for each test. If the casing status results repeat for all tests, the test is considered valid.

CAUTION

To minimize arcing between the carrier pipe and the casing, do not exceed 10 amps circuit current.

1.2.4 Casing Depolarization Test

The casing depolarization test is used to determine if a casing short exists. In this test, an external DC current is applied to the casing and the potential shift between the casing and the carrier pipe is analyzed. A casing depolarization test is shown in *Figure 4*.

Perform the casing depolarization test using the following procedure:

WARNING!

Always obtain and use the appropriate PPE for each task. Only qualified individuals may make electrical connections and operate equipment. Always verify the absence of voltage and current before making connections.

Step 1 Construct a temporary groundbed a minimum of 50 feet (15 meters) from, and perpendicular to, the carrier pipe. Bare steel bars driven into the ground and electrically connected may be used as the temporary groundbed.

Step 2 Connect the positive terminal of a DC source (6- or 12-volt battery) to the casing, and the negative terminal to the temporary groundbed.

Step 3 Insert in the positive lead:

- An On/Off switch to interrupt the circuit
- A variable resistor to limit and adjust the test current
- A shunt to measure the circuit current

NOTE

In this circuit, the casing becomes the anode and the temporary groundbed is the cathode.

Step 4 With the circuit switch in the Off position (no current), measure and record the potential (in volts) of the casing (C1) and of the carrier pipe (P1), with the reference electrode placed at the same location.

CAUTION

Do not use the current-carrying casing vent to measure the casing potentials because this introduces an error in the measurement. If a separate casing test lead is not available, connect the positive of the DC source to the far casing vent (vent A), or use an insulated probe bar to make contact with the casing.

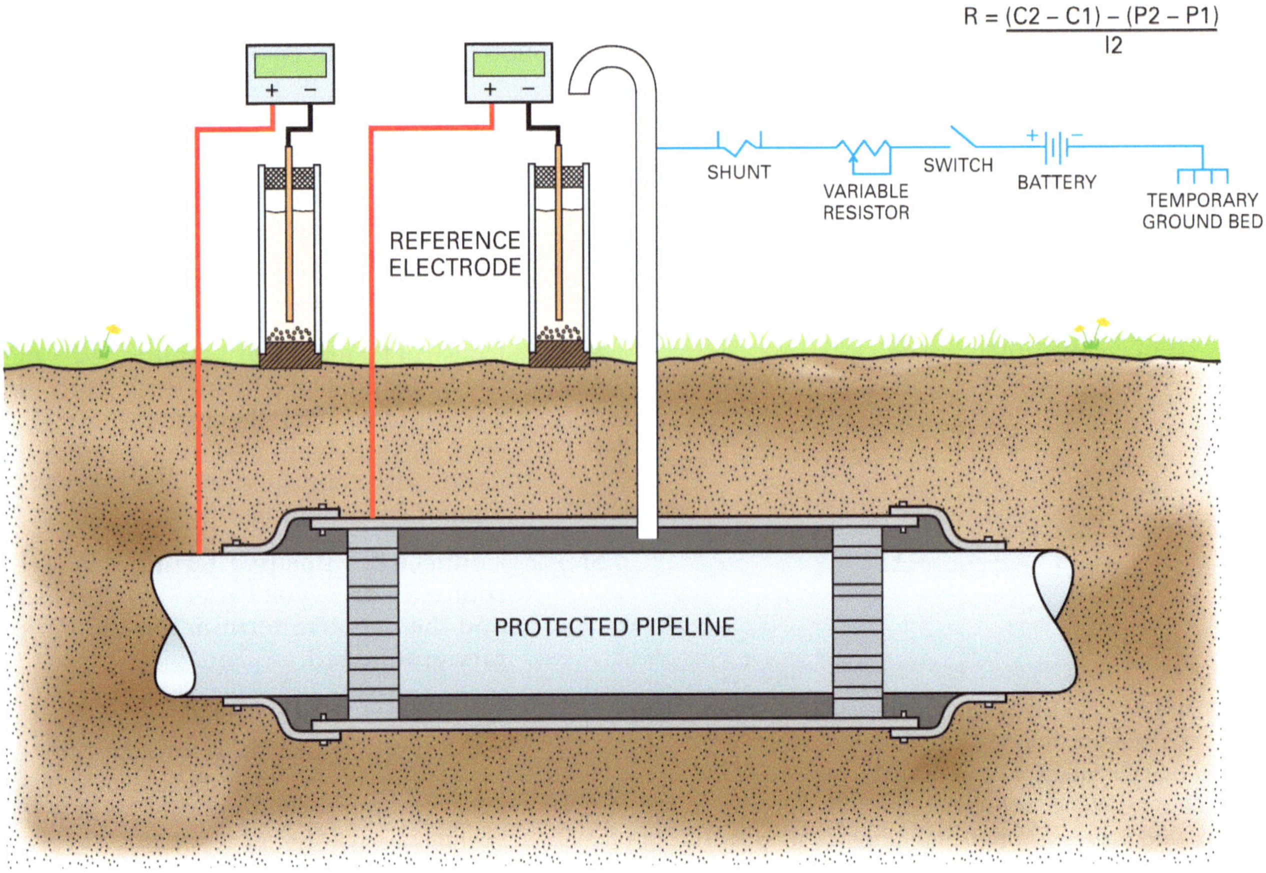

Figure 4 Casing depolarization test.

Step 5 With the circuit switch in the On position, and the current increased approximately 0.5 amp, simultaneously measure and record the new casing (C2) and carrier pipe (P2) potentials in volts and the applied current (I2) in amps. To minimize polarization caused by the increase in current, do not energize the circuit for more than 5 seconds at a time.

Step 6 Obtain the casing potential shift by subtracting the new casing potential (C2) from the initial potential (C1). Obtain the carrier pipe potential shift by subtracting the new carrier potential (P2) from the initial potential (P1).

Step 7 Calculate the effective casing-to-pipe resistance by subtracting the shift in the carrier pipe potential (P2 – P1) from the shift in the casing potential (C2 – C1), and dividing the result by the current applied (I2).

$$R = \frac{(C2 - C1) - (P2 - P1)}{I2}$$

NOTE

If the effective casing-to-pipe resistance is typically less than or equal to 0.08 ohm, a short is indicated. If the resistance is greater than 0.08 ohm, the casing has effective electrical isolation.

Step 8 Repeat the previous steps to obtain a minimum of three casing-to-pipe resistance calculations. Increase the circuit current at least 0.5 amp for each test. If the casing status results repeat for all tests, the test is considered valid.

NOTE

It is usually not necessary to exceed 10 amps from the DC source to the temporary groundbed to develop significant potential shift.

1.2.5 Four-Wire IR Drop Test

The four-wire IR drop test is used to indicate if a casing metallic short exists, as well as the location of the short. The four-wire IR drop test is divided into four parts (Part 1, Part 2, Part 3, and Part 4).

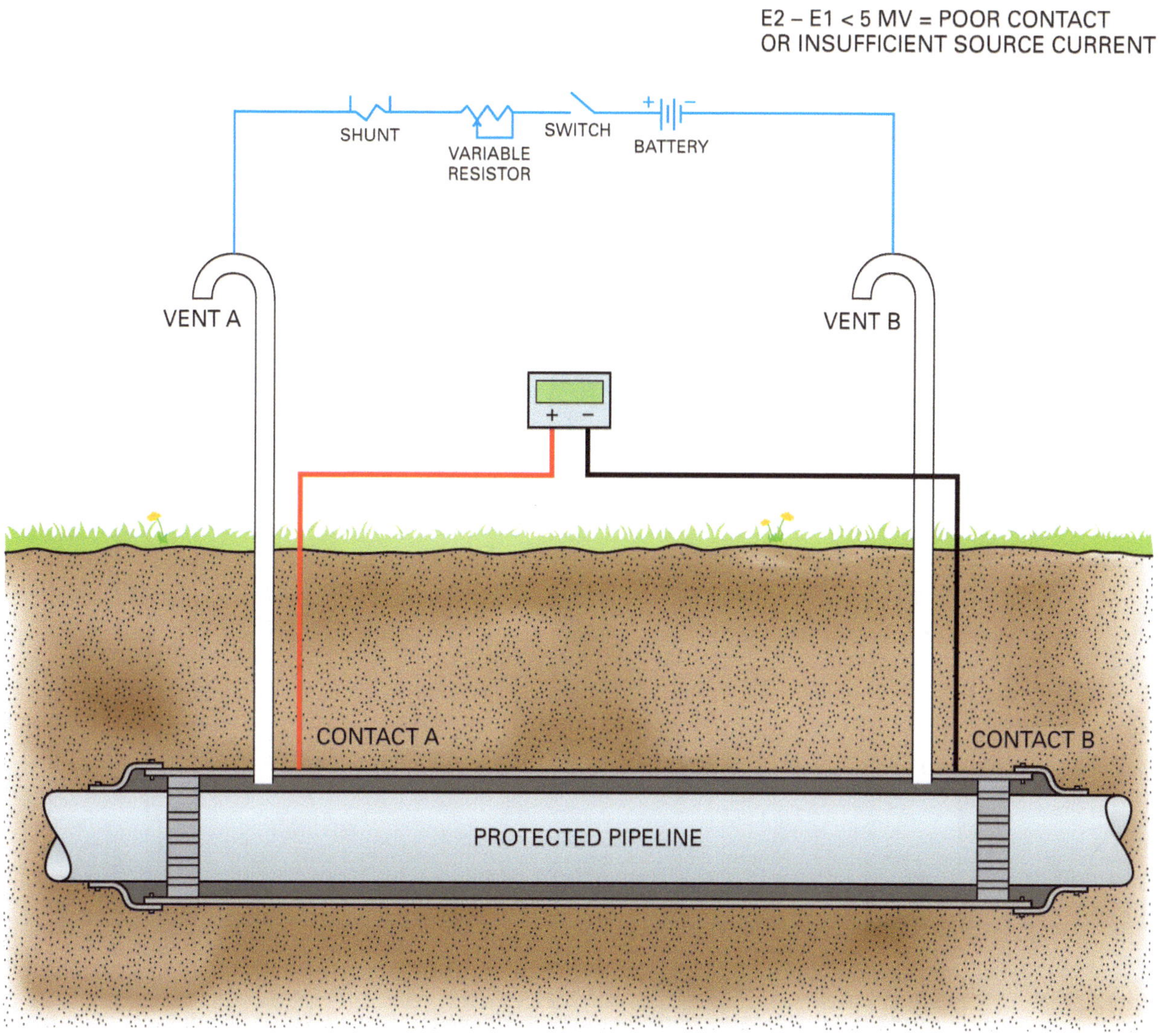

Figure 5 Four-wire IR drop test, Part 1.

> **NOTE**
>
> A $4\frac{1}{2}$ digit voltmeter is recommended for making the voltage drop measurements. Casing vents may be doglegged and not necessarily located at the casing ends.

Part 1 of the four-wire IR drop test is shown in *Figure 5*. Perform Part 1 using the following procedure:

> **WARNING!**
>
> Always obtain and use the appropriate PPE for each task. Only qualified individuals may make electrical connections and operate equipment. Always verify the absence of voltage and current before making connections.

Step 1 Connect the positive terminal from a DC source (6- or 12-volt battery) to casing vent A and the negative terminal to casing vent B.

Step 2 Insert the following in the positive lead:

- An On/Off switch to interrupt the circuit
- A variable resistor to limit and adjust the test current
- A shunt to measure the circuit current

Step 3 With the circuit switch in the Off position (no current), measure and record the voltage drop (E1) in millivolts, and the polarity across the casing length.

Step 4 Contact to the casing may be made with the existing casing test lead and/or insulated probe bar.

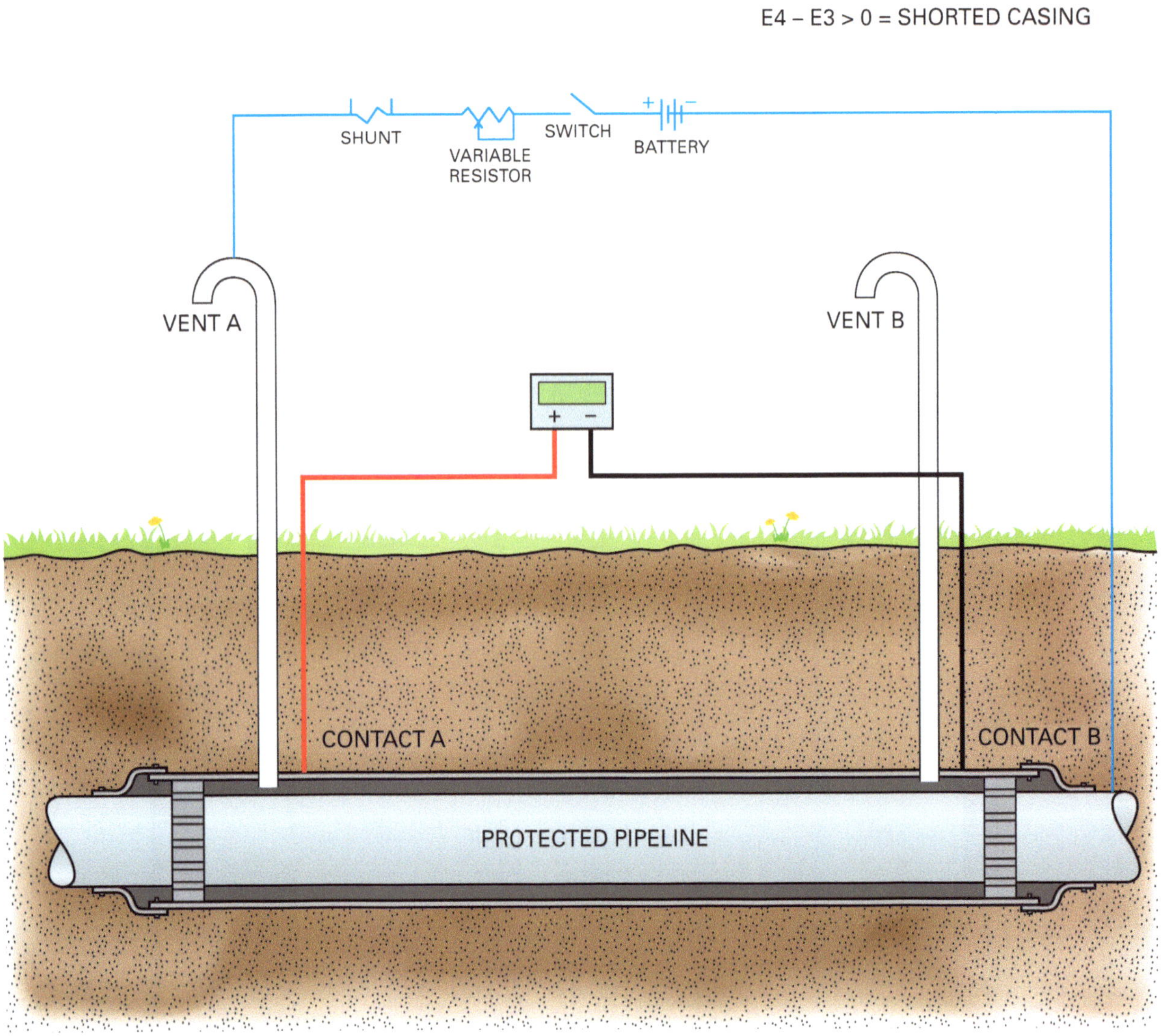

Figure 6 Four-wire IR drop test, Part 2.

> **NOTE**
>
> Do not use the current carrying casing vents as contacts to measure this voltage drop. If insulated probe bars are used, the contact should be made near and inside the casing vents.

Step 5 Label the casing contact point nearest vent A as casing contact A, and the casing contact point nearest vent B as casing contact B.

> **NOTE**
>
> When using insulated probe bars, good metallic contact is vital for accurate measurement interpretation.

Step 6 With the circuit switch in the On position, measure and record the voltage drop (E2) in millivolts, and polarity across the same casing length contact points as in previous steps.

Step 7 Simultaneously measure and record the circuit current (I2) in amps.

Step 8 A casing voltage drop shift (E2 – E1) of less than 5 millivolts (5 mV) may indicate poor contact or insufficient source current. Make the appropriate adjustments before proceeding to Part 2 of the four-wire IR drop test.

> **WARNING!**
>
> Always obtain and use the appropriate PPE for each task. Only qualified individuals may make electrical connections and operate equipment. Always verify the absence of voltage and current before making connections.

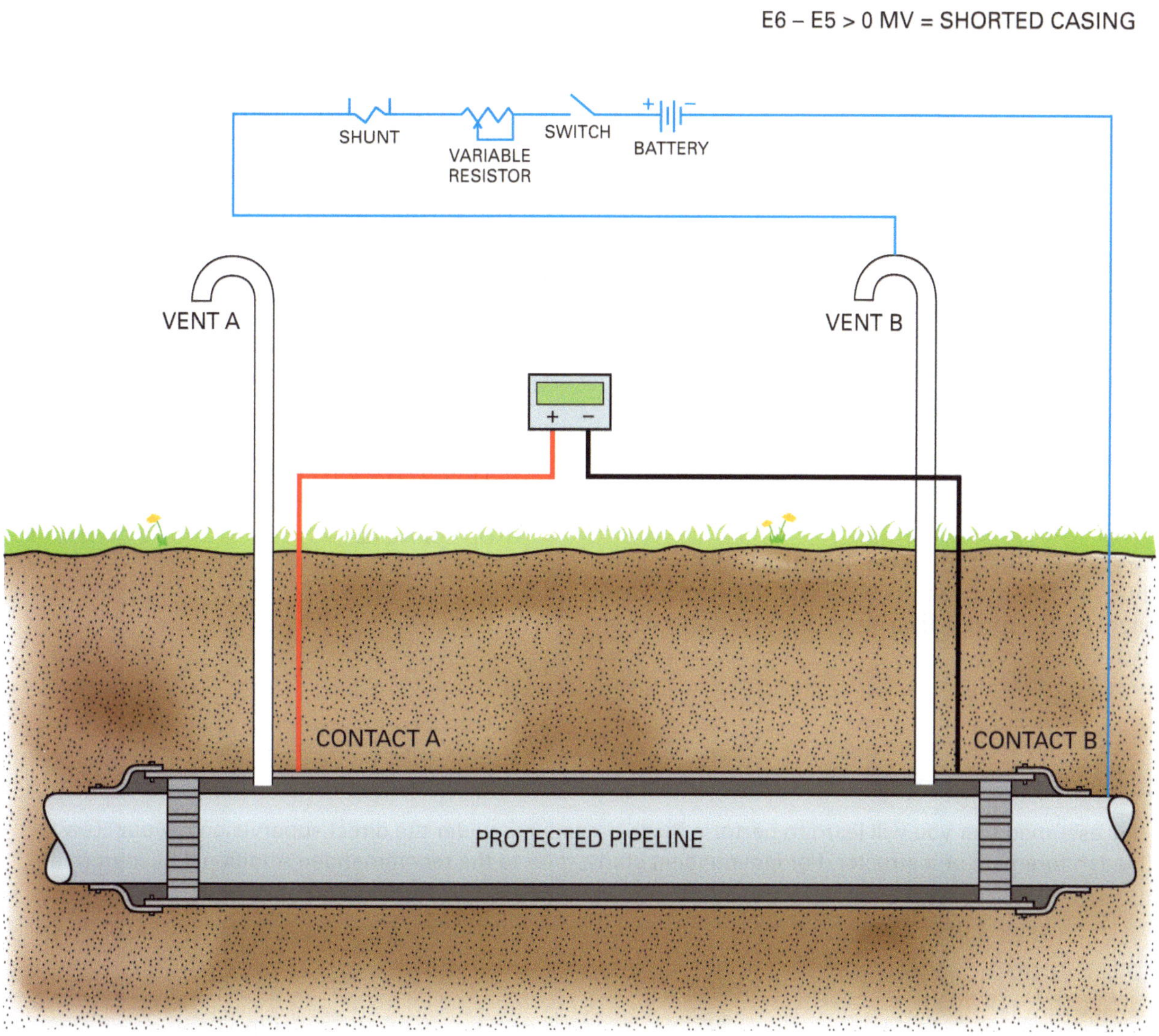

Figure 7 Four-wire IR drop test, Part 3.

Part 2 of the four-wire IR drop test is shown in *Figure 6*. Perform Part 2 using the following procedure:

Step 1 Move the DC source negative terminal from vent B and connect to the carrier pipe while maintaining the positive lead on vent A.

Step 2 With the circuit switch in the Off position (no current), measure and record the voltage drop (E3) in millivolts, and the polarity across the casing length.

Step 3 With the circuit switch in the On position, measure and record the voltage drop (E4) in millivolts and polarity across the same casing length as in Step 2.

Step 4 Calculate and record the voltage shift (E4 – E3).

Step 5 Simultaneously measure the circuit current (I4) in amps and record.

Part 3 of the four-wire IR drop test is shown in *Figure 7*. Perform Part 3 using the following procedure:

WARNING!

Always obtain and use the appropriate PPE for each task. Only qualified individuals may make electrical connections and operate equipment. Always verify the absence of voltage and current before making connections.

Step 1 Move the positive lead from vent A to vent B.

Step 2 With the circuit switch in the Off position (no current), measure and record the voltage drop (E5) in millivolts and the polarity across the casing length.

> **NOTE**
>
> Use the same connecting wire length as in Part 2.

Step 3 With the circuit switch in the On position, adjust the circuit current (if necessary) to the same value as I4 in Part 2.

Step 4 Measure and record the voltage drop (E6) in millivolts across the same casing length as in Step 3.

Step 5 Calculate and record the voltage shift (E6 – E5).

Step 6 Simultaneously measure the circuit current (I6) in amps and record.

Part 4 of the four-wire IR drop test is the analysis of the test results. Use the data collected in each of the preceding tests for the following analysis:

- If E3 = E4 and E5 = E6 (no change in voltage shift), the casing is not shorted.
- If the casing voltage shift (E4 – E3) is greater than zero in Part 2 and/or (E6 – E5) is greater than zero in Part 3, the casing is shorted. If the casing is shorted, the sum of the voltage shift in Part 2 and Part 3 is approximately the voltage shift determined in Part 1 (E2 – E1).
- If the casing is determined shorted and (E4 – E3) is zero and (E6 – E5) is greater than zero, the short is located near vent A.
- If the casing is determined shorted and (E6 – E5) is zero and (E4 – E3) is greater than zero, the short is located near vent B.
- If the casing is determined shorted and (E4 – E3) and (E6 – E5) are both greater than zero, the short is located between the two casing voltage drop contacts.

Additional Resources

It is assumed that you will learn to perform these procedures under the direct supervision of your crew leader, foreman, or instructor. For independent study, refer to the recommended Additional Resources at the back of this module.

1.0.0 Section Review

1. A common cause of shorted casings is _____.
   a. arid soil
   b. anode backfill
   c. high-dielectric fillers
   d. misaligned casing pipes

2. With the potential shift test, the casing is considered electrically isolated if the pipe potential _____.
   a. decreases, while the casing potential remains constant
   b. increases, while the casing potential remains constant
   c. remains constant, while the casing potential decreases
   d. remains constant, while the casing potential increases

SECTION TWO

2.0.0 REPAIR AND REPLACE SHORTED CASINGS

Objective

Describe how to take corrective actions for a shorted casing (CT9_5-17).

a. Identify the prerequisites, competencies, and abnormal operating conditions associated with repairing shorted casings.
b. Describe the procedures for repairing shorted casings.

Performance Task

Repair shorted casings (CT9_5-17).

- Identify potential abnormal operating conditions that may occur, and know the appropriate actions to take in response to them.
- Utilize the appropriate personal protective equipment according to relevant company procedures.
- Clear a workable area and support the pipeline as needed.
- Remove the end seal.
- Visually inspect the ends of the carrier pipe and casing for metallic contact.
- If the pipe has settled, center the carrier pipe within the casing, if possible.
- If the pipeline is coated, ensure the coating is bonded to the carrier pipe (CT5_3-17).
- Install casing insulator and centering cradle while providing adequate support.
- If no metallic contact is found, an electrolyte condition may be the cause of elevated potentials on the casing. This may require removal of the electrolyte material in the casing, if possible, while the end seals are removed.
- Replace the end seal.
- Install the test leads as required (CT1_1-17 and CT2_4-17).
- Conduct a pipe-to-casing potential difference test to determine that the pipe and casing are isolated.
- Complete appropriate documentation as required by operator's procedures.

Trade Terms

Abnormal operating conditions (AOCs): Conditions within a pipeline system, identified by the operator, that may indicate (1) a malfunction of a component or (2) a deviation from normal operations that may indicate a condition exceeding design limits or result in a hazard to persons, property, or the environment.

Casing insulators: Devices attached to carrier pipes that aid in the insertion into the casing and provide electrical isolation from the casing.

Electrolytic short: A current path between the pipe and the casing caused by an electrolytic material such as water or soil.

High dielectric material: A material that has high electrical insulation properties (i.e., a nonconductor of electric current).

Holiday testing: Electrical testing of a coating to locate anomalies or imperfections (holidays).

This section describes some of the basic skills required to repair shorted casings and identifies abnormal operating conditions (AOCs) that personnel may need to respond to when performing this task. It also describes step-by-step procedures for repairing and replacing shorted casings. The task of repairing shorted casings begins when the pipeline casing end(s) have been exposed. The task does not end until the pipeline casing has been successfully tested for electrical isolation.

2.1.0 Prerequisites, Competencies, and Abnormal Operating Conditions

Personnel must possess a variety of skills in order to take corrective actions for a shorted casing. The various operations require a series of steps, which personnel must be competent to perform. Personnel must also be aware of the particular abnormal operating conditions that may occur when repairing and/or replacing shorted casings and know the proper responses to them.

2.1.1 Prerequisites and Competencies

Personnel assigned to repair or replace shorted casings must possess certain basic knowledge and skills prior to performing this task. They must have a basic knowledge of casing systems, including end seals, insulators, and vent connections, and be able to differentiate between a metallic short and an electrolytic short. Personnel must understand the following:

- Casings are oversized pipe required in some instances to reduce external load on the pipeline, such as railroad crossings, interstate highways, etc.
- End seals are kits composed of rubber, vinyl, or other composites to seal the pipeline/casing interface to prohibit water and contaminants from infiltrating the casing.
- Isolating spacers are installed on the pipeline to prevent metallic contact with the casing. Spacers must have sufficient mechanical strength to withstand installation and to maintain isolation.
- Vent connections are made to provide an atmospheric outlet to the casing to prevent pressure buildup and provide access to test the casing atmosphere. One vent is attached on the bottom of the pipe and one is attached on the top to allow insertion of non-metallic material.
- Metallic shorts are caused by metal to metal contact between the pipe and casing.
- Electrolytic shorts are caused by material such as water or soil in the casing that provides a current path between the pipe and the casing.

In addition, individuals assigned to repair or replace shorted casings must be able to:

- Use aboveground methods such as holiday testing as a first step in assessing the coating condition at a cased crossing
- Repair shorted casings after they have been identified
- Measure pipe-to-soil potentials
- Complete testing to ensure effective repair of a shorted casing

2.1.2 Abnormal Operating Conditions

Before repairing or replacing shorted casings, pipeline technicians must be able to identify and properly respond to potential AOCs, should they occur. The following are some potential AOCs, and the appropriate action personnel should take in response to them:

- *Pipeline system damage is found* – Take action, if qualified, and notify personnel of observed condition, as required.
- *A fire or explosion occurs* – Take action, if qualified, and notify personnel of observed condition, as required.
- *Unexplained hydrocarbon is encountered* – Take action, if qualified, and notify personnel of observed condition, as required.

2.2.0 Procedures

The corrective actions that can be taken for shorted casings (a metal-to-metal short or an electrolytic short) are discussed in the following sections. A plan of action and the implementation of the plan should be initiated according to company specifications. The repair of shorted casings consists of the following general procedures:

1. List all known non-repaired shorted casings in the cathodic protection system under test
2. Establish a repair priority to the cathodic protection system under test
3. Take corrective action for the shorted casing, according to the established repair priority
4. Document test results according to company requirements
5. Notify appropriate individuals of shorted casings

2.2.1 Clearing a Shorted Casing

It is important to have a thorough understanding of shorted casings before attempting this procedure. When clearing a casing short, use the following general procedure:

WARNING!

Always obtain and use the appropriate PPE for each task. Only qualified individuals may make electrical connections and operate equipment. Always verify the absence of voltage and current before making connections.

Step 1 Identify potential abnormal operating conditions that may occur, and know the appropriate actions to take in response to them.

Step 2 Utilize the appropriate personal protective equipment according to relevant company procedures.

Step 3 Conduct tests to determine the approximate location of the short.

NOTE

If tests indicate the short is located somewhere other than at the ends of the casing, it may not be practical to clear the short.

Step 4 If tests indicate the short is located at one end of the casing, excavate that end and expose the pipe and casing.

NOTE

Follow current OSHA standards and company policy for excavation and trenching when excavating a ditch or bell hole.

Step 5 Check to see if the test lead wire meets the casing or casing vent. If the wire is the cause of the short, repair the wire insulation or replace the wire. Position the wire so backfilling does not cause a shorted condition.

Step 6 Once exposed, visually inspect the carrier pipe and casing.

Step 7 If the short is caused by the carrier pipe contacting the casing, realign the carrier pipe within the casing using company-approved lifting equipment. If the short is an electrolytic short, flush the casing.

CAUTION

Take care not to damage or cause any unusual stresses on the pipe during this process. Any coating damage to the carrier pipe should be repaired.

Step 8 Maintain clearance by installing new or additional insulators or end seals after clearing the contact.

NOTE

Use sandbags under the end of the casing and under the pipe for support, as the backfill will normally settle.

Step 9 Check for bonding cables or straps connecting the casing to the carrier. If possible, remove the bonding device.

Step 10 Retest the casing before backfilling to verify that the short has been cleared and that there are no other shorts present.

Step 11 Complete appropriate documentation as required by operator's procedures.

2.2.2 Filling a Shorted Casing

It may be necessary to fill the casing with a **high dielectric material**, such as inhibited petroleum-based compounds, if it is impractical to clear a short. Use the following general procedure when filling a shorted casing:

WARNING!

Always obtain and use the appropriate PPE for each task. Only qualified individuals may make electrical connections and operate equipment. Always verify the absence of voltage and current before making connections.

NOTE

While casing fillers may not clear the short, they can mitigate possible corrosion of the carrier pipe by minimizing the corrosive environment of water and oxygen. Follow the manufacturer's recommendations when installing casing fillers.

Step 1 Identify potential abnormal operating conditions that may occur, and know the appropriate actions to take in response to them.

Step 2 Utilize the appropriate personal protective equipment according to relevant company procedures.

Step 3 Ensure that the casing end seals are adequate to contain the filler material during installation and while in service. Rubber link-type end seals or metal end seals that are welded to the casing but insulated from the carrier pipe provide an adequate casing seal, especially during the filling process when high pumping pressure is expected.

Step 4 Install vents at each end of the casing to permit filling of the casing. The opening where the vent is connected to the casing should be adequate in size to permit the filler to flow without requiring excessive pressure. If existing vents are used, the opening to the casing should be checked for adequate size. On the low casing end, consider connecting the vent to the bottom of the casing.

Step 5 Flush the casing with clean water to remove trapped mud, dirt, and other corrosive products. This also ensures that a free flow exists before pumping the casing filler.

Step 6 Test the casing for sealability, as the casing itself may not be watertight.

Step 7 Pump the filler material from the vent on the low end if the casing filler material is heavier than water. Pump from the high end if the casing filler material is lighter than water. Always follow the casing filler manufacturer's recommendations.

Step 8 Calculate the theoretical gallons of filler required to fill the casing.

Step 9 Measure and record the actual gallons of filler pumped.

Step 10 Compare the two amounts to approximate the effectiveness of the filler in the casing.

Step 11 Complete appropriate documentation as required by operator's procedures.

NOTE

Always follow manufacturer's recommendations for determining if casing is properly filled.

2.2.3 Repairing a Shorted Casing

Use the following general procedure when repairing a shorted casing:

WARNING!

Always obtain and use the appropriate PPE for each task. Only qualified individuals may make electrical connections and operate equipment. Always verify the absence of voltage and current before making connections. Follow the manufacturer's instructions for the equipment being installed.

Step 1 Identify potential abnormal operating conditions that may occur, and know the appropriate actions to take in response to them.

Step 2 Utilize the appropriate personal protective equipment according to relevant company procedures.

Step 3 Clear a work area in which to perform testing and repair. Support the pipeline as needed to prevent sagging or damage.

Step 4 Remove the end seal to expose the carrier pipe and casing.

Step 5 Check for any points of physical contact between the pipe and the casing. Pay particular attention to the area around the end seals.

Step 6 If the pipe shows signs of settling, center the carrier pipe within the casing following company policy and procedures.

Step 7 Ensure that all coating is intact and bonded to the carrier pipe (CT5_3-17).

Step 8 Install **casing insulators** (isolating spacers) and centering cradles while providing adequate support.

Step 9 If elevated potentials are measured but there is no metal-to-metal contact, the cause may be an electrolytic condition, such as a leaky casing. If possible, remove the electrolytic material in the casing to resolve the electrolytic condition.

Step 10 Replace the end seal.

Step 11 Install test leads as required (CT2_4-17) and measure the structure-to-soil potential (CT1_1-17).

NOTE

Conduct a pipe-to-casing potential difference test to determine that the pipe and casing are isolated.

Step 12 Document the repair per the company/operator requirements.

2.2.4 Monitoring a Shorted Casing

The procedure for monitoring a shorted casing is described in the following steps:

WARNING!

Always obtain and use the appropriate PPE for each task. Only qualified individuals may make electrical connections and operate equipment. Always verify the absence of voltage and current before making connections.

Step 1 Identify potential abnormal operating conditions that may occur, and know the appropriate actions to take in response to them.

Step 2 Utilize the appropriate personal protective equipment according to relevant company procedures.

Step 3 Monitor the casing vents with leak-detection equipment.

Step 4 Perform the leak-detection test at least twice each calendar year at intervals not exceeding $7^1/_2$ months until the corrective action is performed.

> **NOTE**
> For gas pipelines located in Class 3 and Class 4 population density areas, perform the leak detection test at least four times each calendar year at intervals not exceeding $4^1/_2$ months.

Step 5 If a leak is detected during the monitoring, follow company procedures to repair the leak.

Step 6 If an in-line inspection tool survey indicates no corrosion activity on the carrier pipe within the shorted casing, an evaluation for appropriate corrective action should be taken and a timeframe for corrective action established.

Step 7 Complete appropriate documentation as required by operator's procedures.

2.2.5 Replacing a Shorted Casing

Whenever it is possible and appropriate, replace a cased crossing with an uncased heavy-wall steel pipe or a concrete-coated steel pipe. These types of pipes provide added mechanical strength without having the problems that may be caused by pipelines with casings. Always follow any state and local guidelines when replacing shorted casings.

2.2.6 Replacing the Carrier Pipe Within the Casing

Casing insulating spacers have an insulated skid and are strapped around the pipe. There are many types of casing insulators available. These insulators can be all plastic or can have insulated blocks that are attached to steel bands. Some steel bands are lined with rubber or plastic and are called skids. Skids are placed evenly along the pipeline at a maximum of 10 feet (3 meters) apart. Insulating spacers are also located at the ends of the casing. Always follow manufacturer recommendations for the number, size, and spacing of insulated spacers.

Stresses exerted on spacers during installation are very great, especially when working with large pipe and long-cased crossings. If insulators are snagged and damaged during installation, the insulator should be replaced before continuing the installation. When the installation is complete, make sure that the end insulators are close to the end of the casing. If they are not, install another insulator into the casing and position it near the casing end to provide proper electrical separation at the ends of the casing.

End seals are designed to be flexible and to provide a tight seal between the pipe and the casing. Most seals are watertight. Proper casing preparation is required before the carrier pipe is pulled into it. Check the carrier pipe for the following to make sure it has been prepared properly:

- All welds must be smooth.
- The casing must be straight.
- All debris must be removed.
- All welding and hole burning must be performed before the carrier pipe is installed.
- All vent pipes must be set before the carrier pipe is installed.

Additional Resources

It is assumed that you will learn to perform these procedures under the direct supervision of your crew leader, foreman, or instructor. For independent study, refer to the recommended Additional Resources at the back of this module.

2.0.0 Section Review

1. Before repairing shorted casings, individuals must have basic knowledge of _____.
   a. interference bonds
   b. pit depth measurements
   c. casing systems
   d. pipeline coatings

2. If clearing a short is impractical, it may be necessary to fill the casing with _____.
   a. low dielectric material
   b. high dielectric material
   c. asbestos
   d. concrete

SUMMARY

Pipelines that are run through casings are subject to corrosion. To protect the pipeline and maintain proper electrical isolation, the casings are tested for shorts. Shorted casings are identified by comparing potential readings taken from the carrier and the casing. Tests performed for shorted casings are the potential shift test, the internal resistance test, the casing depolarization test, and the four-wire IR drop test.

Repairing shorted casings involves forming and implementing a plan of corrective action. The short must be cleared and then filled with a high dielectric material. If feasible, the shorted casing should be replaced. This module described the general procedures for repairing and replacing shorted casings.

Review Questions

1. Pipelines are protected against contact between the carrier pipe and the casing through the use of _____.
   a. spacers
   b. shims
   c. O-rings
   d. collars

2. The most common method used for identifying possible shorted casings is the comparison of casing and carrier _____.
   a. alignments
   b. resistance
   c. depolarization
   d. potential readings

3. When performing a recognition test for possible shorted casings, measure the carrier pipe-to-soil potential with respect to a suitable _____.
   a. nearby measurement
   b. landmark
   c. reference electrode
   d. test station

4. The test used to determine if a direct metal-to-metal contact exists between the casing and the carrier pipe is a(n) _____.
   a. internal resistance test
   b. potential shift test
   c. depolarization test
   d. IR drop test

5. Metal-to-metal contact is indicated by an internal casing-to-pipe resistance test reading of less than or equal to _____.
   a. 0.001 ohm
   b. 0.01 ohm
   c. 0.5 ohm
   d. 1 ohm

6. If a fire or explosion occurs when a pipeline technician is repairing a shorted casing, the technician should _____.
   a. immediately call 911
   b. determine the extent and severity of the damage
   c. evacuate all non-essential personnel from the work site
   d. take action, if qualified, and notify personnel as required

7. Casing insulated spacers have a(n) _____.
   a. insulated bushing
   b. insulated skid
   c. electrical fitting
   d. insulated flange

8. Skids are placed evenly along a pipeline at a maximum of _____.
   a. 4 feet (1.2 meters) apart
   b. 6 feet (1.8 meters) apart
   c. 8 feet (2.4 meters) apart
   d. 10 feet (3.0 meters) apart

Trade Terms Quiz

Fill in the blank with the correct term that you learned from your study of this module.

1. The pipe that is installed through the casing is the ________.
2. A sleeve over another pipe is a(n) ________.
3. The change from the open-circuit potential as a result of current across the electrode/electrolyte interface is called ________.
4. Indirect or direct metallic contact between two metallic structures is a(n) ________.
5. Casings in which metallic contact exists between the casing and the carrier pipe are ________.
6. The voltage potential of the casing in the soil or electrolyte as related to a reference electrode is the ________.
7. Water that collects between a pipe and its casing can result in a(n) ________.
8. Conditions within a pipeline system, identified by the operator, that may indicate (1) a malfunction of a component or (2) a deviation from normal operations that may indicate a condition exceeding design limits or result in a hazard to persons, property, or the environment are called ________.
9. A material that has high electrical insulation properties is described as a(n) ________.
10. Devices attached to carrier pipes that aid in the insertion into the casing and provide electrical isolation from the casing are ________.
11. Electrical testing of a coating to locate anomalies or imperfections is ________.
12. A measurement of the voltage or potential difference between a buried or submerged metal structure (pipe) and an electrolyte, such as soil, is the ________.
13. The removal of factors resisting the current in an electrochemical cell is called ________.

Trade Terms

Abnormal operating conditions (AOCs)
Carrier pipe
Casing
Casing insulators
Casing-to-soil potential
Depolarization
Electrolytic short
High dielectric material
Holiday testing
Metallic short
Pipe-to-soil potential
Polarization
Shorted casings

Trade Terms Introduced in This Module

Abnormal operating conditions (AOCs): Conditions within a pipeline system, identified by the operator, that may indicate (1) a malfunction of a component or (2) a deviation from normal operations that may indicate a condition exceeding design limits or result in a hazard to persons, property, or the environment.

Carrier pipe: A pipe that carries the product. The carrier pipe is installed through the casing.

Casing: A sleeve installed under a roadway, railroad tracks, etc., by boring, which will allow another pipe to be slid through it.

Casing insulators: Devices attached to carrier pipes that aid in the insertion into the casing and provide electrical isolation from the casing.

Casing-to-soil potential: The voltage potential of the casing in the soil or electrolyte as related to a reference electrode.

Depolarization: The removal of factors resisting the current in an electrochemical cell.

Electrolytic short: A current path between the pipe and the casing caused by an electrolytic material such as water or soil.

High dielectric material: A material that has high electrical insulation properties (i.e., a non-conductor of electric current).

Holiday testing: Electrical testing of a coating to locate anomalies or imperfections (holidays).

Metallic short: Indirect or direct metallic contact between two metallic structures.

Pipe-to-soil potential: A measurement of the voltage or potential difference between a buried or submerged metal structure (pipe) and an electrolyte, such as soil; also referred to as structure-to-electrolyte potential or structure-to-soil potential.

Polarization: A change from the open-circuit potential as a result of current across the electrode/electrolyte interface.

Shorted casings: Casings in which a metallic contact exists between the casing and the carrier pipe that could cause the casing to shield cathodic protection from the carrier pipe.

Additional Resources

The following recommended resources can provide additional helpful information related to the requirements for this covered task:

API Recommended Practice 1161, Third Edition, © 2014 American Petroleum Institute. API Publishing Services, 1220 L Street NW, Washington, DC 20005. Distributed by Thompson Reuters (Scientific) LLC, **www.techstreet.com**.

49 *CFR* 192, *Subpart I*, **www.ecfr.gov**.

49 *CFR* 195, *Subpart H*, **www.ecfr.gov**.

NCCER Module AOCCC-17, *Abnormal Operating Conditions Control Center*.

NCCER Module AOCFG-17, *Abnormal Operating Conditions Field and Gas*.

NCCER Module 62105-14, *Introduction to Pipeline Documents*.

NCCER Module 64202-02, *Use of Meters and Test Equipment*.

NCCER Module 66101-14, *Introduction to the Pipeline Industry*.

NCCER Module CT1_1-17, *Measure Structure-to-Soil Potential*.

NCCER Module CT1_2-17, *Conduct Close Interval Survey*.

NCCER Module CT1_3-17, *Test and Detect Interference*.

NCCER Module CT1_4-17, *Inspect and Perform Electrical Test of Bonds*.

NCCER Module CT2_2-17, *Repair Damaged Test Leads*.

NCCER Module CT2_4-17, *Install Test Leads by Exothermic Welding Methods*.

NCCER Module CT5_3-17, *Inspect the Condition of External Coating on Buried or Submerged Pipe*.

NCCER Module CT14_1-17, *Locate Line*.

NCCER Module CT32_0-17, *Observation of Excavation Activities*.

NCCER Module CT39_0-17, *Backfilling a Trench Following Maintenance*.

Figure Credits

Tinker & Rasor, Module Opener

Section Review Answer Key

Answer	Section Reference	Objective
Section One		
1. d	1.1.0	1a
2. b	1.2.2	1b
Section Two		
1. c	2.1.1	2a
2. b	2.2.2	2b

NCCER CURRICULA — USER UPDATE

NCCER makes every effort to keep its textbooks up-to-date and free of technical errors. We appreciate your help in this process. If you find an error, a typographical mistake, or an inaccuracy in NCCER's curricula, please fill out this form (or a photocopy), or complete the online form at **www.nccer.org/olf**. Be sure to include the exact module ID number, page number, a detailed description, and your recommended correction. Your input will be brought to the attention of the Authoring Team. Thank you for your assistance.

Instructors – If you have an idea for improving this textbook, or have found that additional materials were necessary to teach this module effectively, please let us know so that we may present your suggestions to the Authoring Team.

NCCER Product Development and Revision

13614 Progress Blvd., Alachua, FL 32615

Email: curriculum@nccer.org
Online: www.nccer.org/olf

❑ Trainee Guide ❑ Lesson Plans ❑ Exam ❑ PowerPoints Other ________________________

Craft / Level: Copyright Date:

Module ID Number / Title:

Section Number(s):

Description:

Recommended Correction:

Your Name:

Address:

Email: Phone:

CT9_6-17

Install Electrical Insulating Device

Overview

The pipe and equipment in a pipeline system must be electrically isolated to keep the system operating safely. Failure to do so compromises system operations and endangers the safety of personnel. This module describes the use of electrical insulating devices designed to isolate pipeline segments and equipment to ensure proper functioning of cathodic protection, lightning protection, and ground fault protection. It explains how to test for electrical isolation and describes procedures for installing various electrical insulating devices.

Trainees with successful module completions may be eligible for credentialing through the NCCER Registry. To learn more, go to **www.nccer.org** or contact us at 1.888.622.3720. Our website has information on the latest product releases and training, as well as online versions of our *Cornerstone* magazine and Pearson's product catalog.

Your feedback is welcome. You may email your comments to **curriculum@nccer.org**, send general comments and inquiries to **info@nccer.org**, or fill in the User Update form at the back of this module.

This information is general in nature and intended for training purposes only. Actual performance of activities described in this manual requires compliance with all applicable operating, service, maintenance, and safety procedures under the direction of qualified personnel. References in this manual to patented or proprietary devices do not constitute a recommendation of their use.

CT9_6-17
Install Electrical Insulating Device

Objectives

When you have completed this module, you will be able to do the following:

1. Describe the use of electrical isolation in CP systems.
   a. Describe the purpose of, and methods used for, electrical insulation/isolation in CP systems.
   b. Describe how to test for electrical isolation.
2. Describe how to install electrical insulating devices (CT9_6-17).
   a. Identify the prerequisites, competencies, and abnormal operating conditions associated with installing electrical insulating devices.
   b. Describe procedures for installing various electrical insulating devices.

Performance Task

Under the supervision of the instructor, you should be able to do the following:

1. Install electrical insulating devices (CT9_6-17).
   - Identify potential abnormal operating conditions that may occur during performance of this CT, and know the appropriate actions to take in response to them.
   - Utilize the appropriate personal protective equipment according to relevant company procedures.
   - Verify the location where the isolation device is needed.
   - Verify the appropriate kit materials.
   - Verify proper flange alignment prior to installation of insulation kit.
   - Install the flange kit and/or dielectric unions according to the manufacturer's specifications. Use precautions to prevent any moisture, soil, or other foreign matter from contacting any portion of the insulating joint prior to its being sealed.
   - Verify isolation using the appropriate insulation tester.
   - Complete appropriate documentation as required by operator's procedures.

Trade Terms

Abnormal operating conditions (AOCs)
Carrier pipe
Casing
Casing-to-soil potential
Depolarization
Dielectric
Metallic short
Pipe-to-soil potential
Polarization

Industry Recognized Credentials

If you are training through an NCCER-accredited sponsor, you may be eligible for credentials from NCCER's Registry. The ID number for this module is CT9_6-17. Note that this module may have been used in other NCCER curricula and may apply to other level completions. Contact NCCER's Registry at 888.622.3720 or go to **www.nccer.org** for more information.

Contents

Figures and Tables

Section One

1.0.0 Electrical Isolation in CP Systems

Objective

Describe the use of electrical isolation in CP systems.

a. Describe the purpose of, and methods used for, electrical insulation/isolation in CP systems.
b. Describe how to test for electrical isolation.

Trade Terms

Carrier pipe: A pipe that carries the product. The carrier pipe is installed through the casing.

Casing: A sleeve installed under a roadway, railroad tracks, etc., by boring, which will allow another pipe to be slid through it.

Casing-to-soil potential: The voltage potential of the casing in the soil or electrolyte as related to a reference electrode.

Depolarization: The removal of factors resisting the current in an electrochemical cell.

Dielectric: An electrically nonconductive material.

Metallic short: Indirect or direct metallic contact between two metallic structures.

Pipe-to-soil potential: The potential difference between the surface of a buried or submerged metallic structure and the electrolyte that is measured with reference to an electrode in contact with the electrolyte; also referred to as structure-to-electrolyte potential or structure-to-soil potential.

Electrical insulating devices are used to separate the pipeline and equipment to ensure the proper operation of ground fault, lightning, and cathodic protection (CP) systems. If metallic contact occurs between the casing and the pipe, protection is drawn away from the pipe which may result in inadequate protection in the cased area.

1.1.0 Electrical Insulation/Isolation Methods

Various methods and electrical insulating devices are designed to isolate pipeline segments and equipment to ensure that proper functioning of cathodic protection, lightning protection, and ground fault protection is not disrupted or compromised.

Pipelines inside casings must have spacers to prevent any electrical contact between the carrier pipe and the casing (*Figure 1*). Casing seals should be installed to prevent water from entering the area between the pipe and the casing. Water in this area can cause corrosion damage.

The following methods/devices for electrical insulation/isolation are covered in this module:

- Insulating flanges or dielectric unions
- Isolation joints
- Lightning and ground fault protection systems
- Electrical isolation from electrical grounding
- Aboveground piping isolation from other buried structures
- Casing isolation/insulation

1.2.0 Isolation Testing

Various tests are used to verify electrical isolation in a pipeline system. Each of these tests involves a procedure that consists of several steps.

> **NOTE**
> Step-by-step procedures for performing the isolation tests mentioned here are covered more fully in NCCER Module CT9_5-17, *Repair Shorted Casings.*

1.2.1 Electrical Isolation Test

The electrical isolation test is a comparison of casing and carrier potential readings. The annual survey should include electrical isolation testing of all casings using the following procedure:

> **WARNING!**
> Always obtain and use the appropriate personal protective equipment (PPE) for each task. Only qualified individuals may make electrical connections and operate equipment. Always verify the absence of voltage and current before making connections.

Step 1 Measure the carrier pipe-to-soil potential (P) with respect to a suitable reference electrode (for example, copper-copper sulfate).

Step 2 With the reference electrode placed at the same location as in Step 1, measure the casing-to-soil potential (C).

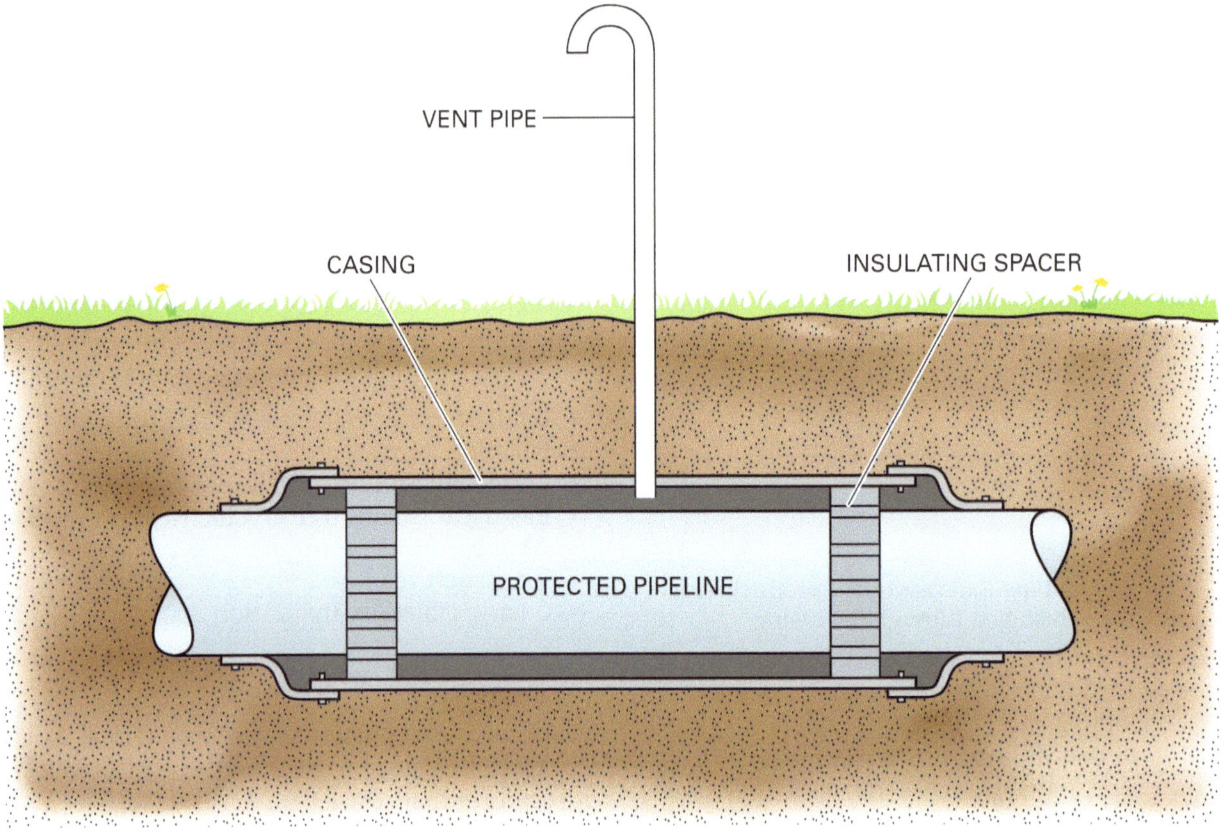

Figure 1 Casing installation requirements.

Step 3 Subtract the casing-to-soil reading from the carrier pipe-to-soil reading (P – C).

Step 4 If the absolute difference (P – C) is equal to or greater than 100 millivolts (0.1 volt), the casing is considered electrically isolated (non-shorted). If the absolute difference (P – C) is less than 100 millivolts (0.1 volt), the casing may be shorted and further testing is required to confirm the possible short.

Step 5 Document all test results per company requirements.

1.2.2 Further Testing

When the absolute difference obtained from an electrical isolation test is less than 100 millivolts, perform one or a combination of the following test(s) to determine if a casing short is present. Document all test results per company requirements.

- *Potential shift test* – The potential shift test (*Figure 2*) is easy to perform and requires minimal equipment for pipelines equipped with impressed current CP systems. The values for carrier pipe-to-soil potential (P) and casing-to-soil potential (C) at different levels of cathodic protection current are measured with respect to a suitable reference electrode. The difference between the two values (P-C) indicates whether the casing may be shorted. If potential shift test results suggest that the casing is shorted, further testing is required to confirm the possible short.
- *Internal resistance test* – The internal resistance test (*Figure 3*) is used to determine if a direct metal-to-metal contact exists between the casing and the carrier pipe. This is done by determining the electrical resistance between the casing and the carrier pipe. The positive terminal of a DC source is connected to the casing vent, and the negative terminal to the carrier pipe. The voltage between the casing and the carrier pipe is then recorded and documented to obtain a minimum of three casing-to-pipe resistance calculations.

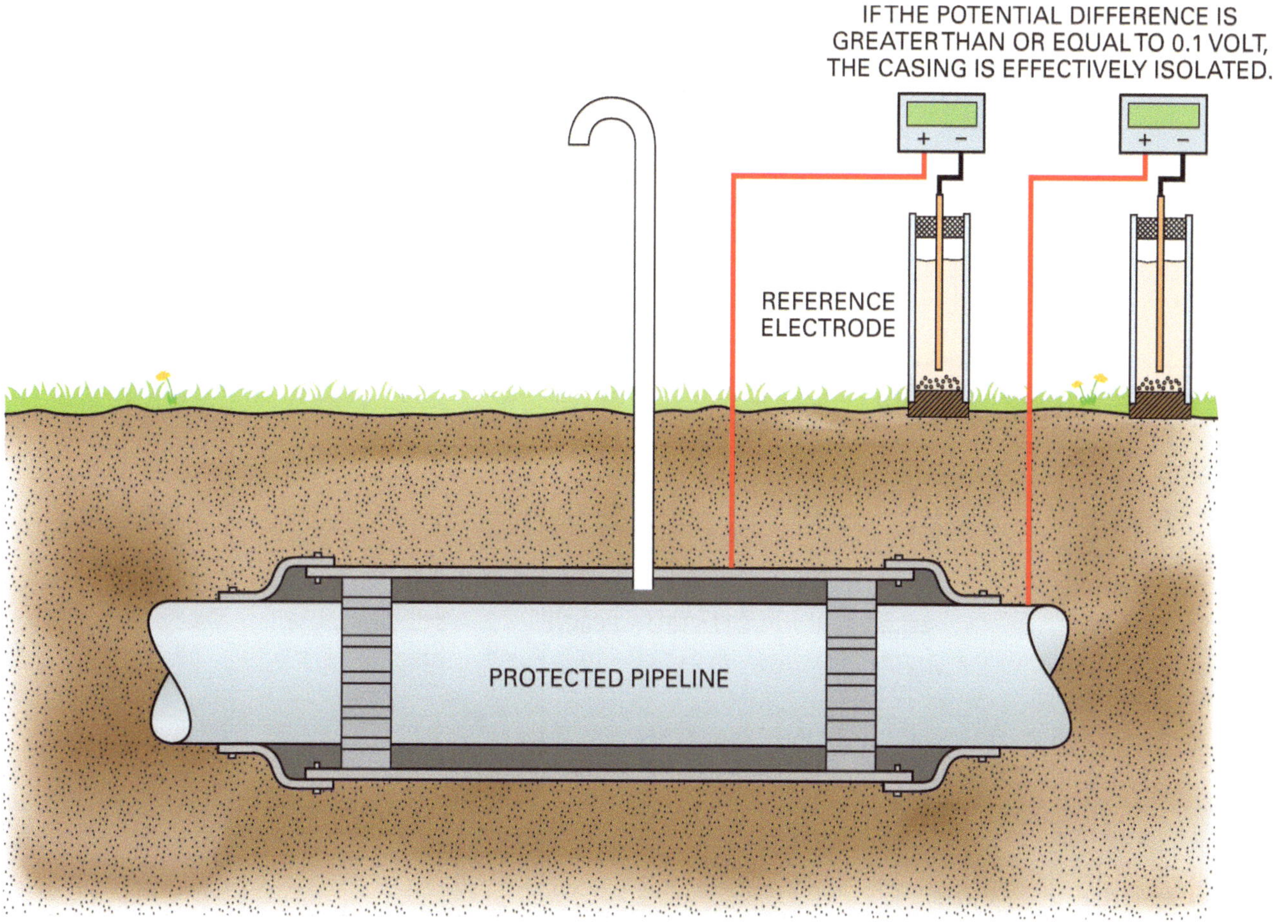

Figure 2 Potential shift test.

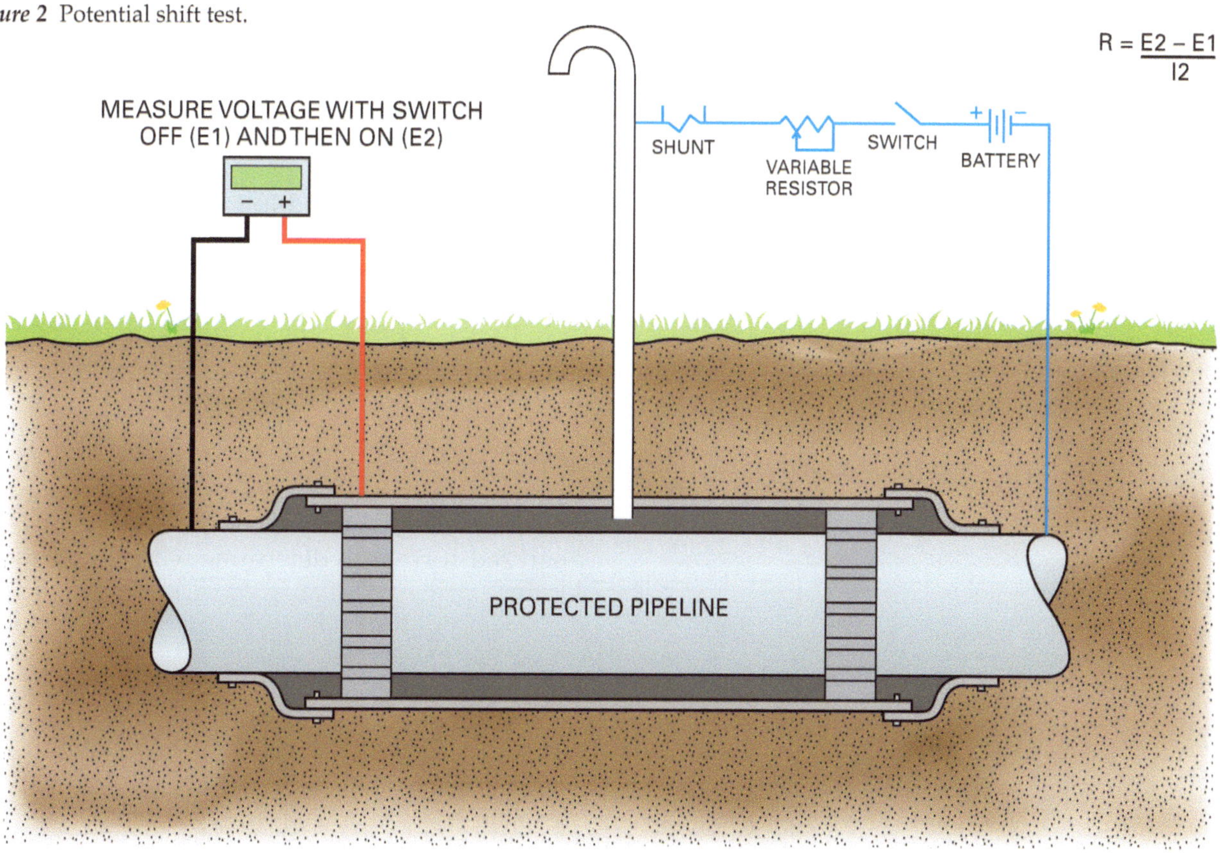

Figure 3 Internal resistance test.

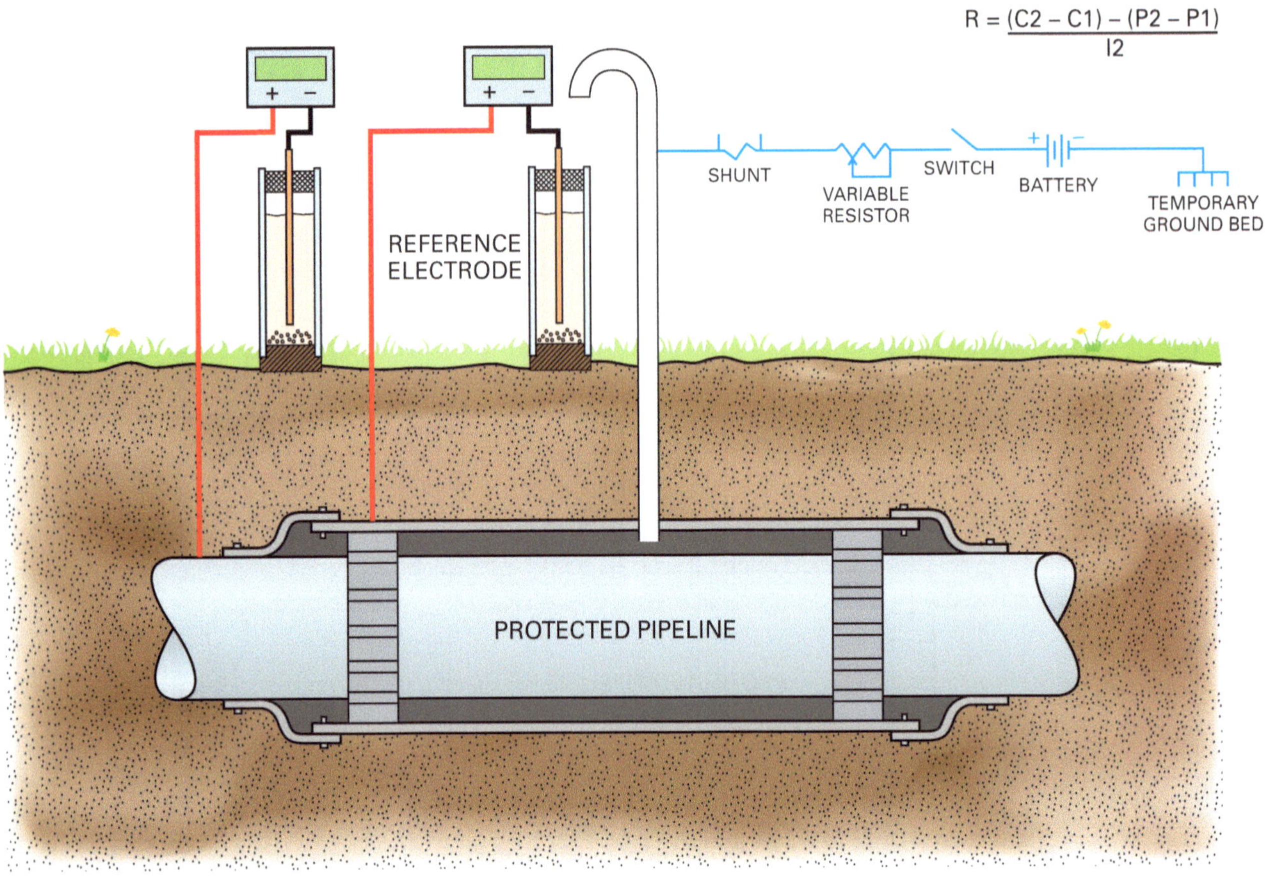

Figure 4 Casing depolarization test.

- *Casing depolarization test* – The casing **depolarization** test (*Figure 4*) is used to determine if a casing short exists. In this test, the positive terminal of a DC source is connected to the casing and the negative terminal is connected to a temporary groundbed. Incremental amounts of external DC current are applied to the casing and the potential shift between the casing and the carrier pipe is calculated and analyzed to obtain a minimum of three casing-to-pipe resistance calculations.
- *Four-wire IR drop test* – The four-wire IR drop test is used to indicate if a casing **metallic short** exists and the location of the short. The four-wire IR drop test is divided into four parts.

In Part 1 (*Figure 5*), the positive terminal of a DC source is connected to casing vent A and the negative terminal to casing vent B. Then the voltage drops at casing contacts A and B are measured and analyzed.

In Part 2 (*Figure 6*), the positive terminal of a DC source is connected to casing vent A and the negative terminal is connected to the carrier pipe. Then the voltage drops at the casing contacts are measured and recorded.

In Part 3 (*Figure 7*), the positive lead is moved to vent B and the voltage shift at the casing contacts is measured and recorded.

In Part 4, the data that was collected in each of the preceding tests is analyzed to determine whether the casing is shorted. If the casing is found to be shorted, then the test results are analyzed further to determine where the short is located.

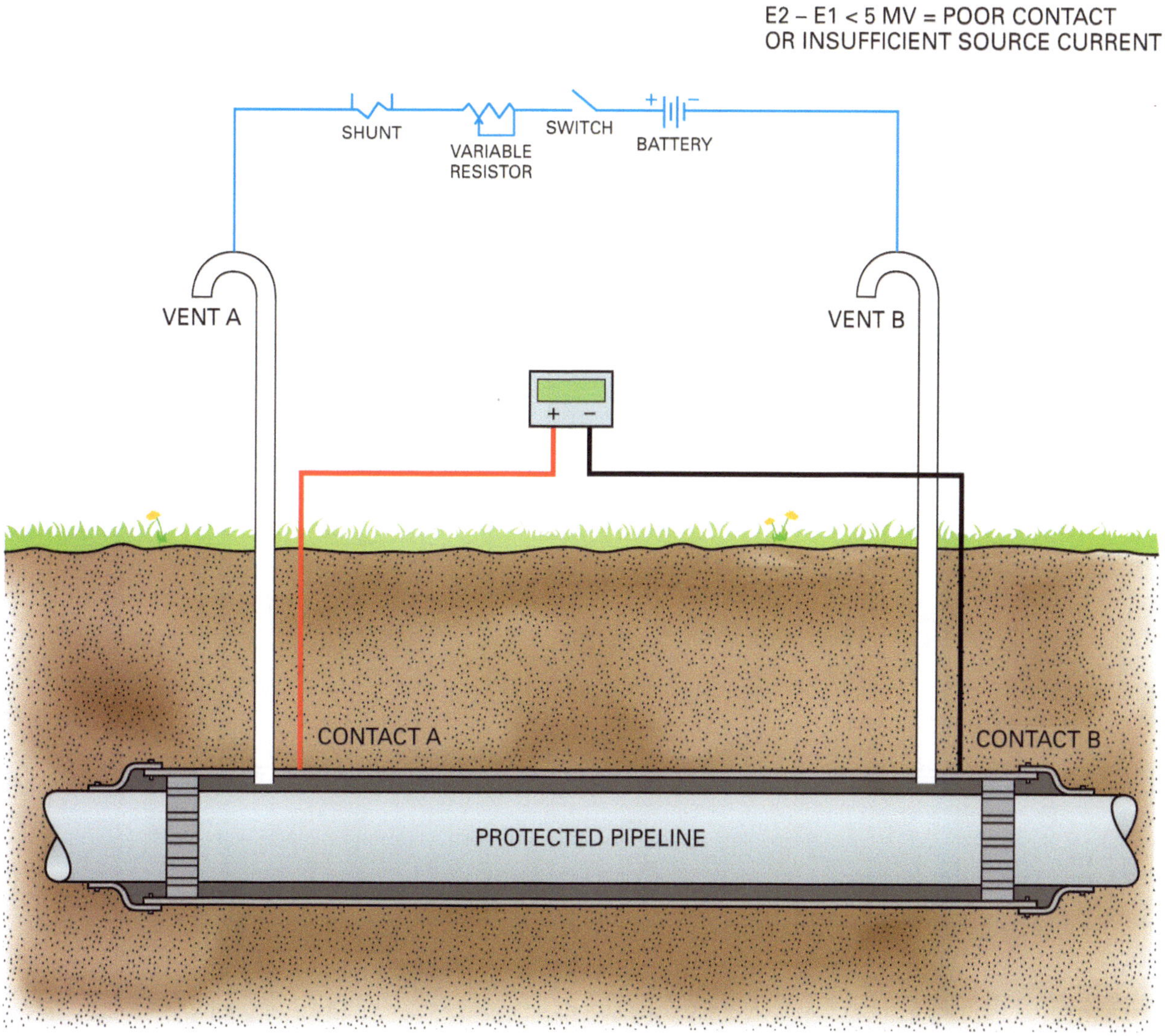

Figure 5 Four-wire IR drop test Part 1.

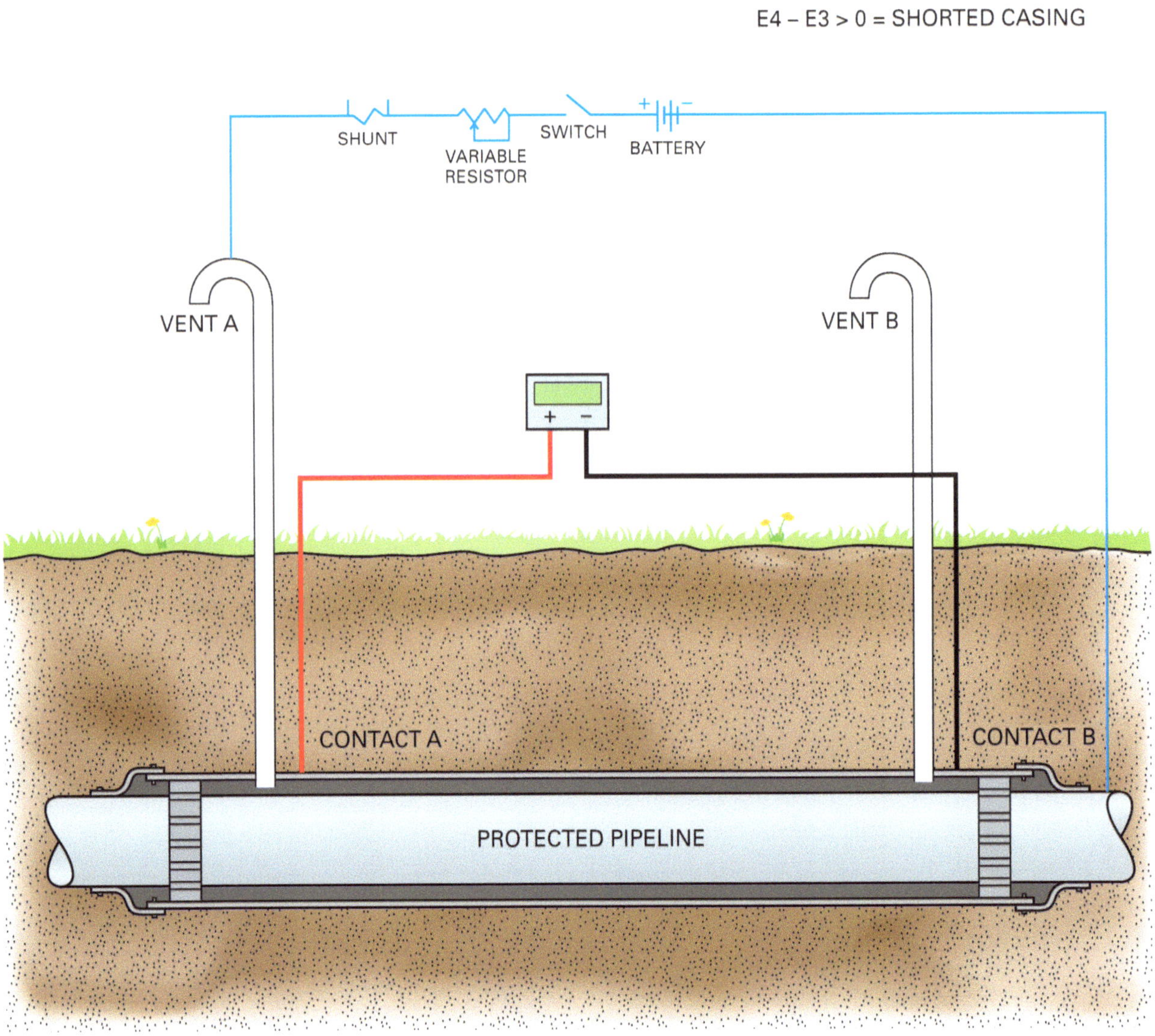

Figure 6 Four-wire IR drop test, Part 2.

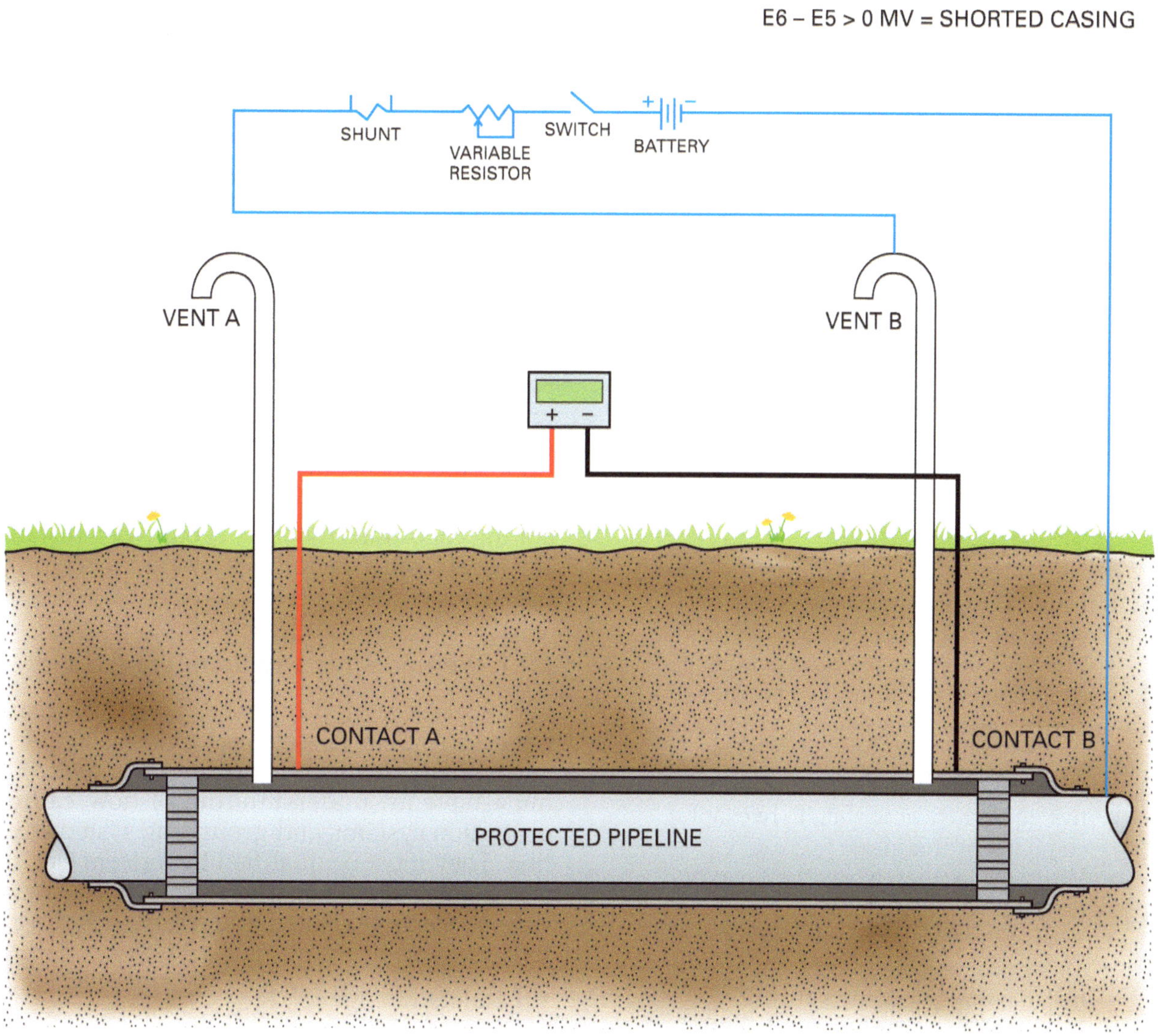

Figure 7 Four-wire IR drop test, Part 3.

Additional Resources

It is assumed that you will learn to perform these procedures under the direct supervision of your crew leader, foreman, or instructor. For independent study, refer to the recommended Additional Resources at the back of this module.

1.0.0 Section Review

3. One method of electrical insulation/isolation is _____.
   a. nitrogen purging
   b. anode depolarization
   c. low-dielectric groundbeds
   d. casing isolation/insulation

4. In the casing depolarization test, a(n) _____.
   a. AC current is applied to the carrier pipe
   b. DC current is applied to the carrier pipe
   c. DC current is applied to the casing
   d. AC current is applied to the casing

SECTION TWO

2.0.0 INSTALL ELECTRICAL INSULATING DEVICES

Objective

Describe how to install electrical insulating devices (CT9_6-17).

a. Identify the prerequisites, competencies, and abnormal operating conditions associated with installing electrical insulating devices.
b. Describe procedures for installing various electrical insulating devices.

Performance Task

- Install electrical insulating devices (CT9_6-17).
- Identify potential abnormal operating conditions that may occur, and know the appropriate actions to take in response to them.
- Utilize the appropriate personal protective equipment according to relevant company procedures.
- Verify the location where the isolation device is needed.
- Verify the appropriate kit materials.
- Verify proper flange alignment prior to installation of insulation kit.
- Install the flange kit and/or dielectric unions according to the manufacturer's specifications. Use precautions to prevent any moisture, soil, or other foreign matter from contacting any portion of the insulating joint prior to its being sealed.
- Verify isolation using the appropriate insulation tester.
- Complete appropriate documentation as required by operator's procedures.

Trade Terms

Abnormal operating conditions (AOCs): Conditions within a pipeline system, identified by the operator, that may indicate (1) a malfunction of a component or (2) a deviation from normal operations that may indicate a condition exceeding design limits or result in a hazard to persons, property, or the environment.

Polarization: A change from the open-circuit potential as a result of current across the electrode/electrolyte interface.

This section describes some of the basic skills required to install electrical insulating devices and identifies **abnormal operating conditions (AOCs)** that personnel may need to respond to when performing this task. It also describes step-by-step procedures for installing various devices that are used for electrical insulation/isolation.

2.1.0 Prerequisites, Competencies, and Abnormal Operating Conditions

Personnel must possess a variety of skills in order to install electrical insulating devices. The various devices require a series of steps, which personnel must be competent to perform. Personnel must also be aware of the particular abnormal operating conditions that may occur when installing electrical insulating devices and know the proper responses to them.

2.1.1 Prerequisites and Competencies

Personnel assigned to install electrical insulating devices must possess certain basic knowledge and skills prior to performing this task. They must have an understanding of how cathodic protection systems and grounding systems function. They must be qualified to perform electrical isolation testing. They must also be familiar with various electrical isolation devices, including the following:

- Insulated flange
- Dielectric union
- Monolithic insulated joint
- Casing isolators/spacers

2.1.2 Abnormal Operating Conditions

Before installing electrical insulating devices, personnel must be able to identify and properly respond to potential abnormal operating conditions (AOCs), should they occur. The following are some potential AOCs, and the appropriate action personnel should take in response to them:

- *Presence of corrosion, pitting, etc. is discovered* – Document as required, and notify appropriate personnel.
- *Pipeline is damaged (dents, gouges, scrapes, etc.)* – Notify appropriate personnel.
- *Unexpected hazardous liquid or carbon dioxide is encountered when preparing surfaces for coating* – Eliminate ignition source and notify appropriate personnel.

2.2.0 Procedures

Installing electrical insulating devices is a task that begins when the need for the device is determined and ends when electrical isolation has been verified and the required information has been properly documented. Use the following general procedure for installing insulating devices:

> **WARNING!**
> Always obtain and use the appropriate PPE for each task. Only qualified individuals may make electrical connections and operate equipment. Always verify the absence of voltage and current before making connections. Follow the manufacturer's instructions for the equipment being installed.

Step 1 Identify potential abnormal operating conditions that may occur, and know the appropriate actions to take in response to them.

Step 2 Utilize the appropriate personal protective equipment according to relevant company procedures.

Step 3 Identify the location where the insulation device will be installed. Ensure that it provides a suitable mounting area/enclosure and is compatible with the pipeline/equipment system.

> **WARNING!**
> Take precautions against arcing when installing in a combustible environment. Couplers can be used to prevent arcing.

Step 4 Verify the appropriate kit materials, such as insulating sleeves on the studs or bolts and insulating washers under the nuts and/or heads of the bolts. Mild steel washers are placed over the top of the insulating washers to prevent damage when tightened.

Step 5 Verify proper flange alignment prior to installation of the insulating kit. Use alignment pins in the flange to prevent damage or destruction of the isolating sleeves that can result from misalignment of the flange faces.

Step 6 Install the insulating device according to the manufacturer's specifications. Ensure that the mounting area and device are free of moisture, soil, and other contaminants. If moisture, soil, or other foreign matter contacts any portion of the insulating joint, the entire joint will need to be disassembled, cleaned with a suitable solvent, and dried prior to re-assembly.

Step 7 Verify isolation using the appropriate isolation tester. Note that a multimeter can give a false indication of isolation.

Step 8 Document the installation per the company/operator procedures.

The subsequent sections of this module present procedural steps for installing several specific means of electrical insulation/isolation.

2.2.1 Installing a Flange or Dielectric Union

The following steps outline a general procedure for installing an insulating flange or dielectric union:

> **WARNING!**
> Always obtain and use the appropriate PPE for each task. Only qualified individuals may make electrical connections and operate equipment. Always verify the absence of voltage and current before making connections. Follow the manufacturer's instructions for the equipment being installed.

Step 1 Identify potential abnormal operating conditions that may occur, and know the appropriate actions to take in response to them.

Step 2 Utilize the appropriate personal protective equipment according to relevant company procedures.

Step 3 Verify the location where the isolation device is needed.

Step 4 If using an insulating kit, verify that the kit materials (insulating sleeves, insulating washers, etc.) are appropriate for the given installation.

> **CAUTION**
> To prevent damage when nuts and/or heads of bolts are tightened, place mild steel washers over top of the insulating washers.

Step 5 Check for proper flange alignment before installing an installation kit. Misaligned flange faces during installation of insulating flanges can damage or destroy isolating sleeves.

NOTE
When installing insulating kits, use alignment pins in the flange to prevent misaligned flange faces.

Step 6 Install the flange kit and/or dielectric unions according to the manufacturer's specifications.

CAUTION
Take care to prevent moisture, soil, or other foreign matter from contacting any portion of the insulating joint before the joint is sealed. Otherwise, the entire joint will have to be disassembled, cleaned with a suitable solvent, and dried prior to re-assembly.

Step 7 Use the appropriate insulation tester to verify electrical isolation. Be aware that a digital multimeter can give a false indication of isolation.

Step 8 Record all required information as specified in the operator's procedures.

2.2.2 Installing an Isolation Joint

The following steps outline a general procedure for installing an isolation joint:

WARNING!
Always obtain and use the appropriate PPE for each task. Only qualified individuals may make electrical connections and operate equipment. Always verify the absence of voltage and current before making connections. Follow the manufacturer's instructions for the equipment being installed.

Step 1 Identify potential abnormal operating conditions that may occur, and know the appropriate actions to take in response to them.

Step 2 Utilize the appropriate personal protective equipment according to relevant company procedures.

Step 3 Verify the location where the isolation device is needed.

Step 4 Verify the type of isolation device required. Only certain types of isolation devices may be used where the isolation point location must be buried or the fluid inside the pipe is electrically conductive. Make sure the isolation joint is adequate for the maximum pressure and temperature conditions that will be encountered.

NOTE
It is recommended that isolation joints be assembled and then tested both hydrostatically and electrically before the joints are installed in a pipeline.

Step 5 Install bond wires/cables on both sides of the encapsulated insulating device. This mitigates interference and enables electrical continuity when necessary.

Step 6 Terminate the wires in a common terminal box at the approved location.

Step 7 Use the appropriate insulation tester to verify electrical isolation. Be aware that a digital multimeter can give a false indication of isolation.

Step 8 Record all required information as specified in the operator's procedures.

2.2.3 Installing Lightning and Ground Fault Protection

The following steps outline a general procedure for installing lightning and ground fault protection.

WARNING!
Always obtain and use the appropriate PPE for each task. Only qualified individuals may make electrical connections and operate equipment. Always verify the absence of voltage and current before making connections. Follow the manufacturer's instructions for the equipment being installed.

Step 1 Identify potential abnormal operating conditions that may occur, and know the appropriate actions to take in response to them.

Step 2 Utilize the appropriate personal protective equipment according to relevant company procedures.

Step 3 Verify the location where the isolation device is needed.

Step 4 Install all insulating devices with the appropriate fault and lightning protection in accordance with the manufacturer's design practice. The equipment typically consists of either a polarization cell replacement (PCR) or a solid state decoupling (SSD) device that is bonded sufficiently and oppositely on both sides of the insulating flange.

Step 5 Use the appropriate insulation tester to verify electrical isolation.

Step 6 Record all required information as specified in the operator's procedures.

2.2.4 Installing Electrical Isolation from Electrical Grounding

The following steps outline a general procedure for installing electrical isolation from electrical grounding:

> **WARNING!**
> Always obtain and use the appropriate PPE for each task. Only qualified individuals may make electrical connections and operate equipment. Always verify the absence of voltage and current before making connections. Follow the manufacturer's instructions for the equipment being installed.

Step 1 Identify potential abnormal operating conditions that may occur, and know the appropriate actions to take in response to them.

Step 2 Utilize the appropriate personal protective equipment according to relevant company procedures.

Step 3 Verify the location where the isolation device is needed.

Step 4 Install the required SSD or PCR to isolate the pipeline from electrical grounding systems.

> **NOTE**
> Wherever structures will be in bare metal contact to the soil electrical isolation must be provided through an approved electrical device (SSD or PCR) that is rated for the service in the given installation.

Step 5 Conduct testing on all utility (electrical, telephone, etc.) supply or feeder cables to identify direct current (DC) flow to the foreign structure.

Step 6 Use the appropriate insulation tester to verify electrical isolation.

Step 7 Record all required information as specified in the operator's procedures.

2.2.5 Installing Aboveground Piping Isolation from Other Buried Structures

The following steps outline a general procedure for installing aboveground piping isolation from other buried structures:

> **WARNING!**
> Always obtain and use the appropriate PPE for each task. Only qualified individuals may make electrical connections and operate equipment. Always verify the absence of voltage and current before making connections. Follow the manufacturer's instructions for the equipment being installed.

Step 1 Identify potential abnormal operating conditions that may occur, and know the appropriate actions to take in response to them.

Step 2 Utilize the appropriate personal protective equipment according to relevant company procedures.

Step 3 Verify the location where the isolation device is needed.

Step 4 Install the piping so that it is not in physical contact with any electrically conductive foreign structure. If this is impractical, cathodic protection must be included.

Step 5 Use non-metallic spacers or shields to align the appropriately sized electrical isolation device between metallic piping and metal pipe supports. This prevents dust and moisture from getting in between the insulation material/pipe interfaces.

Step 6 Install the non-metallic spacers between the insulation material and the pipe interface. This ensures that in addition to having dielectric properties, the insulation material can resist compression and abrasion.

Review Questions

1. Pipelines are protected against contact between the carrier pipe and the casing through the use of _____.
   a. spacers
   b. shims
   c. O-rings
   d. collars

2. Lightning and ground fault protection systems are methods of _____.
   a. cathodic alignment
   b. dielectric resistance
   c. structural depolarization
   d. electrical insulation/isolation

3. If potential shift test results suggest the casing is shorted, _____.
   a. further testing is required
   b. a decoupler should be installed
   c. the casing must be disassembled
   d. the end seals must be reconfigured

4. The test used to determine if a direct metal-to-metal contact exists between the casing and the carrier pipe is a(n) _____.
   e. IR drop test
   a. potential shift test
   b. internal resistance test
   c. depolarization test

5. If a pipeline technician discovers the presence of corrosion when installing an electrical insulating device, the technician should document the condition and _____.
   a. eliminate any ignition sources
   b. notify the appropriate personnel
   c. evacuate non-essential personnel
   d. limit access to the work area

6. To prevent damage when tightening nuts and/or heads of bolts during insulated flange installation, _____.
   a. insert alignment pins in the face of the flange
   b. place mild steel washers on top of the insulating washers
   c. use intrinsically safe, non-sparking hand tools
   d. disassemble, clean, and dry the joint prior to re-assembly

7. The equipment installed for lightning and ground fault protection typically consists of either a(n) _____.
   a. SSD or a GFCI
   b. galvanic anode or an IC anode
   c. DC switch or an AC switch
   d. PCR or an SSD

Step 3 Verify the location where the isolation device is needed.

Step 4 Install all insulating devices with the appropriate fault and lightning protection in accordance with the manufacturer's design practice. The equipment typically consists of either a polarization cell replacement (PCR) or a solid state decoupling (SSD) device that is bonded sufficiently and oppositely on both sides of the insulating flange.

Step 5 Use the appropriate insulation tester to verify electrical isolation.

Step 6 Record all required information as specified in the operator's procedures.

2.2.4 Installing Electrical Isolation from Electrical Grounding

The following steps outline a general procedure for installing electrical isolation from electrical grounding:

WARNING!

Always obtain and use the appropriate PPE for each task. Only qualified individuals may make electrical connections and operate equipment. Always verify the absence of voltage and current before making connections. Follow the manufacturer's instructions for the equipment being installed.

Step 1 Identify potential abnormal operating conditions that may occur, and know the appropriate actions to take in response to them.

Step 2 Utilize the appropriate personal protective equipment according to relevant company procedures.

Step 3 Verify the location where the isolation device is needed.

Step 4 Install the required SSD or PCR to isolate the pipeline from electrical grounding systems.

NOTE

Wherever structures will be in bare metal contact to the soil electrical isolation must be provided through an approved electrical device (SSD or PCR) that is rated for the service in the given installation.

Step 5 Conduct testing on all utility (electrical, telephone, etc.) supply or feeder cables to identify direct current (DC) flow to the foreign structure.

Step 6 Use the appropriate insulation tester to verify electrical isolation.

Step 7 Record all required information as specified in the operator's procedures.

2.2.5 Installing Aboveground Piping Isolation from Other Buried Structures

The following steps outline a general procedure for installing aboveground piping isolation from other buried structures:

WARNING!

Always obtain and use the appropriate PPE for each task. Only qualified individuals may make electrical connections and operate equipment. Always verify the absence of voltage and current before making connections. Follow the manufacturer's instructions for the equipment being installed.

Step 1 Identify potential abnormal operating conditions that may occur, and know the appropriate actions to take in response to them.

Step 2 Utilize the appropriate personal protective equipment according to relevant company procedures.

Step 3 Verify the location where the isolation device is needed.

Step 4 Install the piping so that it is not in physical contact with any electrically conductive foreign structure. If this is impractical, cathodic protection must be included.

Step 5 Use non-metallic spacers or shields to align the appropriately sized electrical isolation device between metallic piping and metal pipe supports. This prevents dust and moisture from getting in between the insulation material/pipe interfaces.

Step 6 Install the non-metallic spacers between the insulation material and the pipe interface. This ensures that in addition to having dielectric properties, the insulation material can resist compression and abrasion.

Step 7 Use the appropriate insulation tester to verify electrical isolation.

Step 8 Record all required information as specified in the operator's procedures.

2.2.6 Installing Casing Isolation/Insulation

The following steps outline a general procedure for installing casing isolation/insulation:

> **WARNING!**
>
> Always obtain and use the appropriate PPE for each task. Only qualified individuals may make electrical connections and operate equipment. Always verify the absence of voltage and current before making connections. Follow the manufacturer's instructions for the equipment being installed.

Step 1 Identify potential abnormal operating conditions that may occur, and know the appropriate actions to take in response to them.

Step 2 Utilize the appropriate personal protective equipment according to relevant company procedures.

Step 3 Verify the location where the isolation device is needed.

Step 4 Make sure that the annular space is clear of any debris and contaminants.

Step 5 Install the non-metallic spacers or shields and seal the ends according to the operator's procedure and the manufacturer's specification.

Step 6 Verify that the annular space is clear of any debris and contaminants and that the spacers are secured.

Step 7 Install the end seals to prevent water or debris from getting in.

> **NOTE**
>
> Always install the casing end seals according to the manufacturer's instructions and in the quantity recommended by the manufacturer or design engineer.

Step 8 Use the appropriate insulation tester to verify electrical isolation.

Step 9 Record all required information as specified in the operator's procedures.

Additional Resources

It is assumed that you will learn to perform these procedures under the direct supervision of your crew leader, foreman, or instructor. For independent study, refer to the recommended Additional Resources at the back of this module.

2.0.0 Section Review

1. Before installing electrical insulating devices, individuals must be qualified to perform _____.
   a. rectifier depolarization
   b. pit depth calculations
   c. electrical isolation testing
   d. pipeline resurfacing

2. Installing electrical insulating devices generally begins _____.
   a. verifying isolation with a multimeter
   b. when the need for the device is determined
   c. calculating the required annular space
   d. documenting the installation of the device

Summary

Electrical insulating devices are installed to separate the pipeline and equipment and ensure the proper operation of protective systems such as lightning, ground fault, and cathodic protection systems. Electrical isolation testing is performed to detect and locate electrical shorts that can occur within pipeline systems. The test results are analyzed to determine if and where shorts are located.

Although the exact procedures vary depending on the specific device, installing any electrical insulating device begins when the need for the device is determined and ends when isolation has been verified and the required information has been documented. Always use the appropriate PPE for each task and follow the manufacturer's instructions for the equipment or device that is being installed.

Review Questions

1. Pipelines are protected against contact between the carrier pipe and the casing through the use of _____.
   a. spacers
   b. shims
   c. O-rings
   d. collars

2. Lightning and ground fault protection systems are methods of _____.
   a. cathodic alignment
   b. dielectric resistance
   c. structural depolarization
   d. electrical insulation/isolation

3. If potential shift test results suggest the casing is shorted, _____.
   a. further testing is required
   b. a decoupler should be installed
   c. the casing must be disassembled
   d. the end seals must be reconfigured

4. The test used to determine if a direct metal-to-metal contact exists between the casing and the carrier pipe is a(n) _____.
   e. IR drop test
   a. potential shift test
   b. internal resistance test
   c. depolarization test

5. If a pipeline technician discovers the presence of corrosion when installing an electrical insulating device, the technician should document the condition and _____.
   a. eliminate any ignition sources
   b. notify the appropriate personnel
   c. evacuate non-essential personnel
   d. limit access to the work area

6. To prevent damage when tightening nuts and/or heads of bolts during insulated flange installation, _____.
   a. insert alignment pins in the face of the flange
   b. place mild steel washers on top of the insulating washers
   c. use intrinsically safe, non-sparking hand tools
   d. disassemble, clean, and dry the joint prior to re-assembly

7. The equipment installed for lightning and ground fault protection typically consists of either a(n) _____.
   a. SSD or a GFCI
   b. galvanic anode or an IC anode
   c. DC switch or an AC switch
   d. PCR or an SSD

Trade Terms Quiz

Fill in the blank with the correct term that you learned from your study of this module.

1. The pipe that is installed through the casing is the ________.
2. A sleeve over another pipe is a(n) ________.
3. The change from the open-circuit potential as a result of current across the electrode/electrolyte interface is called ________.
4. Indirect or direct metallic contact between two metallic structures is a(n) ________.
5. The voltage potential of the casing in the soil or electrolyte as related to a reference electrode is the ________.
6. An electrically nonconductive material is described as ________.
7. Conditions within a pipeline system, identified by the operator, that may indicate (1) a malfunction of a component or (2) a deviation from normal operations that may indicate a condition exceeding design limits or result in a hazard to persons, property, or the environment are called ________.
8. The potential difference between the surface of a buried or submerged metallic structure and the electrolyte that is measured with reference to an electrode that is in contact with the electrolyte is the ________.
9. The removal of factors resisting the current in an electrochemical cell is called ________.

Trade Terms

Abnormal operating conditions (AOCs)
Carrier pipe
Casing
Casing-to-soil potential
Depolarization
Dielectric
Metallic short
Pipe-to-soil potential
Polarization

Trade Terms Introduced in This Module

Abnormal operating conditions (AOCs): Conditions within a pipeline system, identified by the operator, that may indicate (1) a malfunction of a component or (2) a deviation from normal operations that may indicate a condition exceeding design limits or result in a hazard to persons, property, or the environment.

Carrier pipe: A pipe that carries the product. The carrier pipe is installed through the casing.

Casing: A sleeve installed under a roadway, railroad tracks, etc., by boring, which will allow another pipe to be slid through it.

Casing-to-soil potential: The voltage potential of the casing in the soil or electrolyte as related to a reference electrode.

Depolarization: The removal of factors resisting the current in an electrochemical cell.

Dielectric: An electrically nonconductive material.

Metallic short: Indirect or direct metallic contact between two metallic structures.

Pipe-to-soil potential: The potential difference between the surface of a buried or submerged metallic structure and the electrolyte that is measured with reference to an electrode in contact with the electrolyte; also referred to as structure-to-electrolyte potential or structure-to-soil potential.

Polarization: A change from the open-circuit potential as a result of current across the electrode/electrolyte interface.

Additional Resources

The following recommended resources can provide additional helpful information related to the requirements for this covered task:

API Recommended Practice 1161, Third Edition, © 2014 American Petroleum Institute. API Publishing Services, 1220 L Street NW, Washington, DC 20005. Distributed by Thompson Reuters (Scientific) LLC, **www.techstreet.com**.

49 *CFR* 192, *Subpart I*, **www.ecfr.gov**.

49 *CFR* 195, *Subpart H*, **www.ecfr.gov**.

NCCER Module AOCCC-17, *Abnormal Operating Conditions Control Center*.

NCCER Module AOCFG-17, *Abnormal Operating Conditions Field and Gas*.

NCCER Module 62105-14, *Introduction to Pipeline Documents*.

NCCER Module 64202-02, *Use of Meters and Test Equipment*.

NCCER Module 66101-14, *Introduction to the Pipeline Industry*.

NCCER Module CT1_1-17, *Measure Structure-to-Soil Potential*.

NCCER Module CT1_2-17, *Conduct Close Interval Survey*.

NCCER Module CT1_3-17, *Test and Detect Interference*.

NCCER Module CT1_4-17, *Inspect and Perform Electrical Test of Bonds*.

NCCER Module CT2_2-17, *Repair Damaged Test Leads*.

NCCER Module CT2_4-17, *Install Test Leads by Exothermic Welding Methods*.

NCCER Module CT5_3-17, *Inspect the Condition of External Coating on Buried or Submerged Pipe*.

NCCER Module CT9_5-17, *Repair Shorted Casings*.

NCCER Module CT14_1-17, *Locate Line*.

NCCER Module CT32_0-17, *Observation of Excavation Activities*.

NCCER Module CT39_0-17, *Backfilling a Trench Following Maintenance*.

Figure Credits

Tinker & Rasor, Module Opener

Section Review Answer Key

Answer	Section Reference	Objective
Section One		
1. d	1.1.0	1a
2. c	1.2.2	1b
Section Two		
1. c	2.11	2a
2. b	2.2.0	2b

NCCER CURRICULA — USER UPDATE

NCCER makes every effort to keep its textbooks up-to-date and free of technical errors. We appreciate your help in this process. If you find an error, a typographical mistake, or an inaccuracy in NCCER's curricula, please fill out this form (or a photocopy), or complete the online form at **www.nccer.org/olf**. Be sure to include the exact module ID number, page number, a detailed description, and your recommended correction. Your input will be brought to the attention of the Authoring Team. Thank you for your assistance.

Instructors – If you have an idea for improving this textbook, or have found that additional materials were necessary to teach this module effectively, please let us know so that we may present your suggestions to the Authoring Team.

NCCER Product Development and Revision

13614 Progress Blvd., Alachua, FL 32615

Email: curriculum@nccer.org
Online: www.nccer.org/olf

❑ Trainee Guide ❑ Lesson Plans ❑ Exam ❑ PowerPoints Other ________________________

Craft / Level: Copyright Date:

Module ID Number / Title:

Section Number(s):

Description:

Recommended Correction:

Your Name:

Address:

Email: Phone:

CT10_1-17

Insert and Remove Coupons

Overview

Pipeline corrosion is a problem that can occur anywhere there is an electrolyte in the pipeline. Controlling corrosion is a critical part of safe pipeline operations. Pipeline corrosion coupons are the most widely used corrosion monitoring devices. This module discusses factors that contribute to internal pipeline corrosion and describes procedures for inserting and removing corrosion coupons.

Trainees with successful module completions may be eligible for credentialing through the NCCER Registry. To learn more, go to **www.nccer.org** or contact us at 1.888.622.3720. Our website has information on the latest product releases and training, as well as online versions of our *Cornerstone* magazine and Pearson's product catalog.

Your feedback is welcome. You may email your comments to **curriculum@nccer.org**, send general comments and inquiries to **info@nccer.org**, or fill in the User Update form at the back of this module.

This information is general in nature and intended for training purposes only. Actual performance of activities described in this manual requires compliance with all applicable operating, service, maintenance, and safety procedures under the direction of qualified personnel. References in this manual to patented or proprietary devices do not constitute a recommendation of their use.

CT10_1-17
Insert and Remove Coupons

Objectives

When you have completed this module, you will be able to do the following:

1. Describe the use of coupons to monitor corrosion.
   a. Understand the factors affecting internal corrosion.
   b. Describe how coupons are used for corrosion monitoring.
2. Describe how to insert and remove corrosion coupons (CT10_1-17).
   a. Identify the prerequisites, competencies, and abnormal operating conditions associated with using corrosion coupons.
   b. Describe the procedure used to insert and remove corrosion coupons.

Performance Task

Under the supervision of the instructor, you should be able to do the following:

1. Insert and remove corrosion coupons (CT10_1-17).
   - Identify potential abnormal operating conditions that may occur during performance of this CT, and know the appropriate actions to take in response to them.
   - Utilize the appropriate personal protective equipment according to relevant company procedures.
   - Retract coupon according to manufacturer and company procedures. Perform this function slowly, monitoring for any leaks or change in pressure.
   - Close isolation valve. Follow company procedures to isolate coupon.
   - Package and label coupon. Document the coupon's visual characteristics. Confirm that the coupon is placed in its original shipping envelope.
   - Prepare new coupon. Properly record the new coupon's information. Thoroughly clean and prepare the coupon holder and ancillary equipment for coupon installation.
   - Install the new coupon on insulating assembly. Install coupon holder according to manufacturer and company procedures. Perform this function slowly to prevent damage to the coupon and plug assembly.
   - Open the isolation valve after the coupon holder assembly is properly installed on the valve.
   - Reinsert the coupon.
   - Perform the necessary notifications upon completion of the inspection.
   - Document inspection results according to company procedure.

Trade Terms

Abnormal operating conditions (AOCs)
Coupon
Mils per year (MPY)
Partial pressure
Scale
Water of condensation

Industry Recognized Credentials

If you are training through an NCCER-accredited sponsor, you may be eligible for credentials from NCCER's Registry. The ID number for this module is CT10_1-17. Note that this module may have been used in other NCCER curricula and may apply to other level completions. Contact NCCER's Registry at 888.622.3720 or go to **www.nccer.org** for more information.

Contents

Figures

SECTION ONE

1.0.0 CORROSION COUPONS

Objective

Describe the use of coupons to monitor corrosion.

a. Understand the factors affecting internal corrosion.
b. Describe how coupons are used for corrosion monitoring.

Trade Terms

Abnormal operating conditions (AOCs): A condition within a pipeline system, identified by the operator, that may indicate (1) a malfunction of a component or (2) a deviation from normal operations that may indicate a condition exceeding design limits or result in a hazard to persons, property, or the environment.

Coupon: A clean, pre-weighed metal sample inserted into a pipeline system to monitor the rate of corrosion.

Mils per year (MPY): The rate of corrosion measured in 1/1,000 of an inch/year.

Partial pressure: The pressure exerted by a single gas in a mixture of gases.

Scale: The formation of mineral deposits in a pipeline system.

Water of condensation: Condensation that forms when water is present in the product and the temperature of the exterior of the pipeline drops significantly below the product temperature.

Corrosion in a pipeline system can lead to expensive maintenance, product damage, environmental problems, and safety issues. Corrosion can occur anywhere there is an electrolyte (such as water) in the pipeline. Water in a pipeline system can lead to both internal corrosion and bacterial growth. Corrosion can also occur under deposits, in heat-affected zones around welds, and at locations where there are significant changes in temperature or pressure.

Mineral deposits of scale in a pipeline system can lead to operational problems, including decreased flow, loss of heat transfer capacity, and corrosion. Analysis of the pipeline contents can help identify possible causes of corrosion and scale so they can be controlled before widespread damage occurs.

1.1.0 Factors Affecting Internal Corrosion

Internal corrosion can be caused and affected by a variety of factors. This section discuss each of these influencing factors in detail. These factors include the following:

- Acid gases
- pH
- Chlorides
- Temperature
- Water content
- Partial pressure
- Velocity
- Bacterial growth

1.1.1 Acid Gases

The major acid gases are carbon dioxide, hydrogen sulfide, and oxygen. These gases react with water to form acids. Carbon dioxide (CO_2) combines with water to form carbonic acid. This acid causes a type of localized corrosion referred to as Mesa pitting (*Figure 1*).

Hydrogen sulfide (H_2S) combines with water to form a weak acid with a characteristic rotten-egg smell. Hydrogen sulfide corrosion is usually identified by the etching in the bottom of the pits. An example of etching in pits is shown in *Figure 2*.

Oxygen (O_2) is a strong depolarizer and can greatly accelerate the action of other corrosive agents. Corrosion caused by oxygen is usually found on surface equipment where air has an entry point. Oxygen corrosion typically forms deep pits (*Figure 3*).

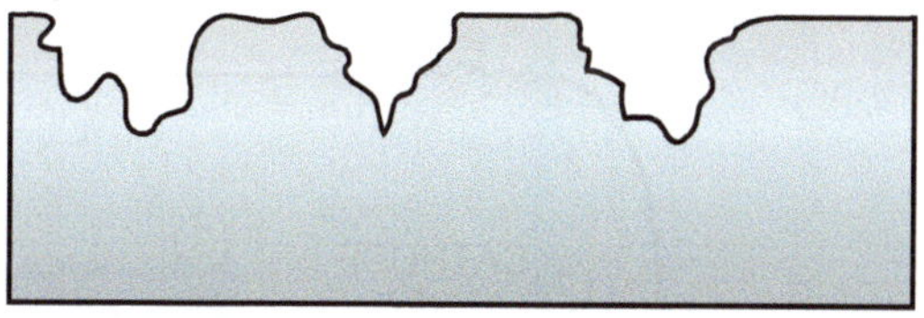

Figure 1 Mesa pitting.

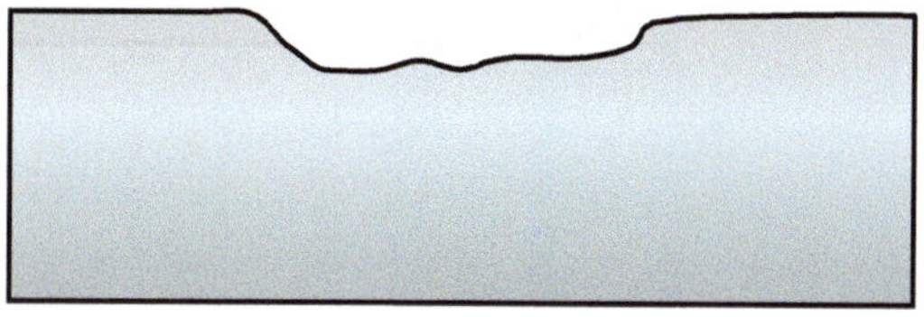

Figure 2 Etching in pits.

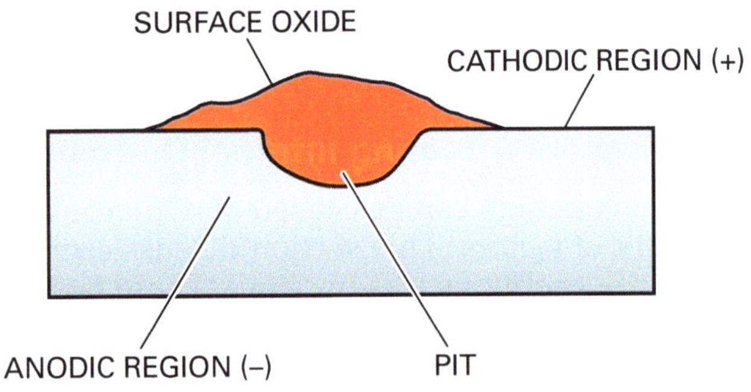

Figure 3 Deep pits created by oxygen corrosion.

1.1.2 pH Level

The normal range for pH is from 0 to 12. A pH of 0 is very acidic, a pH of 7 is neutral, and a pH of 12 is very basic. Refer to the pH scale in *Figure 4*.

The pH of water indicates the tendency of the water to be corrosive or to form scale. The lower the pH reading, the more acidic the water, and the greater its tendency to be corrosive. The higher the pH value of the water, the more likely it is to form scale.

1.1.3 Water Content

The water content of a liquid or gas sample is the amount of water available in the sample. Water of condensation forms when the temperature of the exterior of the pipeline drops below the temperature of the product. This water serves as the electrolyte needed to complete a corrosion cell. A typical corrosion cell is shown in *Figure 5*.

The following are the four elements in a corrosion cell:

- *Anode* – The electrode of a cell where oxidation occurs. Electrons flow away from the anode.
- *Cathode* – The cell electrode at which reduction occurs. Electrons flow toward the cathode.
- *Metallic path* – A conductor (such as metal piping) that electrically connects the anode with the cathode.
- *Electrolyte* – An electrically conductive fluid (such as water) in which the anode and cathode must be immersed.

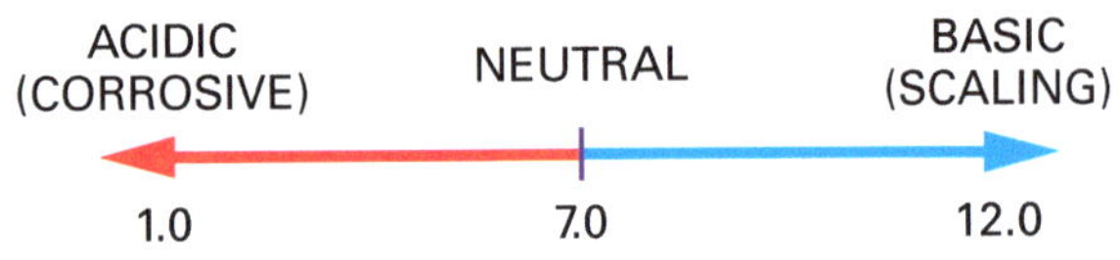

Figure 4 pH scale.

1.1.4 Temperature

Corrosion is accelerated in warm environments. The pipeline temperature may contribute to internal corrosion when it exceeds approximately 110°F (43°C).

1.1.5 Partial Pressure

Partial pressure is the pressure exerted by a single gas in a mixture of gases. The percentage of each gas in the total volume is the same as the percentage of each partial pressure. For example, if there are 100 parts of CO_2 in a total of 1,000 parts of gas, the partial pressure is $^{100}/_{1,000} = {}^{1}/_{10}$ or 10 percent. High partial pressures can result in increased levels of corrosion.

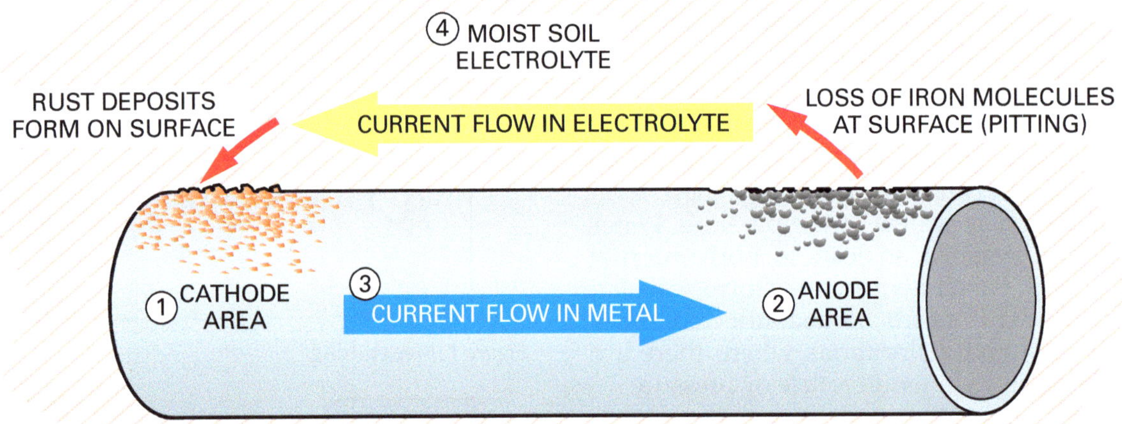

Figure 5 Corrosion cell.

1.1.6 Velocity

Velocity is the speed of fluids or gases as they move through a pipeline. Erosion occurs as these fluids or gases flow over the surface and wear away minute amounts of material. Liquids and gases passing through a pipeline at high velocities can cause turbulence and increase corrosion rates by stripping the protective layer of inhibitor from the metal surface, exposing bare metal. The movement of a fluid through a pipeline must be monitored and adjusted to prevent high velocities and turbulence.

1.2.0 Using Coupons for Corrosion Monitoring

A pipeline corrosion coupon is the most widely used corrosion monitoring device. Coupons are made of steel that is calibrated at a specific weight and designed to represent the interior wall of the pipe. Coupons are inserted into the system, exposed to the fluids or gas in the pipeline for a specific period of time, and then removed, cleaned, and weighed. The change in overall mass is recorded as weight loss in milligrams and the coupons are visually inspected for obvious signs of corrosion. The weight loss during the exposure time is extrapolated into weight loss for one year and normally expressed in mils per year (MPY). One mil equals $^{1}/_{1000}$ inch (0.001 inch).

Corrosion coupon documentation typically includes the following:

- Coupon number
- Coupon location
- Coupon initial weight
- Date installed
- Date removed
- Visual observations

Always follow company policies and procedures when removing and replacing corrosion coupons. A corrosion coupon mailing pouch is shown in *Figure 6*.

1.2.1 Weight Loss and Scale Coupons

There are two main types of coupons, weight loss coupons and scale coupons. Weight loss coupons are used to measure metal loss due to corrosion, while scale coupons are used to measure weight gain due to scale deposits.

Weight loss coupons can be made in many shapes and sizes, including rods, flat strips, or discs (*Figure 7*). Weight loss coupons are usually made of a single material (normally steel), although specialty coupons may include other materials. The test results from weight loss coupons are reported as weight loss over time.

Scale coupons, also called *deposition coupons*, are designed to create a pressure differential between the upstream and downstream sides of the coupon. This pressure differential encourages the

Figure 6 Corrosion coupon mailing pouch.

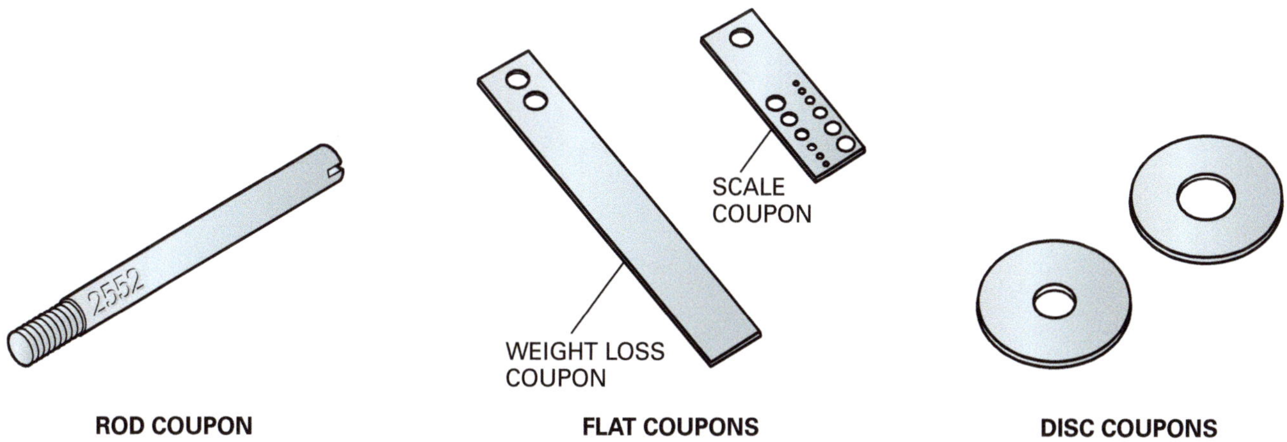

Figure 7 Coupon types.

formation of scale. Scale coupons contain multiple holes of increasing diameter. Scale will form fastest in the smallest hole. The extent of scaling can be determined by the largest hole with accumulated scale. The test results from scale coupons are reported as weight gain over time. The deposits are also analyzed in order to select the appropriate chemical treatment or other remediation.

1.2.2 Coupon Mounting Devices

Coupons are held in place by a variety of mounting devices which fall into two categories: fixed and retractable. Fixed mounting devices (*Figure 8*) typically can only be removed when the system is de-pressurized. A retractable device can be removed and replaced while the pipeline remains in service. All mounting devices include a nonconductive insulator to electrically isolate the coupon from the mounting device.

It is important to orient flat coupons properly when they are installed in the path of product flow. Flat coupons should be fixed so product flows over their narrow profile. Because of their shape, the placement of rod and disc coupons is not as critical. *Figure 9* shows correct and incorrect placement of a coupon within a pipe.

ROD COUPONS

FLAT COUPONS

Figure 8 Fixed coupon mounting devices.

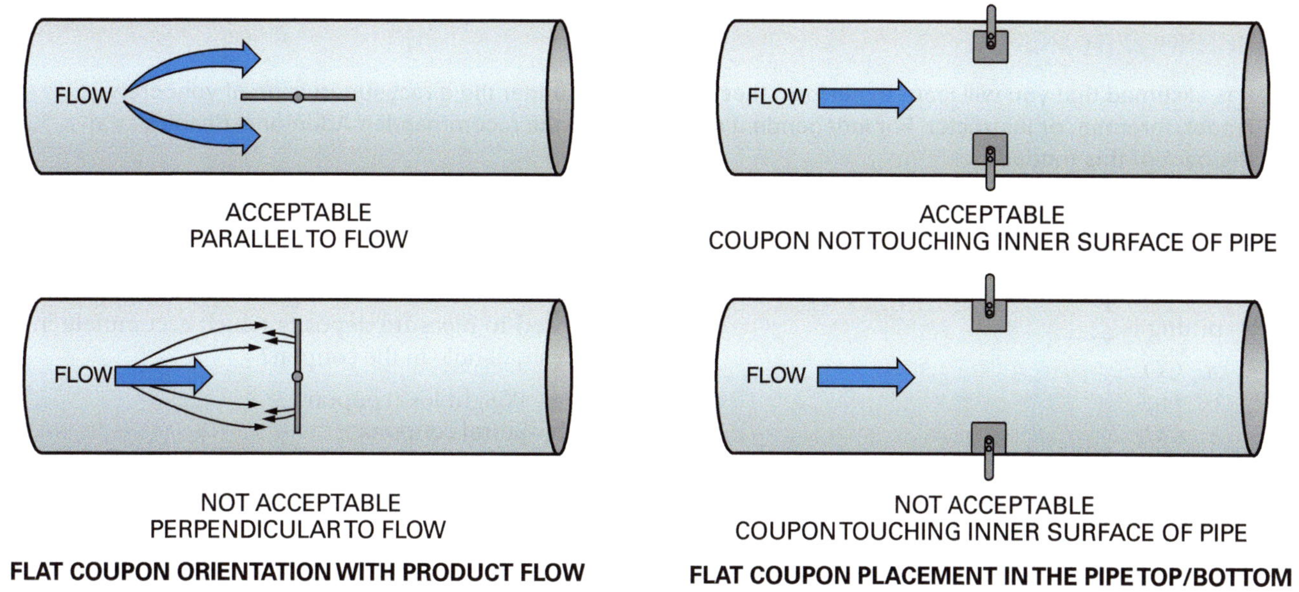

Figure 9 Coupon orientation.

Sidestream Assemblies

Sidestream assemblies are modular systems that can be used to perform a variety of system measurements. The system shown here can be customized to provide measurements of various system values such as metal loss, corrosion rate, erosion rate, pressure, temperature, flow, and bacterial activity.

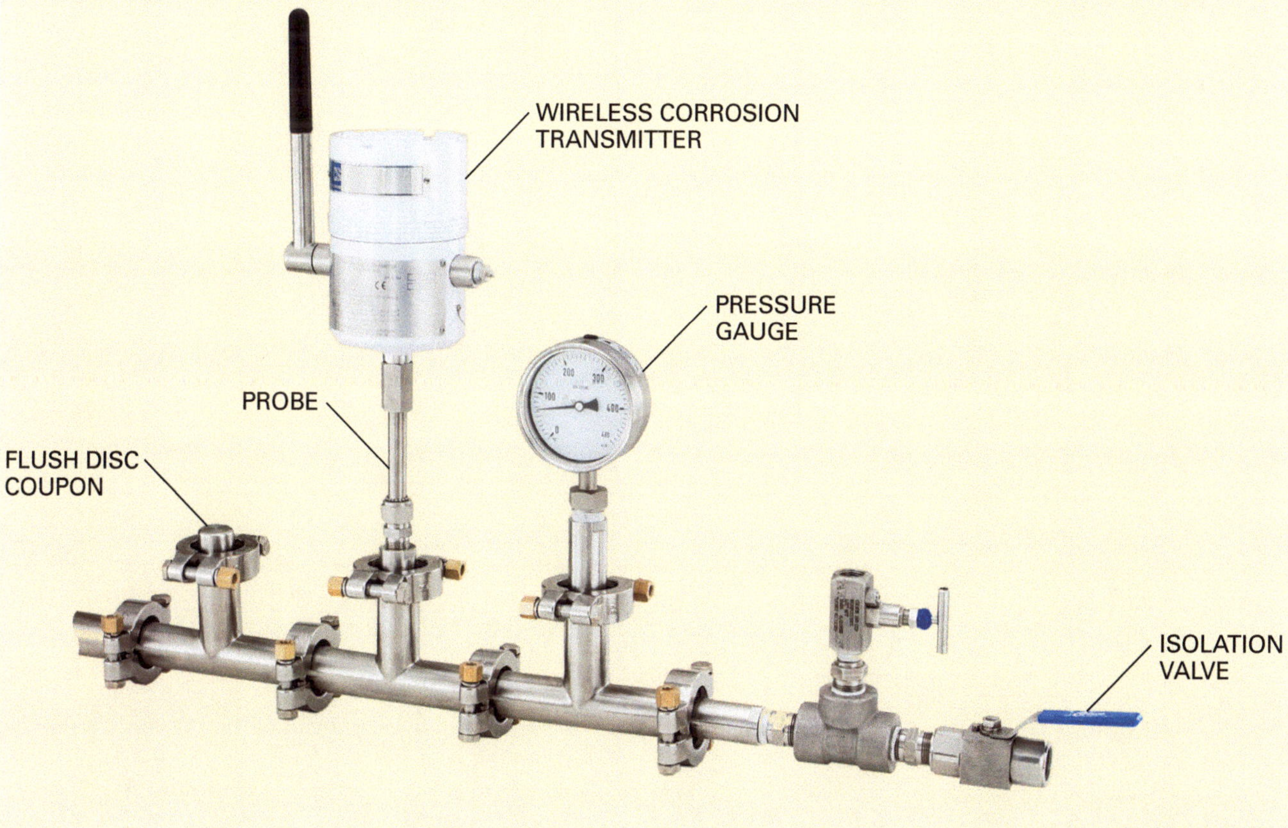

Additional Resources

It is assumed that you will learn to perform these procedures under the direct supervision of your crew leader, foreman, or instructor. For independent study, refer to the recommended Additional Resources at the back of this module.

1.0.0 Section Review

1. The acid gas that is most likely to result in deep pitting is _____.
   a. CO
   b. H_2S
   c. CO_2
   d. O_2

2. Which of the following coupons is typically used to measure deposits which accumulate in a small hole in the coupon?
   a. Weight loss coupons
   b. Spiral coupons
   c. Disc coupons
   d. Scale coupons

SECTION TWO

2.0.0 INSERTING AND REMOVING CORROSION COUPONS

Objective

Describe how to insert and remove corrosion coupons (CT10_1-17).

a. Identify the prerequisites, competencies, and abnormal operating conditions associated with using corrosion coupons.
b. Describe the procedure used to insert and remove corrosion coupons.

Performance Task

1. Insert and remove corrosion coupons (CT10_1-17).
   - Identify potential abnormal operating conditions that may occur during performance of this CT, and know the appropriate actions to take in response to them.
   - Utilize the appropriate personal protective equipment according to relevant company procedures.
   - Retract coupon according to manufacturer and company procedures. Perform this function slowly, monitoring for any leaks or change in pressure.
   - Close isolation valve. Follow company procedures to isolate coupon.
   - Package and label coupon. Document the coupon's visual characteristics. Confirm that the coupon is placed in its original shipping envelope.
   - Prepare new coupon. Properly record the new coupon's information. Thoroughly clean and prepare the coupon holder and ancillary equipment for coupon installation.
   - Install the new coupon on insulating assembly. Install coupon holder according to manufacturer and company procedures. Perform this function slowly to prevent damage to the coupon and plug assembly.
   - Open the isolation valve after the coupon holder assembly is properly installed on the valve.
   - Reinsert the coupon.
   - Perform the necessary notifications upon completion of the inspection.
   - Document inspection results according to company procedure.

Trade Terms

Abnormal operating conditions (AOCs): A condition within a pipeline system, identified by the operator, that may indicate (1) a malfunction of a component or (2) a deviation from normal operations that may indicate a condition exceeding design limits or result in a hazard to persons, property, or the environment.

Corrosion coupons are to be inserted and removed following company and manufacturer specifications. Pipeline technicians are required to have some basic competencies in order to accurately insert and remove corrosion coupons. This section presents the general requirements and procedures for inserting and removing coupons.

2.1.0 Prerequisites, Competencies, and Abnormal Operating Conditions

Personnel must possess a variety of skills in order to insert and remove coupons. The various operations require a series of steps, which a pipeline technician must be competent to perform. Personnel must also be aware of the particular abnormal operating conditions (AOCs) that may occur during the procedure and know the proper responses to them.

2.1.1 Prerequisites and Competencies

Personnel assigned to insert and remove coupons must understand that the task begins when the isolation valve has been closed in preparation for removal, and ends when the old coupon has been submitted for testing and a new coupon has been installed.

Personnel must understand the electrochemical reactions that cause corrosion, and be able to recognize pitting, crevice, cavitation, and impingement types of localized corrosion occurring on the pipe or pipeline component surfaces. They must also be familiar with the design and function of coupons and coupon holders.

Personnel must know how to isolate the valve, safely extract a coupon using a retrieval tool, correctly insert a new one, and document and submit the vital information. They must also be familiar with the company safety policies, understand the correct energy isolation procedures, and be able to select and use the proper PPE.

2.1.2 Abnormal Operating Conditions

Personnel must be familiar with the AOCs that may be encountered during the procedure and know the proper response should one occur. Potential AOCs may include the following:

- *Isolation/service valve malfunctions* – Stop the procedure and make the proper notifications.
- *Damaged or malfunctioning coupon holder assembly causing a leak or preventing the proper insertion or retraction of the coupon is discovered* – Stop the procedure and make the proper notifications.
- *Severe corrosion on the coupon and/or coupon holder is discovered* – Notify the supervisor immediately.
- *Mechanical damage to the threads or sealing surfaces of the coupon holder is discovered* – Notify the supervisor immediately.
- *An unexpected leak of hazardous liquid or gas from the valve or coupon holder is discovered* – Isolate the leak and ventilate or clean the area if it is safe to do so. Proper notifications must be made.
- *Fire or explosion occurs due to the ignition of hydrocarbons* – Isolate and extinguish the fire if it is safe to do so. If necessary, make the proper emergency notifications, including calling 911.

2.2.0 Procedure for Inserting and Removing Coupons

The coupon insertion and retrieval procedures vary by coupon type, holder type, and manufacturer. The instructions contained in this section are intended to serve as general guidelines only. Always refer to the instructions provided by the manufacturer for the coupon and holder in use. Shows a typical retractable coupon holder is shown in *Figure 10*.

WARNING!

Only trained, qualified individuals may perform pipeline monitoring procedures. Wear all appropriate personal protective equipment (PPE) and stay alert to the possibility of unexpected leaks. Safely eliminate all ignition sources and immediately notify the appropriate personnel in the event of a leak.

CAUTION

Oils and acids from hands can affect the weight loss of the coupon. Do not touch the coupon with bare hands. Protect the coupon against contact with any contaminants.

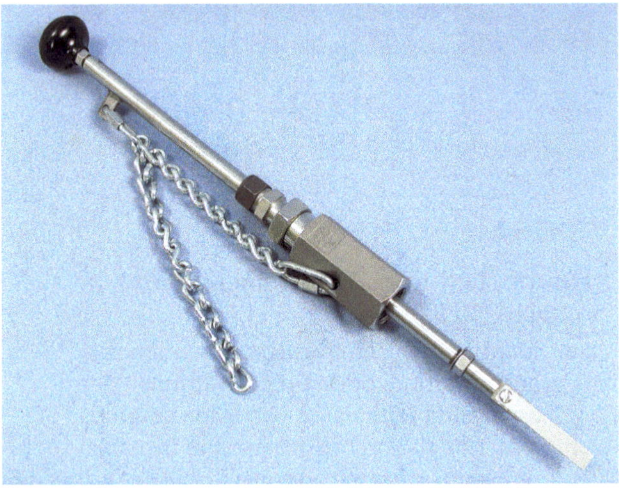

Figure 10 Retractable coupon holder.

The following steps outline the general procedure for inserting and removing coupons:

Step 1 Identify potential abnormal operating conditions (AOCs) that may occur, and know the appropriate actions to take in response to them.

Step 2 Utilize the appropriate personal protective equipment according to relevant company procedures.

Step 3 Verify that the isolation valve is closed and secure, in order to isolate the coupon and plug assembly from the pipeline contents.

Step 4 Remove the fitting cap slowly and carefully (if so equipped). This provides access to the coupon. Follow the manufacturer and company procedure for this action.

Step 5 Remove the coupon/holder assembly and conduct a visual inspection per manufacturer and company procedures. Retrieval procedures vary by holder assembly type and manufacturer.

CAUTION

Use caution when handling coupons. Touching the coupon with bare hands or allowing it to come into contact with external contaminants could affect the weight loss measure of the coupon.

Step 6 Remove the coupon from the holder and conduct a visual inspection per manufacturer and company procedures. If obvious corrosion is present, make the required notifications. A corroded coupon is shown in *Figure 11*.

Figure 11 Corroded coupon.

Step 7 Place the removed coupon directly in its original protective packaging. Record the visual characteristics, date, and remover's name.

Step 8 Properly document the new coupon's serial number or identification, associated protective packaging, date of installation, location of installation, and installer's name.

NOTE

Save the packaging from the new coupon for future use.

Step 9 Clean and inspect the coupon holder and any ancillary equipment. Follow manufacturer and company procedures to avoid contamination.

Step 10 Replace the coupon in the coupon holder per manufacturer and company procedures. Installation procedures vary by coupon and holder type and manufacturer. Make sure the coupon is secure and electrically isolated from the holder.

Step 11 Reinstall the coupon holder in the pipeline per manufacturer and company procedures. Perform this function slowly to prevent damage to the coupon and/or plug assembly. Installation procedures vary by holder type and manufacturer.

Step 12 Replace the fitting cap.

Step 13 Open the isolation valve. Only individuals who are qualified to open the isolation valve may do so. If the coupon installer is not qualified to open the valve, then someone who is qualified must be notified.

Step 14 After the isolation valve is opened, check for leaks.

Step 15 Submit the removed coupon to the appropriate facility for analysis. Properly store the protective packaging from the new coupon for later use.

Step 16 Complete appropriate documentation as required by operator's procedures.

Additional Resources

It is assumed that you will learn to perform these procedures under the direct supervision of your crew leader, foreman, or instructor. For independent study, refer to the recommended Additional Resources at the back of this module.

2.0.0 Section Review

1. Localized corrosion may include pitting, crevice, impingement, and _____.
   a. exfoliation
   b. cavitation
   c. gouging
   d. cracking

2. The first step in coupon removal is to _____.
   a. remove the fitting cap
   b. remove the fitting/holder assembly
   c. verify that the coupon is isolated
   d. inspect the fitting/holder assembly

SUMMARY

Controlling internal corrosion is an integral part of safe pipeline operations. Some of the factors that may contribute to internal corrosion include acid gases, pH, chlorides, temperature, water content, partial pressure, velocity, and bacterial growth. All of these variables must be controlled to the extent possible in order to minimize corrosion.

Pipeline corrosion coupons are the most widely used corrosion monitoring devices. Coupons are made of steel calibrated at a specific weight and designed to represent the interior wall of the pipe. Coupons are inserted into the system, exposed to the fluids or gas in the pipeline for a specific period of time, and then removed, cleaned, and weighed. The change in overall mass is recorded and the coupons are inspected for signs of corrosion.

Review Questions

1. The type of corrosion that is usually identifiable by etching in the bottom of the pits is called _____.
   a. general corrosion
   b. erosion corrosion
   c. localized corrosion
   d. hydrogen sulfide corrosion

2. The type of acid gas most likely to form Mesa pitting is _____.
   a. carbon monoxide
   b. oxygen
   a. carbon dioxide
   b. hydrogen sulfide

3. Of the following values, the pH value that is most acidic is _____.
   a. 3
   b. 6
   c. 9
   d. 12

4. The ambient temperature increases the degree of internal corrosion when it exceeds _____.
   a. 80°F (27°C)
   b. 90°F (32°C)
   c. 100°F (38°C)
   d. 110°F (43°C)

5. All corrosion coupon mounting devices are _____.
   a. electrically isolated from the coupon
   b. fixed in place
   c. mounted on the top half of the pipe
   d. retractable to prevent pig damage

6. The main difference between fixed and retractable coupon holders is that _____.
   a. fixed coupon holders do not require an insulating device
   b. retractable coupon holders can only be used on scale coupons
   c. fixed coupon holders cannot be replaced while the pipe remains in service
   d. retractable devices can only be removed with the system de-pressurized

7. When inserting a retractable coupon holder, refer to the detailed procedure contained in _____.
   a. 49*CFR*195
   b. DOT standards for corrosion testing
   c. the manufacturer's instructions
   d. company safety procedures

Trade Terms Quiz

Fill in the blank with the correct term that you learned from your study of this module.

1. A clean, pre-weighed metal sample inserted into a pipeline system to monitor the rate of corrosion is called a(n) ________.
2. The formation of mineral deposits in a pipeline system is called ________.
3. Conditions within a pipeline system that may indicate (1) a malfunction of a component or (2) a deviation from normal operations that may indicate a condition exceeding design limits or result in a hazard to persons, property, or the environment are called ________.
4. The pressure exerted by a single gas in a mixture of gases is called the ________.
5. Condensation that forms when water is present in the product and the temperature of the exterior of the pipeline drops significantly below the product temperature is called ________.
6. The rate of corrosion measured in 1/1,000 of an inch/year is called ________.

Trade Terms

Abnormal operating conditions (AOCs)
Coupon
Mils per year (MPY)
Partial pressure
Scale
Water of condensation

Trade Terms Introduced in This Module

Abnormal operating conditions (AOCs): A condition within a pipeline system, identified by the operator, that may indicate (1) a malfunction of a component or (2) a deviation from normal operations that may indicate a condition exceeding design limits or result in a hazard to persons, property, or the environment.

Coupon: A clean, pre-weighed metal sample inserted into a pipeline system to monitor the rate of corrosion.

Mils per year (MPY): The rate of corrosion measured in 1/1,000 of an inch/year.

Partial pressure: The pressure exerted by a single gas in a mixture of gases.

Scale: The formation of mineral deposits in a pipeline system.

Water of condensation: Condensation that forms when water is present in the product and the temperature of the exterior of the pipeline drops significantly below the product temperature.

Additional Resources

The following recommended resources can provide additional helpful information related to the requirements for this covered task:

API Recommended Practice 1161, Third Edition, © 2014 American Petroleum Institute. API Publishing Services, 1220 L Street, NW, Washington, DC 20005. Distributed by Thompson Reuters (Scientific) LLC, **www.techstreet.com**.

49 *CFR* 192, Subpart I, **www.ecfr.gov**

49 *CFR* 195, Subpart H, **www.ecfr.gov**

MNL20-2ND, Corrosion Tests and Standards: Application and Interpretation. 2005. Available from ASTM International at **www.astm.org**

Oil and Gas Pipelines: Integrity and Safety Handbook. R. Winston Revie. 2015. Hoboken, NJ: John Wiley and Sons, Inc.

NCCER Module CT8_1-17, *Measure Pit Depth with Pit Gauge*

NCCER Module CT8_2-17, *Measure Wall Thickness with Handheld Ultrasonic Meter*

NCCER Module CT8_3-17, *Measure Corroded Area*

NCCER Module CT10_2-17, *Monitor Probes(Online)*

Figure Credits

Technical ToolBoxes, Inc., Module Opener

Courtesy of Metal Samples Company, Figures 6, 8, 10, 11

Rohrback Cosasco Systems, Inc., SA01

Section Review Answer Key

Answer	Section Reference	Objective
Section One		
1. d	1.1.1	1a
2. d	1.2.1	1b
Section Two		
1. b	2.1.1	2a
2. c	2.2.0	2b

NCCER CURRICULA — USER UPDATE

NCCER makes every effort to keep its textbooks up-to-date and free of technical errors. We appreciate your help in this process. If you find an error, a typographical mistake, or an inaccuracy in NCCER's curricula, please fill out this form (or a photocopy), or complete the online form at **www.nccer.org/olf**. Be sure to include the exact module ID number, page number, a detailed description, and your recommended correction. Your input will be brought to the attention of the Authoring Team. Thank you for your assistance.

Instructors – If you have an idea for improving this textbook, or have found that additional materials were necessary to teach this module effectively, please let us know so that we may present your suggestions to the Authoring Team.

NCCER Product Development and Revision

13614 Progress Blvd., Alachua, FL 32615

Email: curriculum@nccer.org

Online: www.nccer.org/olf

❑ Trainee Guide ❑ Lesson Plans ❑ Exam ❑ PowerPoints Other ____________

Craft / Level: Copyright Date:

Module ID Number / Title:

Section Number(s):

Description:

Recommended Correction:

Your Name:

Address:

Email: Phone:

CT10_2-17

Monitor Probes (Online)

Overview

Controlling internal corrosion is an integral part of safe pipeline operations. Probes are sensing elements that extend into gas or liquid streams to measure electrical potential and other factors that influence corrosion. Corrosion detection probes must be monitored on a regular basis to ensure that the probe data measurements are functioning properly. This module identifies probes used to monitor conditions in a pipeline system, and describes a typical general procedure for verifying that a probe is functioning correctly.

Trainees with successful module completions may be eligible for credentialing through the NCCER Registry. To learn more, go to **www.nccer.org** or contact us at 1.888.622.3720. Our website has information on the latest product releases and training, as well as online versions of our *Cornerstone* magazine and Pearson's product catalog.

Your feedback is welcome. You may email your comments to **curriculum@nccer.org**, send general comments and inquiries to **info@nccer.org**, or fill in the User Update form at the back of this module.

This information is general in nature and intended for training purposes only. Actual performance of activities described in this manual requires compliance with all applicable operating, service, maintenance, and safety procedures under the direction of qualified personnel. References in this manual to patented or proprietary devices do not constitute a recommendation of their use.

CT10_2-17

Monitor Probes (Online)

Objectives

When you have completed this module, you will be able to do the following:

1. Describe the function of online probes.
   a. Understand the factors affecting internal corrosion.
   b. Identify various probes and their applications.
2. Describe how to read a probe (CT10_2-17).
   a. Identify the prerequisites, competencies, and abnormal operating conditions associated with reading probes.
   b. Describe the procedure used to read a probe.

Performance Task

Under the supervision of the instructor, you should be able to do the following:

1. Monitor online probes (CT10_2-17).
   - Identify potential abnormal operating conditions that may occur during performance of this CT, and know the appropriate actions to take in response to them.
   - Utilize the appropriate personal protective equipment according to relevant company procedures.
   - Locate the probe site and remove any secondary containment covers to gain access to the probe.
   - Confirm the probe terminals are acceptable for use.
   - Connect the data cords from the data logger to the appropriate terminal of the probe.
   - Turn the data logger on and obtain the reading (follow the manufacturer's operating procedures).
   - Document the reading.
   - Disconnect the data logger leads from the probe.
   - Dress the secondary container cover with anti-seize compound and place the cap back on probe adapter.
   - Confirm data logger readings are appropriate (the data logger is functioning properly).
   - Complete appropriate documentation as required by operator's procedures.

Trade Terms

Abnormal operating conditions (AOCs)
Microbiologically influenced corrosion (MIC)
Partial pressure
Planktonic bacteria
Probe element
Resistance
Scale
Sessile bacteria
Sulfate-reducing bacteria (SRB)
Water of condensation

Industry Recognized Credentials

If you are training through an NCCER-accredited sponsor, you may be eligible for credentials from NCCER's Registry. The ID number for this module is CT10_2-17. Note that this module may have been used in other NCCER curricula and may apply to other level completions. Contact NCCER's Registry at 888.622.3720 or go to **www.nccer.org** for more information.

Contents

Figures

Section One

1.0.0 Online Probes

Objective

Describe the function of online probes.

a. Understand the factors affecting internal corrosion.
b. Identify various probes and their applications.

Trade Terms

Microbiologically influenced corrosion (MIC): Electrochemical corrosion accelerated by the metabolic activities of certain microorganisms, such as sulfate-reducing bacteria (SRB).

Partial pressure: The pressure exerted by a single gas in a mixture of gases.

Planktonic bacteria: Bacteria carried by the process fluid in a pipeline system.

Probe element: A device used to monitor corrosion through the changes in electrical resistance of the element material in an electrolyte.

Resistance: In this context, the ability of a gasket to withstand the effects of chemicals and other substances without damage or change.

Scale: The formation of mineral deposits in a pipeline system.

Sessile bacteria: Bacteria that attach to the pipeline walls and may result in microbiologically influenced corrosion (MIC).

Sulfate-reducing bacteria (SRB): A group of bacteria commonly associated with microbiologically influenced corrosion (MIC) in petroleum pipelines. SRBs reduce sulfates to sulfides, releasing sulfuric acid and hydrogen sulfide.

Water of condensation: Condensation that forms when water is present in the product and the temperature of the exterior of the pipeline drops significantly below the product temperature.

Conditions in the product flow stream that may create, or contribute to, a corrosive environment can be monitored by the use of probes which are continuously exposed to the fluid. These probes are sensing elements that are placed in gas or liquid pipelines to measure various system values, such oxygen level, corrosion rate, or bacteria level. The data generated by the probes can be accessed online.

1.1.0 Factors Affecting Internal Corrosion

Internal corrosion can be caused and affected by a variety of factors. This section discuss each of these influencing factors in detail. These factors include the following:

- Acid gases
- pH
- Chlorides
- Temperature
- Water content
- Partial pressure
- Velocity
- Bacterial growth

1.1.1 Acid Gases

The major acid gases are carbon dioxide, hydrogen sulfide, and oxygen. These gases react with water to form acids. Carbon dioxide (CO_2) combines with water to form carbonic acid. This acid causes a type of localized corrosion referred to as Mesa pitting (*Figure 1*).

Hydrogen sulfide (H_2S) combines with water to form a weak acid with a characteristic rotten-egg smell. Hydrogen sulfide corrosion is usually identified by the etching in the bottom of the pits. An example of etching in pits is shown in *Figure 2*.

Oxygen (O_2) is a strong depolarizer and can greatly accelerate the action of other corrosive agents. Corrosion caused by oxygen is usually found on surface equipment where air has an entry point. Oxygen corrosion typically forms deep pits (*Figure 3*).

1.1.2 pH Level

The normal range for pH is from 0 to 12. A pH of 0 is very acidic, a pH of 7 is neutral, and a pH of 12 is very basic. Refer to the pH scale in *Figure 4*.

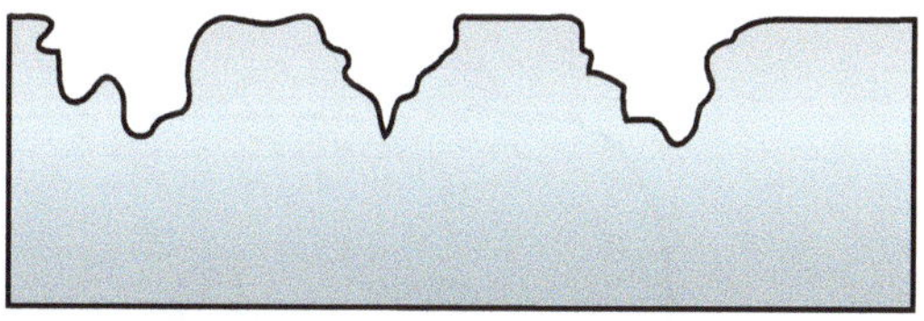

Figure 1 Mesa pitting.

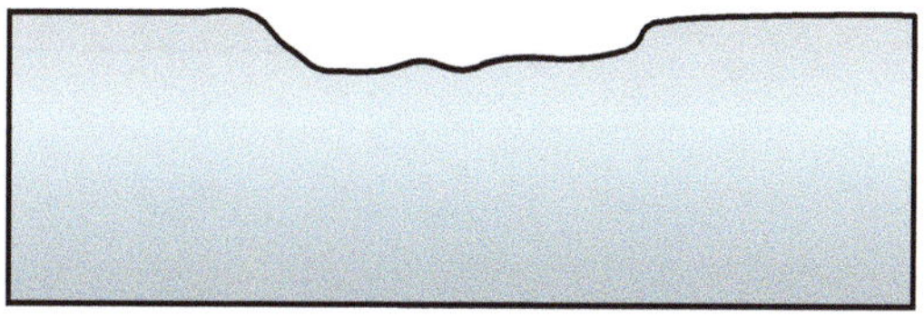

Figure 2 Etching in pits.

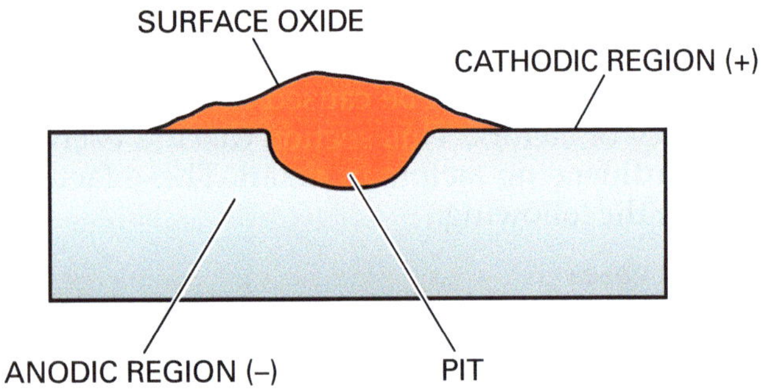

Figure 3 Deep pits created by oxygen corrosion.

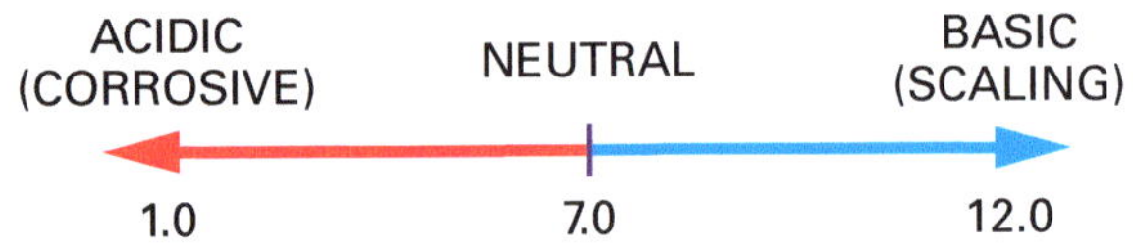

Figure 4 pH scale.

The pH of water indicates the tendency of the water to be corrosive or to form scale. The lower the pH reading, the more acidic the water, and the greater its tendency to be corrosive. The higher the pH value of the water, the more likely it is to form scale.

1.1.3 Water Content

The water content of a liquid or gas sample is the amount of water available in the sample. Water of condensation forms when the temperature of the exterior of the pipeline drops below the temperature of the product. This water serves as the electrolyte needed to complete a corrosion cell. A typical corrosion cell is shown in *Figure 5*.

The following are the four elements in a corrosion cell:

- *Anode* – The electrode of a cell where oxidation occurs; electrons flow away from the anode.
- *Cathode* – The cell electrode at which reduction occurs; electrons flow toward the cathode.
- *Metallic path* – A conductor (such as metal piping) that electrically connects the anode with the cathode.
- *Electrolyte* – An electrically conductive fluid (such as water) in which the anode and cathode must be immersed.

1.1.4 Temperature

The pipeline temperature becomes a factor for internal corrosion when it exceeds approximately 110°F (43°C). Corrosion is accelerated in warm environments.

1.1.5 Partial Pressure

Partial pressure is the pressure exerted by a single gas in a mixture of gases. The percentage of each gas in the total volume is the same as the percentage of each partial pressure. For example, if there are 100 parts of CO_2 in a total of 1,000 parts of gas, the partial pressure is $^{100}/_{1,000} = {}^{1}/_{10}$ or 10 percent. High partial pressures can result in increased levels of corrosion.

1.1.6 Velocity

Velocity is the speed of fluids or gases as they move through a pipeline. Erosion occurs as these

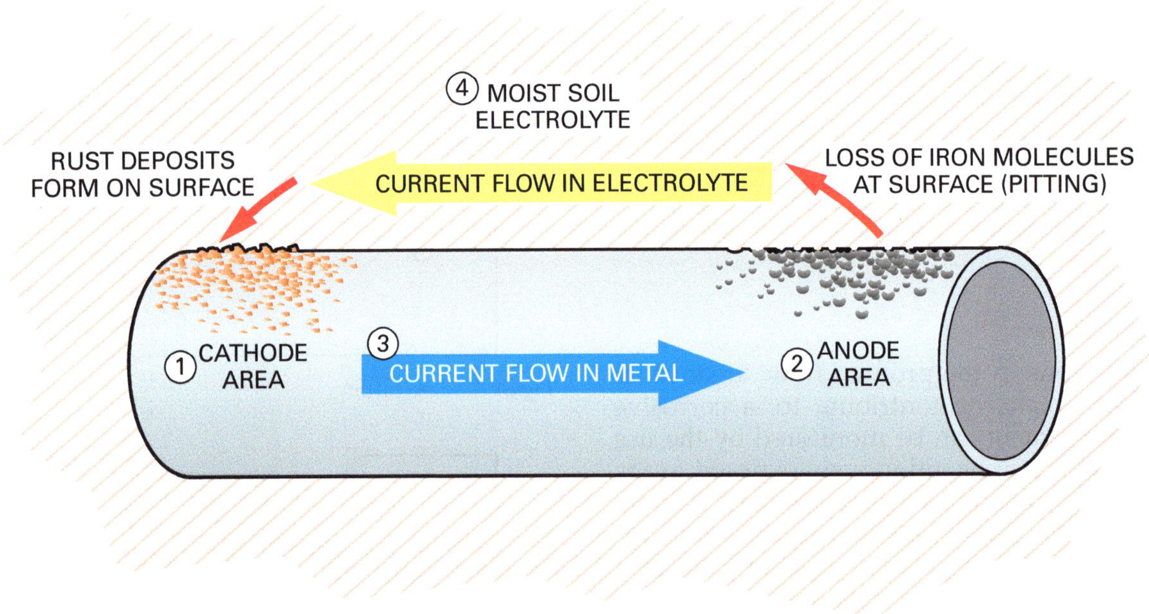

Figure 5 Corrosion cell.

fluids or gases flow over the surface and wear away minute amounts of material. Liquids and gases passing through a pipeline at high velocities can cause turbulence and increase corrosion rates by stripping the protective layer of inhibitor from the metal surface, exposing bare metal. The movement of a fluid through a pipeline must be monitored and adjusted to prevent high velocities and turbulence.

1.2.0 Types of Probes and Their Applications

Probes are available in a variety of configurations, and can be fixed or retractable. Common types of probes include electrical **resistance**, linear polarization resistance, galvanic, hydrogen, and biological.

1.2.1 Electrical Resistance (ER) Probes

Electrical resistance probes work as automatic coupons. They function on the principle that as the diameter of a conductor decreases, the resistance to current flow increases. ER probes pass a current through an element exposed to the fluid or gas product in a pipeline. The element can be manufactured from almost any conductive metal. As the element corrodes, there is a corresponding increase in resistance to current flow. This change in resistance is measured and recorded over a period of time. Unlike other types of probes, ER probes do not have to be immersed in an electrolyte to function.

The surface area of the exposed element is calculated at the time of insertion into the product stream. As the change in resistance occurs, the change in surface area (metal loss) over time can be determined. This metal loss is then extrapolated into weight loss in mils per year. These probes are available in either fixed or retractable styles.

Figure 6 shows a typical ER probe, remote data logger, and handheld reader. The reader shown can be used to retrieve stored data from remote data loggers and can also collect readings directly from the probes. The data is uploaded to a personal computer (PC) for use with the manufacturer's corrosion software for graphing and analysis.

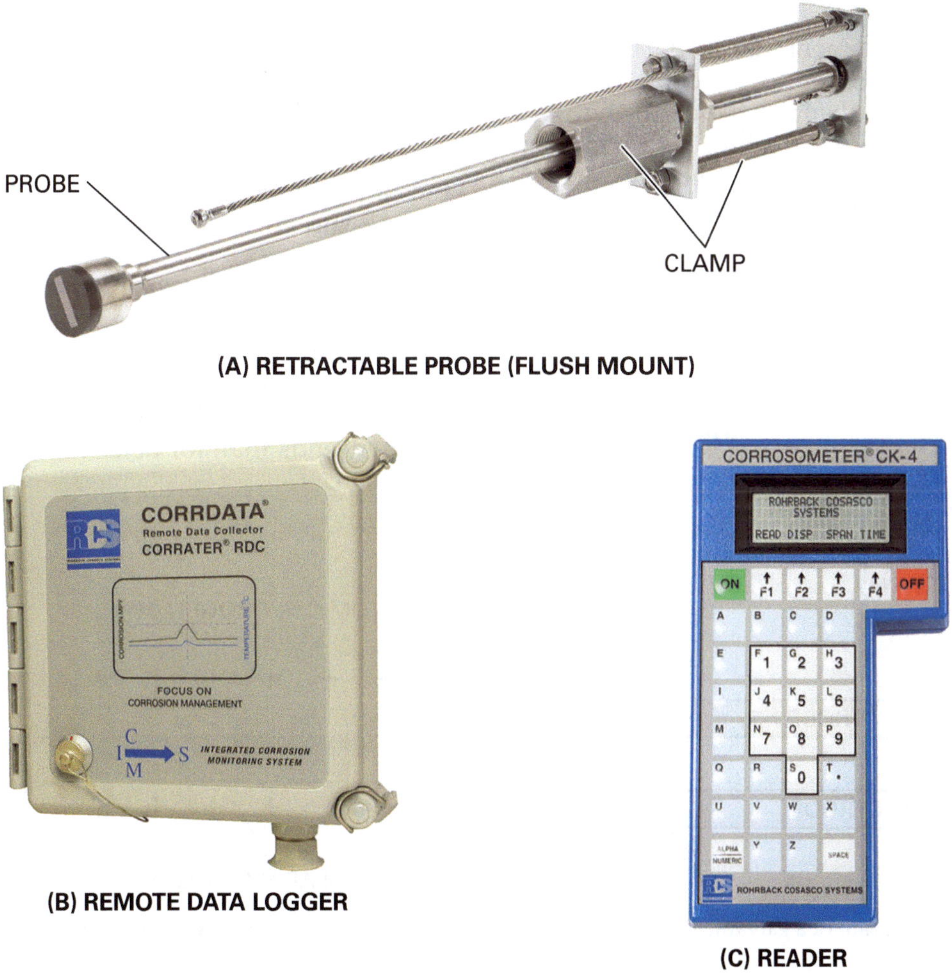

Figure 6 Retractable ER probe, data logger, and reader.

1.2.2 Linear Polarization Resistance (LPR) Probes

Linear polarization resistance probes are capable of providing instantaneous corrosion rate information. LPR probes function on the principle of polarization decay. These probes are available in two- and three-probe configurations. LPR probes must be immersed in an electrolyte to function properly. They work best in a very clean electrolyte.

LPR probes require either a handheld reader or a combination transmitter and data logger to provide readings. A fixed LPR probe is shown in *Figure 7*.

1.2.3 Galvanic Probes

A galvanic probe functions using the inherent differences in potential between two dissimilar metals. A **probe element** is commonly made of brass and steel. If there is an electrolyte between the two metal elements, the potential difference between the two metals causes a corrosion current to flow. This current flow is measured using a meter or data logger connected to the probe. A schematic diagram of a galvanic probe is shown in *Figure 8*.

Galvanic probe elements must be in contact with electrolyte in order to function properly. Any change in the makeup of the electrolyte will result in a change in the reading obtained from the probes. Galvanic probes (*Figure 9*) may be used to detect oxygen infiltration into systems and to predict corrosion inhibitor film life.

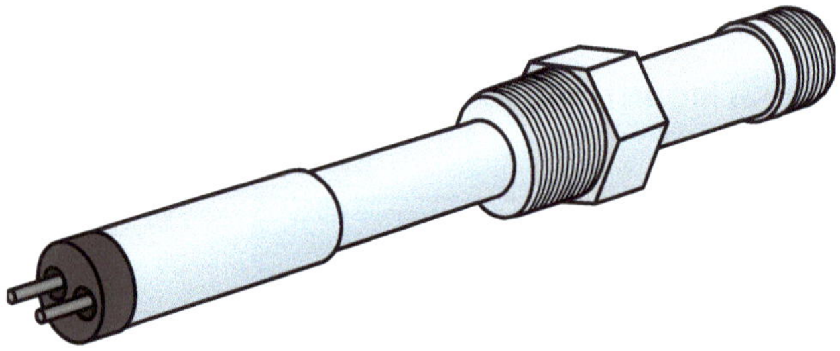

Figure 7 Fixed LPR probe.

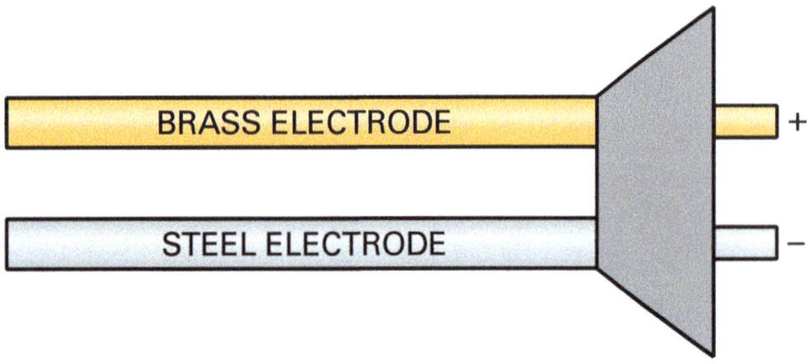

Figure 8 Schematic diagram of a galvanic probe.

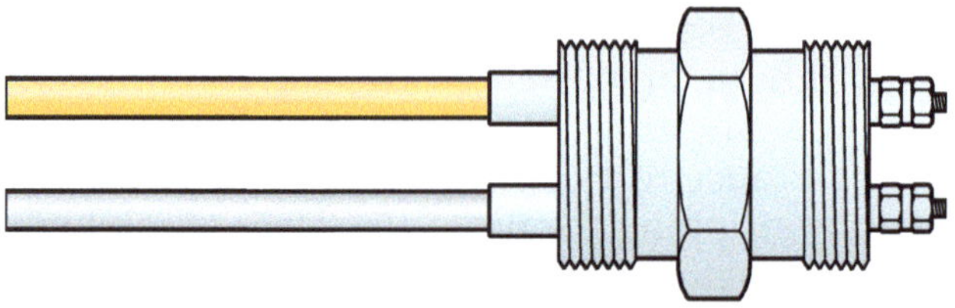

Figure 9 Galvanic probe.

1.2.4 Hydrogen Probes

Hydrogen probes function by reacting to the hydrogen film formed by the corrosion process. As the hydrogen film is formed, some of the atomic hydrogen enters the probe element and combines with other atomic hydrogen atoms to form molecular hydrogen. A hydrogen probe does not need to be immersed in electrolyte to function properly. A schematic diagram of a hydrogen probe is shown in *Figure 10*.

When two hydrogen atoms combine to form molecular hydrogen, the molecule is too large to migrate back through the probe element. As additional hydrogen molecules are formed, the space starts to fill, causing a corresponding increase in pressure inside the element. This change in pressure is recorded over time.

There are two basic types of hydrogen probes. One type consists of a thin carbon steel tube with a solid rod inside it and a small space around the rod. This tube is inserted into the flow stream. Hydrogen atoms migrate from the product into the flow stream. As hydrogen molecules collect, the pressure increases and is displayed by a pressure gauge.

The second type, the patch probe, operates on the same principle. In this case, the patch is sealed to the exterior surface of the pipeline and collects hydrogen atoms that penetrate the wall. Hydrogen probes are more suited for identifying trends than for determining exact rates of corrosion.

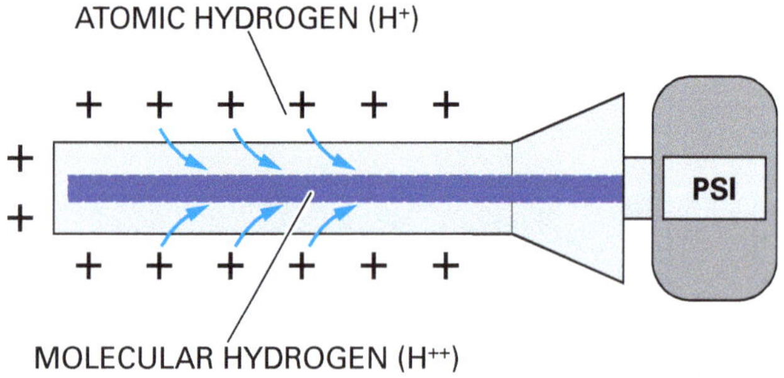

Figure 10 Schematic diagram of a hydrogen probe.

1.2.5 Biological Probes

Microbiologically influenced corrosion (MIC) is electrochemical corrosion accelerated by the metabolic activities of microorganisms such as **sulfate-reducing bacteria (SRB)**. SRBs reduce sulfates to sulfides, releasing sulfuric acid and hydrogen sulfide, which then react with iron to form a black precipitate (iron sulfide). Hydrogen sulfide gas is toxic, flammable, and reduces the quality of the petroleum in the process line.

Biological probes are used to collect samples of bacteria in pipeline systems. Bacteria may be present in a system as either **sessile bacteria** (attached to the metal surface), or **planktonic bacteria** (present in the process fluid). The unrestricted growth of sessile bacteria is responsible for a large percentage of corrosion failures in pipeline systems. It must be controlled with chemical treatment before it creates a bio-film on the interior surface of the pipe. Once the microorganisms have formed a bio-film, they become more resistant to biocides and grow rapidly. *Figure 11* shows a biological probe designed for flush installation.

Figure 11 Biological probe.

Sidestream Assemblies

Sidestream assemblies are modular systems that can be used to perform a variety of system measurements. The system shown here can be customized to provide measurements of various system values such as metal loss, corrosion rate, erosion rate, pressure, temperature, flow, and bacterial activity.

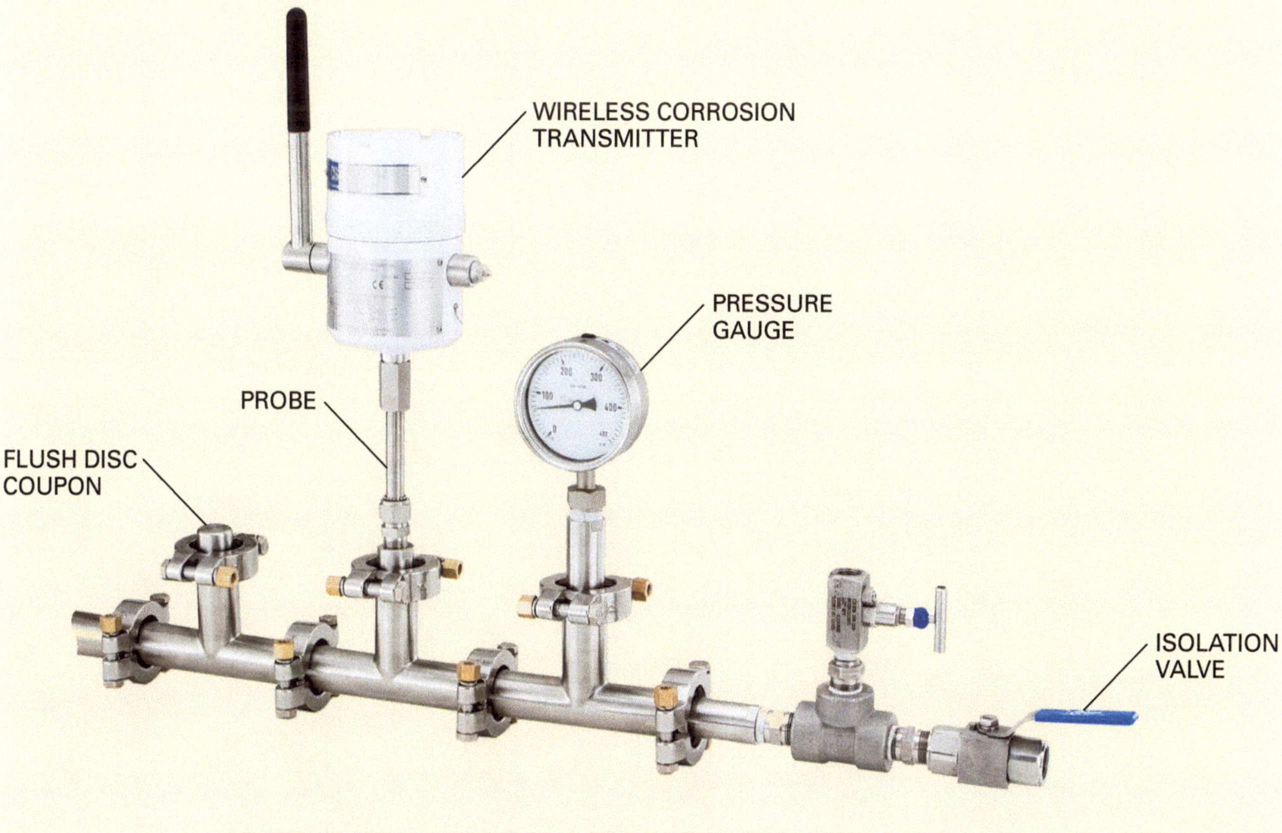

Additional Resources

It is assumed that you will learn to perform these procedures under the direct supervision of your crew leader, foreman, or instructor. For independent study, refer to the recommended Additional Resources at the back of this module.

1.0.0 Section Review

1. The acid gas that is most likely to result in deep pitting is _____.
   a. CO
   b. H_2S
   c. CO_2
   d. O_2

2. The type of probe often used to detect the entrance of oxygen into a system is a(n) _____.
   a. electrical resistance probe
   b. linear polarization resistance probe
   c. hydrogen probe
   d. galvanic probe

Section Two

2.0.0 Reading Online Probes

Objective

Describe how to read a probe (CT10_2-17).

a. Identify the prerequisites, competencies, and abnormal operating conditions (AOCs) associated with reading probes.
b. Describe the procedure used to read a probe.

Performance Task

1. Monitor online probes (CT10_2-17).
    - Identify potential abnormal operating conditions (AOCs) that may occur during performance of this CT, and know the appropriate actions to take in response to them.
    - Utilize the appropriate personal protective equipment according to relevant company procedures.
    - Locate the probe site and remove any secondary containment covers to gain access to the probe.
    - Confirm the probe terminals are acceptabe for use.
    - Connect the data cords from the data logger to the appropriate terminal of the probe.
    - Turn the data logger on and obtain the reading (follow the manufacturer's operating procedures).
    - Document the reading.
    - Disconnect the data logger leads from the probe.
    - Dress the secondary container cover with anti-seize compound and place the cap back on probe adapter.
    - Confirm data logger readings are appropriate (the data logger is functioning properly).
    - Complete appropriate documentation as required by operator's procedures.

Trade Term

Abnormal operating conditions (AOCs): Conditions within a pipeline system, identified by the operator, that may indicate (1) a malfunction of a component or (2) a deviation from normal operations that may indicate a condition exceeding design limits or result in a hazard to persons, property, or the environment.

Personnel assigned to monitor online probes are required to have some basic competencies. They must have the ability to carefully follow the required procedures for reading probes and recording the results. This section describes some of the basic skills needed by pipeline technicians and identifies abnormal operating conditions (AOCs) that personnel may need to respond to when monitoring online probes. It also explains the basic procedures for reading online probes.

2.1.0 Prerequisites, Competencies, and Abnormal Operating Conditions

Personnel must possess a variety of skills in order to safely and successfully monitor online probes. The various operations require a series of steps, which a pipeline technician must be competent to perform. Personnel must also be aware of the particular abnormal operating conditions (AOCs) that may occur during the procedure and know the proper responses to them.

2.1.1 Prerequisites and Competencies

Personnel must be familiar with different types of probes, including ER probes, LPR probes, galvanic probes, hydrogen probes, and biological probes. They must know how to access a probe and connect a data logger to it, make and verify the reading, determine the condition of the probe, and document and submit the vital information. They may also need to know how to extract a probe and insert a replacement probe, if necessary. Personnel must follow company safety policies, understand the correct energy isolation procedures, and be able to select and use the proper PPE. Personnel assigned to monitor probes must understand that the task begins when the secondary containment cover is opened, and ends when the probe data has been obtained, verified, and documented.

2.1.2 Abnormal Operating Conditions

Personnel monitoring probes online must be able to identify potential abnormal operating conditions (AOCs) and know the proper response to each, should they occur. The following are some potential AOCs, and the appropriate action personnel should take in response to them:

- *An interruption of communications, the control system, or the power occurs* – Begin troubleshooting procedures in an attempt to determine the cause and restore the power or communications.

- *A component such as a probe lead is damaged or broken* – Repair or replace the lead or other component.
- *An unexpected leak of hazardous liquid or gas occurs* – Isolate the leak and ventilate or clean the area if it is safe to do so. Proper notifications must be made.
- *Fire or explosion due to the ignition of hydrocarbons occurs* – Call the fire department and isolate and extinguish the fire if it is safe to do so.

2.2.0 Procedure

The procedure for reading a probe varies according to the type of probe. Always refer to the instructions provided by the manufacturer for the probe and data logger/recording instrument in use. This procedure is intended to serve as a general guideline only.

The following steps outline the general procedure for reading a probe:

WARNING!

Only trained, qualified individuals may perform pipeline monitoring procedures. Make sure the test equipment and leads are in working order before proceeding. Wear all appropriate personal protective equipment (PPE) and stay alert to the possibility of unexpected leaks. Safely eliminate all ignition sources and immediately notify the appropriate personnel in the event of a leak.

Step 1 Identify potential abnormal operating conditions (AOCs) that may occur, and know the appropriate actions to take in response to them.

Step 2 Notify the pipeline control center of the planned activity.

Step 3 Use company-prescribed PPE and clothing such as hard hat, safety glasses, and hard- toe shoes.

Step 4 Locate the probe site and remove any secondary containment covers to gain access to the probe.

Step 5 Inspect the probe elements to ensure they are not damaged or corroded.

Step 6 Connect the leads to the recording instrument (data logger) and the appropriate probe terminal. Follow the manufacturer's instructions to turn the data logger on and obtain the reading. If a data logger reading is unattainable and all external components are in good working order, there may be a problem with the probe itself.

Step 7 Document the reading on the appropriate company form. Depending on the type of data logger used, if may be necessary to immediately document the reading indicated on the LED if the data is not stored in the data logger for later retrieval.

Step 8 Determine if the readings indicate that the probe needs to be inspected and/or replaced. If replacement is indicated, seek assistance from the appropriate personnel for additional investigative or corrective action.

Step 9 Disconnect the data logger leads from the probe.

Step 10 Apply anti-seize compound to the secondary container cover and replace the cap on the probe adapter.

Step 11 Check the readings to ensure they are consistent with the expected readings supplied by the probe manufacturer. If they are outside the expected range, verify the operation of the data logger and/or check the probe.

Step 12 Complete proper documentation and make required notifications that the activity is complete.

Additional Resources

It is assumed that you will learn to perform these procedures under the direct supervision of your crew leader, foreman, or instructor. For independent study, refer to the recommended Additional Resources at the back of this module.

2.0.0 Section Review

1. If an interruption of power occurs, the proper response is to _____.
   a. notify the control center
   b. attempt to determine the cause and restore the power
   c. manually record the probe data
   d. disconnect the leads from the logger

2. Which of the following is true when reading a probe?
   a. All probes are read in the same way.
   b. The cover must stay in place during a reading.
   c. The instrument will automatically record required documentation.
   d. The leads are connected to the recording instrument and the probe terminal.

SUMMARY

Controlling internal corrosion is an integral part of safe pipeline operations. Some of the factors that may contribute to internal corrosion include acid gases, pH, chlorides, temperature, water content, partial pressure, velocity, and bacterial growth. All of these variables must be controlled to the extent possible in order to minimize corrosion.

Probes are sensing elements that extend into gas or liquid streams to measure electrical potential and other factors that influence corrosion. Corrosion detection probes must be monitored on a regular basis to ensure that the probe data measurements are functioning properly. Technicians use data loggers to obtain readings that are then used for corrosion monitoring and documentation. This module identified types of probes and described the general procedure for safely monitoring probes online.

Review Questions

1. The type of corrosion that is usually identifiable by etching in the bottom of the pits is called _____.
   a. general corrosion
   b. erosion corrosion
   c. localized corrosion
   d. hydrogen sulfide corrosion

2. The type of acid gas most likely to form Mesa pitting is _____.
   a. carbon monoxide
   b. oxygen
   c. carbon dioxide
   d. hydrogen sulfide

3. Of the following values, the pH value that is most acidic is _____.
   a. 3
   b. 6
   c. 9
   d. 12

4. The ambient temperature increases the degree of internal corrosion when it exceeds _____.
   a. 80°F
   b. 90°F
   c. 100°F
   d. 110°F

5. Which type of probe works as an automatic coupon?
   a. Hydrogen probes
   b. LPR probes
   c. Galvanic probes
   d. Electrical resistance probes

6. Galvanic probes function by detecting the difference in potential between _____.
   a. any corrosion and the pipe wall
   b. two dissimilar metals
   c. two parallel probes
   d. the anodic and cathodic areas on the pipe

7. Hydrogen atoms migrate through the outer skin of the hydrogen probe element and combine with other hydrogen atoms to form a measurable _____.
   a. corrosion
   b. film
   c. pressure increase
   d. barrier to corrosion

8. Which of the following statements is true?
   a. Probe data is always stored on the data logger.
   b. Probe data is only provided by an LED display.
   c. Probe data is not always stored on the data logger.
   d. Probe data is transmitted from the logger to the control center.

Trade Terms Quiz

Fill in the blank with the correct term that you learned from your study of this module.

1. Electrochemical corrosion accelerated by the metabolic activities of certain microorganisms, such as sulfate-reducing bacteria, is called ________.
2. A device used to monitor corrosion through the changes in electrical resistance of the element material in an electrolyte is called a(n) ________.
3. Conditions within a pipeline system, identified by the operator, that may indicate (1) a malfunction of a component or (2) a deviation from normal operations that may indicate a condition exceeding design limits or result in a hazard to persons, property, or the environment are called ________.
4. The formation of mineral deposits in a pipeline system is called ________.
5. The pressure exerted by a single gas in a mixture of gases is called the ________.
6. Condensation that forms when water is present in the product and the temperature of the exterior of the pipeline drops significantly below the product temperature is called ________.
7. Bacteria that attach to the pipeline walls and may result in microbiologically influenced corrosion (MIC) are called ________.
8. A group of bacteria commonly associated with microbiologically influenced corrosion (MIC) in petroleum pipelines is called ________.
9. Bacteria carried by the process fluid in a pipeline system are called ________.
10. The ability of a gasket to withstand the effects of chemicals and other substances without damage or change is called ________.

Trade Terms

Abnormal operating conditions (AOCs)
Microbiologically influenced corrosion (MIC)
Partial pressure
Planktonic bacteria
Probe element
Resistance
Scale
Sessile bacteria
Sulfate-reducing bacteria (SRB)
Water of condensation

Trade Terms Introduced in This Module

Abnormal operating conditions (AOCs): Conditions within a pipeline system, identified by the operator, that may indicate (1) a malfunction of a component or (2) a deviation from normal operations that may indicate a condition exceeding design limits or result in a hazard to persons, property, or the environment.

Microbiologically influenced corrosion (MIC): Electrochemical corrosion accelerated by the metabolic activities of certain microorganisms, such as sulfate-reducing bacteria (SRB).

Partial pressure: The pressure exerted by a single gas in a mixture of gases.

Planktonic bacteria: Bacteria carried by the process fluid in a pipeline system.

Probe element: A device used to monitor corrosion through the changes in electrical resistance of the element material in an electrolyte.

Resistance: In this context, the ability of a gasket to withstand the effects of chemicals and other substances without damage or change.

Scale: The formation of mineral deposits in a pipeline system.

Sessile bacteria: Bacteria that attach to the pipeline walls and may result in microbiologically influenced corrosion (MIC).

Sulfate-reducing bacteria (SRB): A group of bacteria commonly associated with microbiologically influenced corrosion (MIC) in petroleum pipelines. SRBs reduce sulfates to sulfides, releasing sulfuric acid and hydrogen sulfide.

Water of condensation: Condensation that forms when water is present in the product and the temperature of the exterior of the pipeline drops significantly below the product temperature.

Additional Resources

The following recommended resources can provide additional helpful information related to the requirements for this covered task:

API Recommended Practice 1161, Third Edition, © 2014 American Petroleum Institute. API Publishing Services, 1220 L Street, NW, Washington, DC 20005. Distributed by Thompson Reuters (Scientific) LLC, **www.techstreet.com**.

49 *CFR* 192, Subpart I, **www.ecfr.gov**

49 *CFR* 195, Subpart H, **www.ecfr.gov**

NCCER Module AOCCC-17, *Abnormal Operating Conditions Control Center*

NCCER Module AOCFG-17, *Abnormal Operating Conditions Field and Gas*

NCCER Module CT8_1-17, *Measure Pit Depth with Pit Gauge*

NCCER Module CT8_2-17, *Measure Wall Thickness with Handheld Ultrasonic Meter*

NCCER Module CT8_3-17, *Measure Corroded Area*

NCCER Module CT10_1-17, *Insert and Remove Coupons*

Electrical Resistance E/RTM Monitoring. Metal Samples Company. Munford, AL.

Linear Polarization Resistance (LPR). Metal Samples Company. Munford, AL.

MNL20-2ND, Corrosion Tests and Standards: Application and Interpretation. 2005, available from ASTM International at **www.astm.org**.

Oil and Gas Pipelines: Integrity and Safety Handbook. R. Winston Revie. 2015. Hoboken, NJ: John Wiley and Sons, Inc.

Figure Credits

Technical ToolBoxes, Inc., Module Opener

Rohrback Cosasco Systems, Inc., Figures 6, 11, SA01

Section Review Answer Key

Answer	Section Reference	Objective
Section One		
1. d	1.1.1	1a
2. d	1.2.3	1b
Section Two		
1. b	2.1.2	2a
2. d	2.2.0	2b

NCCER CURRICULA — USER UPDATE

NCCER makes every effort to keep its textbooks up-to-date and free of technical errors. We appreciate your help in this process. If you find an error, a typographical mistake, or an inaccuracy in NCCER's curricula, please fill out this form (or a photocopy), or complete the online form at **www.nccer.org/olf**. Be sure to include the exact module ID number, page number, a detailed description, and your recommended correction. Your input will be brought to the attention of the Authoring Team. Thank you for your assistance.

Instructors – If you have an idea for improving this textbook, or have found that additional materials were necessary to teach this module effectively, please let us know so that we may present your suggestions to the Authoring Team.

NCCER Product Development and Revision

13614 Progress Blvd., Alachua, FL 32615

Email: curriculum@nccer.org

Online: www.nccer.org/olf

❏ Trainee Guide ❏ Lesson Plans ❏ Exam ❏ PowerPoints Other ____________________

Craft / Level: Copyright Date:

Module ID Number / Title:

Section Number(s):

Description:

Recommended Correction:

Your Name:

Address:

Email: Phone:

CT11_0-17

Monitoring and Controlling the Injection Rate of the Corrosion Inhibitor

Overview

When corrosion inhibitors are used to mitigate internal corrosion, the operator must inject the inhibitor in sufficient quantities to ensure design coverage of the inhibitor. This module describes the chemical injection systems used to inject corrosion inhibitors into a pipeline system and explains how to monitor and adjust the inhibitor injection rates.

Trainees with successful module completions may be eligible for credentialing through the NCCER Registry. To learn more, go to **www.nccer.org** or contact us at 1.888.622.3720. Our website has information on the latest product releases and training, as well as online versions of our *Cornerstone* magazine and Pearson's product catalog.

Your feedback is welcome. You may email your comments to **curriculum@nccer.org**, send general comments and inquiries to **info@nccer.org**, or fill in the User Update form at the back of this module.

This information is general in nature and intended for training purposes only. Actual performance of activities described in this manual requires compliance with all applicable operating, service, maintenance, and safety procedures under the direction of qualified personnel. References in this manual to patented or proprietary devices do not constitute a recommendation of their use.

CT11_0-17
Monitoring and Controlling the Injection Rate of the Corrosion Inhibitor

Objectives

When you have completed this module, you will be able to do the following:

1. Identify the components of a typical chemical injection system, and describe how to monitor and adjust the inhibitor injection rates.
   a. Identify the components of a typical chemical injection system.
   b. Describe how to monitor and adjust the inhibitor injection rates.
2. Describe how to monitor and control the injection rate of the corrosion inhibitor (CT11_0-17).
   a. Identify the prerequisites, competencies, and abnormal operating conditions (AOCs) associated with monitoring and controlling the injection rate of the corrosion inhibitor.
   b. Describe procedures for monitoring and controlling the injection rate of the corrosion inhibitor.

Performance Task

Under the supervision of the instructor, you should be able to do the following:

1. Monitor and control the injection rate of the corrosion inhibitor (CT11_0-17).
   - Identify potential abnormal operating conditions (AOCs) that may occur during performance of this CT, and know the appropriate actions to take in response to them.
   - Utilize the appropriate personal protective equipment according to relevant company procedures.
   - Visually inspect the tank and injection system to verify that the injection system is operating. Check for indications of leakage.
   - Ensure that the volume of inhibitor is sufficient to last until the next inspection.
   - Monitor the flow rate of the inhibitor to determine if it is correct to the established flow rate.
   - Adjust the injection rate to meet the established flow rate, if necessary.
   - Complete appropriate documentation as required by operator's procedures.

Trade Terms

Abnormal operating conditions (AOCs)
Atomizing nozzle
Inhibitor
Nipple
Parts per million (ppm)
Quill
Viscosity

Industry Recognized Credentials

If you are training through an NCCER-accredited sponsor, you may be eligible for credentials from NCCER's Registry. The ID number for this module is CT11_0-17. Note that this module may have been used in other NCCER curricula and may apply to other level completions. Contact NCCER's Registry at 888.622.3720 or go to **www.nccer.org** for more information.

Contents

Figures and Tables

SECTION ONE

1.0.0 CHEMICAL INJECTION SYSTEMS

Objective

Identify the components of a typical chemical injection system, and describe how to monitor and adjust the inhibitor injection rates.

a. Identify the components of a typical chemical injection system.
b. Describe how to monitor and adjust the inhibitor injection rates.

Trade Terms

Atomizing nozzle: A type of tubing termination designed to spray an arc of fine mist; typically used in gas pipelines.

Inhibitor: A chemical substance or combination of substances that, when in the proper concentration, forms an environment that prevents or reduces corrosion.

Nipple: A straight end tubing termination often used for drip injection into liquid pipeline systems.

Parts per million (ppm): Concentration of chemical/inhibitor per volume treated as calculated to control internal corrosion.

Quill: A type of tubing termination with a beveled end similar to a fountain pen. Quills are used to deliver liquid product in a single direction such as along a pipe wall.

Viscosity: The thickness of a fluid or a measure of its resistance to fluid flow.

In order for corrosion to occur, four elements must be present: an anode, a cathode, an electrolyte, and a metal pathway for current flow. The most common electrolyte is liquid water, which may enter the system through a variety of means, including condensation, chemicals introduced into the system, or when the system is opened for maintenance or repair. Special precautions must be taken to avoid the entry of water vapor into a pipeline system. For example, all equipment must be properly dehydrated prior to installation.

Even when all precautions are taken, some water will inevitably enter the system. Chemical inhibition can be used to control internal corrosion. An inhibitor is added to a fluid to change the characteristics of the fluid or to form a protective coat on the inner surface of the pipeline. Some inhibitors, instead of forming a film on the entire surface, deposit a film only on anodic or cathodic locations along the interior surface. Others work by dissolving in the water, thereby preventing it from becoming an effective electrolyte. Chemical inhibitors are injected into the pipeline at a specific rate using an injection system.

1.1.0 Chemical Injection System Components

The components in a typical chemical injection system include the corrosion inhibitor, a tank, pump, power source, containment basin, and injection tubing. Corrosion inhibitors may be injected using either a continuous or a batch injection. *Figure 1* shows a typical chemical injection system.

1.1.1 Corrosion Inhibitors

Corrosion inhibitors are highly concentrated in order to minimize the frequency of tank refills. These chemicals may damage components if not rated for use in the system. Soft materials such as seals are particularly vulnerable to chemical degradation.

In addition to chemical compatibility, the inhibitor must be supplied at the correct viscosity for the distribution system. A chemical that is too thick or too thin will not be distributed at the anticipated rate. Excess viscosity will also result in premature wearing of pumps and other components.

The inhibitor must also be matched to the delivery system type, either continuous flow or batch injection. Some chemicals are designed for delivery via continuous flow, while others are designed for intermittent or batch injection. Using the wrong type of delivery system will affect system performance and may allow corrosion to occur.

Special safety precautions are required when using inhibitors. These chemicals are flammable and may be fatal if ingested by mouth or through airways. Inhibitors are also toxic to wildlife. Always handle them with care to prevent personal injury and destructive environmental impacts. Refer to the safety data sheet (SDS) for the chemical in use, and wear the appropriate personal protective equipment when working with or near these chemicals. The SDS for a typical corrosion inhibitor is provided in the *Appendix* of this module.

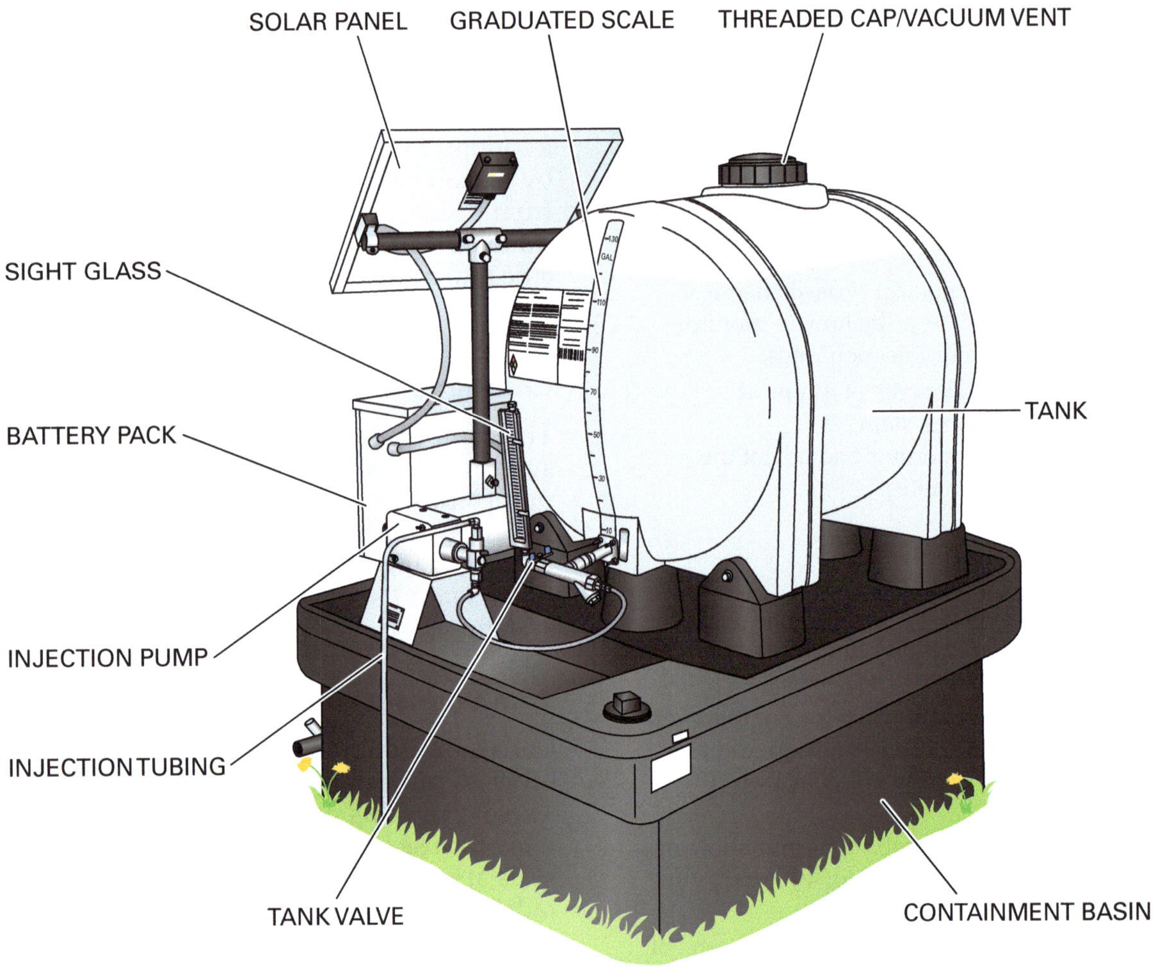

Figure 1 Typical chemical injection system.

1.1.2 Tanks

Chemical injection system tanks can be made of polyethylene, stainless steel, or fiberglass. These tanks are available in various sizes from 65 gallons (246 liters) up to 335 gallons (1,268 liters) or more.

Tanks must be sturdy and designed to resist chemical and ultraviolet (UV) damage. Polyethylene tanks offer the advantage of being lightweight and economical. Unlike steel or fiberglass, polyethylene tanks are seamless, which reduces the chance of leaks.

Tanks may be vertical or horizontal in configuration, and typically include graduated fill markings on the front or side. Indents may be provided on horizontal units for secure banding to the tank stand (*Figure 2*). Tank stands may be steel or polyethylene. Polyethylene offers the advantage of being resistant to weather, chemicals, and corrosion.

1.1.3 Pumps

All pumps must be sized for the expected range of injection rates and a minimum discharge pressure. This pressure is determined by taking the system operating pressure and adding the pressure limitations of any system components. A gauge is used to regulate the discharge pressure and make adjustments as necessary.

When continuous injection is used, a pulsation dampener may be added to reduce pump pressure spikes and excess wear. All pump components and fittings must be compatible with the type of chemical(s) used in the system.

1.1.4 Power Sources

The pumps used to power injection systems can be pneumatic, electrical, or solar-operated.

Pneumatic systems can be operated using natural gas or instrument air. Chemical injectors are often powered by natural gas since it is readily available on many sites. However, care must be taken to ensure that the gas supply is compatible with the pump and its components. Many natural gas supplies contain concentrations of hydrogen sulfide or carbon dioxide, which can be harmful to system components. All natural gas compo-

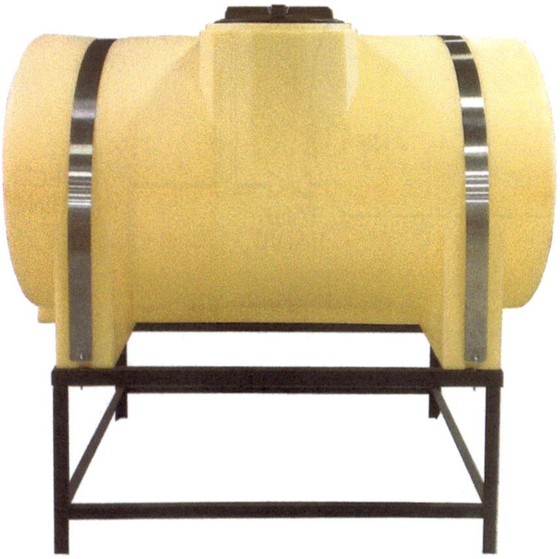

Figure 2 Polyethylene tank on steel stand with banding.

nents must meet the requirements of *ANSI/NACE MR0175 /ISO 15156*, Part 3.

Electrical pumps operate on either 120 or 240VAC. They require the use of utility power and may be more expensive to operate than pneumatic or solar-powered equipment. *Figure 3* shows a tank system with two pumps.

Solar-powered pumps offer the advantage of energy efficiency and allow for use in remote areas where electrical hookups are not available. However, they require regular battery checks and maintenance, and are not suitable for use in all climates.

1.1.5 Containment Basins

All tanks require a containment basin to protect the environment in the event of a spill. The containment basin capacity must equal or exceed the tank capacity, and it must be enclosed to prevent wildlife from ingesting the chemicals.

Containment basins may be made of stainless steel, fiberglass, or polyethylene. Polyethylene is preferred as it is lightweight, corrosion-proof, and resists both chemical and UV damage.

1.1.6 Injection Tubing

The tubing used with chemical injection systems is normally a narrow, rigid steel tube available in various lengths and diameters, with pressure ratings up to 1,500 pounds/square inch (psi) or 10.3 megapascals (MPa). Tubing must be sized to match the pump fittings and the maximum anticipated discharge pressure. The tubing may terminate in a **quill**, **nipple**, or an **atomizing nozzle**.

Figure 3 Tank system with electrically powered pumps.

A quill is a type of tubing termination with a beveled end, similar to the tip of a fountain pen. Quills are used to inject relatively small amounts of inhibitor under pressure to provide even dispersal throughout the pipeline product.

A nipple or open-end termination is the simplest type of termination. It can be used for drip injection of small quantities or larger batch injections.

An atomizing nozzle sprays an arc of fine mist to provide even dispersal of the inhibitor. Atomizers are precision components and must be carefully matched to the system pressure, the inhibitor used, and the desired injection rate. The chemical viscosity is important with all tubing terminations, but it is particularly critical with the use of atomizers. Atomizers must be equipped with a filter to avoid clogging the nozzle, preventing delivery of the inhibitor.

Generally, atomizers are used for gas pipelines, and quills or nipples are used for liquid pipelines. All terminations must be sized to match the required tubing insertion diameter and the maximum anticipated discharge pressure. A check valve (often a simple ball valve) is used to prevent the backflow of process fluids into the injection system. *Figure 4* shows various tubing terminations.

The tubing material must be compatible with the type of chemical used in the system. All

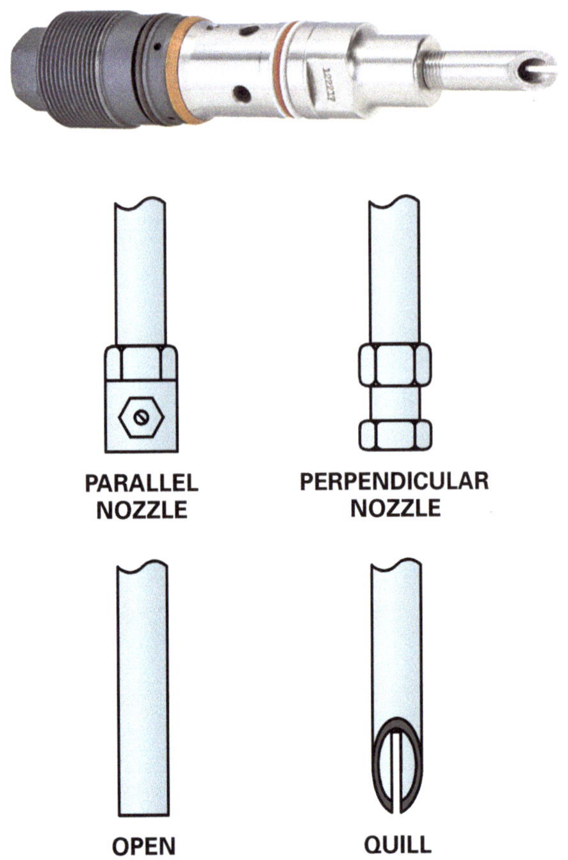

Figure 4 Tubing terminations.

system connections must be tight in order to prevent leaks and protect against the entry of moisture. *Figure 5* shows a chemical injection system attached to a pipeline.

1.2.0 Monitoring and Adjusting Inhibitor Injection Rates

Chemical injection systems require regular maintenance to ensure that the system is operating properly and delivering the inhibitor at the expected injection rate. Injection rates are determined by qualified individuals who take into consideration the volume, flow rate, and parts per million (ppm) required to protect the line. Too much inhibitor, or too little, can result in corrosion. The injection chemicals and rates may also be adjusted based on the requirements of the specific location in the pipeline.

All injection system components must be tested for proper operation and replaced if damaged. The flow rate must be monitored and adjusted

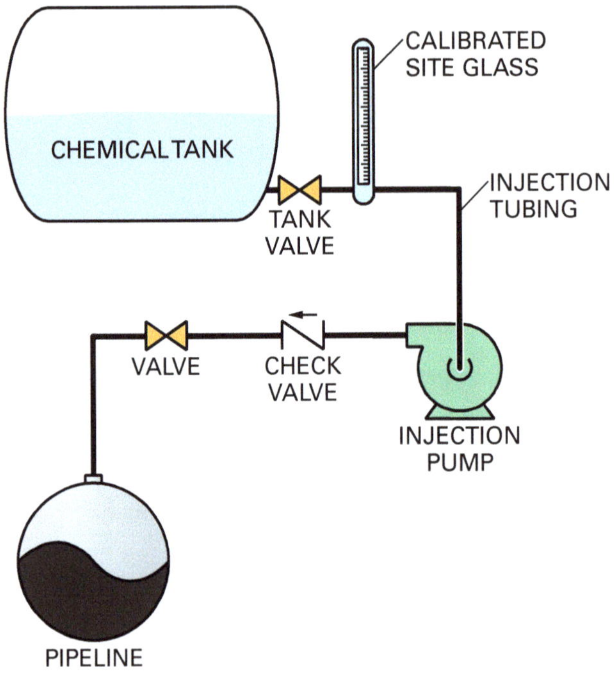

Figure 5 Chemical injection system attached to a pipeline.

as necessary. Possible items to document during inspections include the following:

- Amount of chemical in tank
- Injection rate
- Physical condition of system components
- Change in injection rate
- Follow-up items for next inspection
- Repairs done/spare parts used

Types of advanced testing include monitoring the downstream product to determine if it contains the anticipated concentration of inhibitor, and coupon monitoring to check for evidence of corrosion. Always follow company policies and procedures when performing any testing.

1.2.1 Monitoring the Injection Rate Using a Sight Glass

The injection rate is the amount of chemical to be injected within a specified time. It is defined in units of measure over a specified period, such as quarts per day (QPD) or liters per day (LPD). Chemical injection rates are normally determined by designated company personnel and verified using sight glass measurements.

On most sight glasses, there will be an inside set of marks and an outside set of marks. The inside marks represent the injection rate; the marks

Pigging

Prior to the use of chemical inhibitors it is important to start with a clean line, such as can be achieved through the use of an effective pigging program. Examples of pigs used for pipeline cleaning are shown here.

Figure Credit: Girard Industries

on the left are usually graduated in quarts per 24 hours, and the marks on the right are in liters per 24 hours. The outside marks usually indicate the total volume of chemical available; the marks on the left indicate gallons and the marks on the right indicate liters.

1.2.2 Adjusting the Injection Rate

After monitoring the flow rate of the inhibitor, adjust the injection rate as necessary to conform to the injection rate required to ensure that the proper amount of inhibitor is preventing corrosion activity. Always follow company policies and procedures when adjusting the injection rate.

SCADA Monitoring

Inhibitor injection systems may be designed for use with supervisory control and data acquisition (SCADA) monitoring. These systems are often used for remote monitoring of data such as flow rates and discharge pressures. A pneumatically powered chemical injection package for use with SCADA monitoring is shown here. This data helps to predict system problems and is used to pinpoint repairs, but does not replace the annual on-site survey.

Figure Credit: CheckPoint Pumps & Systems®

Additional Resources

It is assumed that you will learn to perform these procedures under the direct supervision of your crew leader, foreman, or instructor. For independent study, refer to the recommended Additional Resources at the back of this module.

1.0.0 Section Review

1. The viscosity of a fluid refers to its _____.
   a. specific gravity
   b. toxicity
   c. resistance to flow
   d. operating pressure

2. Injection rates are commonly measured on the sight glass in _____.
   a. quarts per 24 hours
   b. quarts per 8 hours
   c. quarts per hour
   d. quarts per minute

Section Two

2.0.0 Monitor and Control the Injection Rate of the Corrosion Inhibitor

Objective

Describe how to monitor and control the injection rate of the corrosion inhibitor (CT11_0-17).

a. Identify the prerequisites, competencies, and abnormal operating conditions (AOCs) associated with monitoring and controlling the injection rate of the corrosion inhibitor.
b. Describe procedures for monitoring and controlling the injection rate of the corrosion inhibitor.

Performance Task

Monitor and control the injection rate of the corrosion inhibitor (CT11_0-17).

- Identify potential abnormal operating conditions (AOCs) that may occur during performance of this CT, and know the appropriate actions to take in response to them.
- Utilize the appropriate personal protective equipment according to relevant company procedures.
- Visually inspect the tank and injection system to verify that the injection system is operating. Check for indications of leakage.
- Ensure that the volume of inhibitor is sufficient to last until the next inspection.
- Monitor the flow rate of the inhibitor to determine if it is correct to the established flow rate.
- Adjust the injection rate to meet the established flow rate, if necessary.
- Complete appropriate documentation as required by operator's procedures.

Trade Term

Abnormal operating conditions (AOCs): Conditions within a pipeline system, identified by the operator, that may indicate (1) a malfunction of a component or (2) a deviation from normal operations that may indicate a condition exceeding design limits or result in a hazard to persons, property, or the environment.

This section describes the procedures for monitoring and controlling the injection rate of the corrosion inhibitor. It also provides the prerequisites and competencies required for this covered task, and describes the **abnormal operating conditions (AOCs)** that may be encountered. This task begins with a visual observation of the injection system, and ends when proper documentation and notifications are completed.

2.1.0 Prerequisites, Competencies, and Abnormal Operating Conditions

Personnel must possess a variety of skills in order to monitor and control the injection rate of the corrosion inhibitor safely and efficiently. The various operations require a series of steps, which personnel must be competent to perform. Personnel must also be aware of the particular abnormal operating conditions (AOCs) that may occur during the installation procedure and know the proper responses to them.

2.1.1 Prerequisites and Competencies

Personnel assigned to monitor and control the injection rate of the corrosion inhibitor must possess the following basic knowledge and skills prior to performing this task:

- Familiarity with the product in the pipeline
- Familiarity with the testing equipment for injection system components
- Access to and familiarity with the Safety Data Sheet (SDS) for the inhibitor used
- Knowledge of company procedures and appropriate PPE to be used when performing this task

Personnel authorized to adjust the chemical injection rate must have a thorough knowledge of the injection system and of the permissible range for injection rates and volumes over time.

2.1.2 Abnormal Operating Conditions

When monitoring and controlling the injection rate of the corrosion inhibitor, a pipeline technician must be able to identify and properly respond to abnormal operating conditions (AOCs) should they occur. The following are potential AOCs, and the appropriate action personnel should take:

- *Damage or malfunction to injection system occurs, causing leak* – Stop the process and make proper notifications.

- *Unexpected hazardous liquid or carbon dioxide leak is encountered* – Identify and isolate the source of the leak if possible and if trained to do so. Proper notifications should be made as soon as possible, including calling 911 if necessary.
- *Fire or explosion occurs* – Limit access to the area and make the proper notifications, including calling 911 if necessary.

2.2.0 Procedures

Personnel assigned to monitor and control the injection rate of the corrosion inhibitor must always follow company policy and procedures. The basic steps for this task are outlined in this section.

> **WARNING!**
>
> Refer to the inhibitor SDS for the protective equipment to be worn when working on or near the injection system. Stay alert to the possibility of unexpected leaks. In the event of a leak, safely eliminate all ignition sources and immediately notify the appropriate personnel.

To monitor the injection rate using a sight glass, use the following general procedure:

Step 1 Identify potential abnormal operating conditions (AOCs) that may occur, and know the appropriate actions to take in response to them.

Step 2 Utilize the appropriate personal protective equipment according to relevant company procedures.

Step 3 Ensure that there is power to the injection unit.

Step 4 Allow the sight glass to fill to a convenient level using either the built-in fill valve or by closing the pump discharge valve between the tank and the sight glass.

Step 5 Mark the sight glass at the existing level.

Step 6 Open the pump discharge valve to the pipeline.

Step 7 Turn on the pump and allow it to run for one minute, and then turn it off again.

Step 8 Read the change in the fluid level on the sight glass.

Step 9 Re-calibrate two more times, then average the results and record the rate.

Step 10 Make sure to leave the tank valve in the Open position.

Step 11 Complete appropriate documentation as required by operator's procedures.

To adjust the injection rate, use the following general procedure:

Step 1 Identify potential abnormal operating conditions (AOCs) that may occur, and know the appropriate actions to take in response to them.

Step 2 Utilize the appropriate personal protective equipment according to relevant company procedures.

Step 3 Perform a visual inspection of the tank and injection system for damage and signs of leaking. Verify that the injection system is operating correctly.

Step 4 Verify that the volume of inhibitor is sufficient to last until the next inspection.

Step 5 Monitor the flow rate of the inhibitor. This may include measuring the change in volume of the inhibitor using a sight glass, as described previously.

Step 6 Adjust the injection rate as necessary to conform to the required injection rate.

Step 7 Document all measurements and/or changes and make notifications per company procedures.

Additional Resources

It is assumed that you will learn to perform these procedures under the direct supervision of your crew leader, foreman, or instructor. For independent study, refer to the recommended Additional Resources at the back of this module.

2.0.0 Section Review

1. If there is a leak due to damage or malfunction of the injection system, personnel must _____.
   a. stop the process and make notifications
   b. immediately evacuate the area
   c. eliminate any source of ignition
   d. attempt to repair the damage

2. After monitoring the injection rate using a sight glass, _____.
   a. recalibrate five times and average the results
   b. leave the tank valve in the Open position
   c. make appropriate notifications
   d. make sure the tank valve is closed

SUMMARY

Even when all precautions are taken, some water will inevitably enter a pipeline system. Chemical inhibition can be used to control internal corrosion. The components in a typical chemical injection system include the corrosion inhibitor, a tank, pump, power source, containment basin, and injection tubing. Corrosion inhibitors may be injected using either a continuous or a batch injection. All chemical injection systems must be monitored and controlled for correct injection rates.

Review Questions

1. The most common electrolyte is _____.
   a. salt
   b. water
   c. soil
   d. oil

2. According to Section 2 of the SDS in the *Appendix*, which of the following is *true* regarding the hazards of the product?
   a. It is extremely flammable.
   b. It is only flammable when mixed with petroleum products.
   c. It does not present an explosion hazard.
   d. It is stable when heated.

3. According to Section 5 of the SDS in the *Appendix*, the type of extinguisher that is likely to spread a fire involving this product is a(n) _____.
   a. carbon dioxide extinguisher
   b. foam
   c. water fog
   d. water jet

4. According to Section 8 of the SDS in the *Appendix*, the use of this product requires safety glasses with side shields (or goggles), protective gloves, and chemical resistant clothing. What type of gloves are recommended?
   a. canvas
   b. latex
   c. cotton
   d. nitrile/neoprene

5. According to Section 9 of the SDS in the *Appendix*, an inhibitor leak is most likely to _____.
   a. smell like sulfur
   b. smell like petroleum
   c. smell like ammonia
   d. be odorless

6. Section 12 of the SDS in the *Appendix* contains ecological information about the effects of this product on _____.
   a. birds
   b. beneficial insects
   c. aquatic life
   d. deer

7. Continuous injection systems subject to pressure spikes may benefit from the addition of a _____.
   a. larger pump
   b. double-walled pipe
   c. pulsation dampener
   d. second nozzle

8. Which of the following tubing terminations is most likely to be used in a gas pipeline system?
   a. Nipple
   b. Atomizing nozzle
   c. Quill
   d. Open end

9. Which of the following tubing terminations is most likely to be used to deliver a batch injection of inhibitor?
   a. Perpendicular atomizing nozzle
   b. Parallel atomizing nozzle
   c. Notched quill
   d. Open end

10. Chemical inhibitor injection rates are normally set by _____.
    a. the inhibitor vendor
    b. management procedure
    c. OSHA standards
    d. designated individuals in the company

Trade Terms Quiz

Fill in the blank with the correct term that you learned from your study of this module.

1. A chemical substance or combination of substances that, when in the proper concentration, forms an environment that prevents or reduces corrosion is called a(n) ________.
2. The thickness of a fluid or a measure of its resistance to fluid flow is its ________.
3. A straight end tubing termination often used for drip injection into liquid pipeline systems is called a(n) ________.
4. Conditions within a pipeline system, identified by the operator, that may indicate (1) a malfunction of a component or (2) a deviation from normal operations that may indicate a condition exceeding design limits or result in a hazard to persons, property, or the environment are called ________.
5. A type of tubing termination designed to spray an arc of fine mist, typically used in gas pipelines is a(n) ________.
6. A type of tubing termination with a beveled end similar to a fountain pen which is used to deliver liquid product in a single direction such as along a pipe wall is a(n) ________.
7. The concentration of chemical/inhibitor per volume treated as calculated to control internal corrosion is referred to in terms of ________

Trade Terms

Abnormal operating conditions (AOCs)
Atomizing nozzle
Inhibitor
Nipple
Parts per million (ppm)
Quill
Viscosity

Trade Terms Introduced in This Module

Abnormal operating conditions (AOCs): Conditions within a pipeline system, identified by the operator, that may indicate (1) a malfunction of a component or (2) a deviation from normal operations that may indicate a condition exceeding design limits or result in a hazard to persons, property, or the environment.

Atomizing nozzle: A type of tubing termination designed to spray an arc of fine mist; typically used in gas pipelines.

Inhibitor: A chemical substance or combination of substances that, when in the proper concentration, forms an environment that prevents or reduces corrosion.

Nipple: A straight end tubing termination often used for drip injection into liquid pipeline systems.

Parts per million (ppm): Concentration of chemical/inhibitor per volume treated as calculated to control internal corrosion.

Quill: A type of tubing termination with a beveled end similar to a fountain pen. Quills are used to deliver liquid product in a single direction such as along a pipe wall.

Viscosity: The thickness of a fluid or a measure of its resistance to fluid flow.

Corrosion Inhibitor Safety Data Sheet

SAFETY DATA SHEET

1. Identification

Product identifier	**Heavy Duty Corrosion Inhibitor**
Other means of identification	
Product code	06026
Recommended use	Corrosion inhibitor
Recommended restrictions	None known.

Manufacturer/Importer/Supplier/Distributor information

Manufactured or sold by:

Company name	CRC Industries, Inc.
Address	885 Louis Dr. Warminster, PA 18974 US
Telephone	
General Information	215-674-4300
Technical Assistance	800-521-3168
Customer Service	800-272-4620
24-Hour Emergency (CHEMTREC)	800-424-9300 (US) 703-527-3887 (International)
Website	www.crcindustries.com

2. Hazard(s) identification

Physical hazards	Flammable aerosols	Category 1
	Gases under pressure	Liquefied gas
Health hazards	Skin corrosion/irritation	Category 2
	Serious eye damage/eye irritation	Category 2A
	Reproductive toxicity (fertility)	Category 2
	Specific target organ toxicity, single exposure	Category 3 narcotic effects
	Specific target organ toxicity, repeated exposure	Category 2
	Aspiration hazard	Category 1
Environmental hazards	Hazardous to the aquatic environment, acute hazard	Category 2
	Hazardous to the aquatic environment, long-term hazard	Category 2
OSHA defined hazards	Not classified.	

Label elements

Signal word Danger

Hazard statement Extremely flammable aerosol. Contains gas under pressure; may explode if heated. May be fatal if swallowed and enters airways. Causes skin irritation. Causes serious eye irritation. May cause drowsiness or dizziness. Suspected of damaging fertility. May cause damage to organs (central nervous system, respiratory system, eyes, skin) through prolonged or repeated exposure. Toxic to aquatic life. Toxic to aquatic life with long lasting effects.

Material name: Heavy Duty Corrosion Inhibitor
06026 Version #: 01 Issue date: 12-02-2014

SDS US
1 / 10

Precautionary statement

Prevention	Obtain special instructions before use. Do not handle until all safety precautions have been read and understood. Keep away from heat/sparks/open flames/hot surfaces. - No smoking. Do not spray on an open flame or other ignition source. Do not apply while equipment is energized. Pressurized container: Do not pierce or burn, even after use. Extinguish all flames, pilot lights and heaters. Vapors will accumulate readily and may ignite. Use only with adequate ventilation; maintain ventilation during use and until all vapors are gone. Open doors and windows or use other means to ensure a fresh air supply during use and while product is drying. If you experience any symptoms listed on this label, increase ventilation or leave the area. Do not breathe gas. Do not breathe mist or vapor. Wash thoroughly after handling. Wear protective gloves/protective clothing/eye protection/face protection. Avoid release to the environment.
Response	If swallowed: Immediately call a poison center/doctor. Do NOT induce vomiting. If on skin: Wash with plenty of water. If skin irritation occurs: Get medical attention. Take off contaminated clothing and wash before reuse. If inhaled: Remove person to fresh air and keep comfortable for breathing. Call a poison center/doctor if you feel unwell. If in eyes: Rinse cautiously with water for several minutes. Remove contact lenses, if present and easy to do. Continue rinsing. If eye irritation persists: Get medical attention. If exposed or concerned: Get medical attention. Collect spillage.
Storage	Store in a well-ventilated place. Store locked up. Protect from sunlight. Do not expose to temperatures exceeding 50°C/122°F. Exposure to high temperature may cause can to burst.
Disposal	Dispose of contents/container in accordance with local/regional/national regulations.
Hazard(s) not otherwise classified (HNOC)	Static accumulating flammable liquid can become electrostatically charged even in bonded and grounded equipment. Sparks may ignite liquid and vapor. May cause flash fire or explosion.

3. Composition/information on ingredients

Mixtures

Chemical name	Common name and synonyms	CAS number	%
Liquefied Petroleum Gas		68476-86-8	20 - 30
2-Methylpentane		107-83-5	10 - 20
Naphtha (petroleum), hydrotreated light		64742-49-0	10 - 20
Stoddard Solvent		8052-41-3	10 - 20
Dipropylene glycol monomethyl ether		34590-94-8	3 - 5
Distillates (petroleum), hydrotreated light		64742-47-8	3 - 5
n-Hexane		110-54-3	< 1
Petrolatum, Micro Soft Wax		8009-03-8	< 0.2

Specific chemical identity and/or percentage of composition has been withheld as a trade secret.

4. First-aid measures

Inhalation	Remove victim to fresh air and keep at rest in a position comfortable for breathing. Call a POISON CENTER or doctor/physician if you feel unwell.
Skin contact	Remove contaminated clothing. Rinse skin with water/shower. If skin irritation occurs: Get medical advice/attention. Wash contaminated clothing before reuse.
Eye contact	Immediately flush eyes with plenty of water for at least 15 minutes. Remove contact lenses, if present and easy to do. Continue rinsing. Get medical attention if irritation develops and persists.
Ingestion	Call a physician or poison control center immediately. Rinse mouth. Do not induce vomiting. If vomiting occurs, keep head low so that stomach content doesn't get into the lungs.
Most important symptoms/effects, acute and delayed	Vapors have a narcotic effect and may cause headache, fatigue, dizziness and nausea. Aspiration may cause pulmonary edema and pneumonitis. Severe eye irritation. Symptoms may include stinging, tearing, redness, swelling, and blurred vision. Skin irritation. May cause redness and pain. Prolonged exposure may cause chronic effects.
Indication of immediate medical attention and special treatment needed	Provide general supportive measures and treat symptomatically. Keep victim under observation. Symptoms may be delayed.
General information	IF exposed or concerned: Get medical advice/attention. Ensure that medical personnel are aware of the material(s) involved, and take precautions to protect themselves. Show this safety data sheet to the doctor in attendance.

5. Fire-fighting measures

Suitable extinguishing media	Water fog. Foam. Carbon dioxide (CO_2). Dry chemical powder, carbon dioxide, sand or earth may be used for small fires only.
Unsuitable extinguishing media	Do not use water jet as an extinguisher, as this will spread the fire.
Specific hazards arising from the chemical	Contents under pressure. Pressurized container may rupture when exposed to heat or flame. This product is a poor conductor of electricity and can become electrostatically charged. If sufficient charge is accumulated, ignition of flammable mixtures can occur. Static electricity accumulation may be significantly increased by the presence of small quantities of water or other contaminants. Material will float and may ignite on surface of water. During fire, gases hazardous to health may be formed.
Special protective equipment and precautions for firefighters	Firefighters must use standard protective equipment including flame retardant coat, helmet with face shield, gloves, rubber boots, and in enclosed spaces, SCBA.
Fire-fighting equipment/instructions	In case of fire: Stop leak if safe to do so. Move containers from fire area if you can do so without risk. Containers should be cooled with water to prevent vapor pressure build up.
General fire hazards	Extremely flammable aerosol. Contents under pressure. Pressurized container may rupture when exposed to heat or flame.

6. Accidental release measures

Personal precautions, protective equipment and emergency procedures	Keep unnecessary personnel away. Keep people away from and upwind of spill/leak. Keep out of low areas. Remove all possible sources of ignition in the surrounding area. Many gases are heavier than air and will spread along ground and collect in low or confined areas (sewers, basements, tanks). Wear appropriate protective equipment and clothing during clean-up. Do not breathe mist or vapor. Do not breathe gas. Emergency personnel need self-contained breathing equipment. Do not touch damaged containers or spilled material unless wearing appropriate protective clothing. Ventilate closed spaces before entering them. Use appropriate containment to avoid environmental contamination. Local authorities should be advised if significant spillages cannot be contained. For personal protection, see section 8 of the SDS.
Methods and materials for containment and cleaning up	Eliminate all ignition sources (no smoking, flares, sparks, or flames in immediate area). Keep combustibles (wood, paper, oil, etc.) away from spilled material. Stop the flow of material, if this is without risk. Wipe up with absorbent material (e.g. cloth, fleece). Clean surface thoroughly to remove residual contamination. For waste disposal, see section 13 of the SDS. Prevent entry into waterways, sewer, basements or confined areas.
Environmental precautions	Avoid release to the environment. Prevent further leakage or spillage if safe to do so. Avoid discharge into drains, water courses or onto the ground. Inform appropriate managerial or supervisory personnel of all environmental releases. Use appropriate containment to avoid environmental contamination.

7. Handling and storage

Precautions for safe handling	Obtain special instructions before use. Do not handle until all safety precautions have been read and understood. Minimize fire risks from flammable and combustible materials (including combustible dust and static accumulating liquids) or dangerous reactions with incompatible materials. Pressurized container: Do not pierce or burn, even after use. Do not use if spray button is missing or defective. Do not spray on a naked flame or any other incandescent material. Do not smoke while using or until sprayed surface is thoroughly dry. Do not cut, weld, solder, drill, grind, or expose containers to heat, flame, sparks, or other sources of ignition. Use caution around energized equipment. The metal container will conduct electricity if it contacts a live source. This may result in injury to the user from electrical shock and/or flash fire. Do not breathe mist or vapor. Do not breathe gas. Avoid contact with eyes, skin, and clothing. Pregnant or breastfeeding women must not handle this product. Should be handled in closed systems, if possible. Use only in well-ventilated areas. Wear appropriate personal protective equipment. Wash hands thoroughly after handling. Avoid release to the environment. Observe good industrial hygiene practices. For product usage instructions, please see the product label.
Conditions for safe storage, including any incompatibilities	Level 3 Aerosol. Pressurized container. Protect from sunlight and do not expose to temperatures exceeding 50°C/122 °F. Do not puncture, incinerate or crush. Do not handle or store near an open flame, heat or other sources of ignition. This material can accumulate static charge which may cause spark and become an ignition source. Avoid spark promoters. These alone may be insufficient to remove static electricity. Store in a well-ventilated place. Store away from incompatible materials (see Section 10 of the SDS).

8. Exposure controls/personal protection

Occupational exposure limits

US. OSHA Table Z-1 Limits for Air Contaminants (29 CFR 1910.1000)

Components	Type	Value	Form
Dipropylene glycol monomethyl ether (CAS 34590-94-8)	PEL	600 mg/m3	
		100 ppm	
n-Hexane (CAS 110-54-3)	PEL	1800 mg/m3	
		500 ppm	
Petrolatum, Micro Soft Wax (CAS 8009-03-8)	PEL	5 mg/m3	Mist.
Stoddard Solvent (CAS 8052-41-3)	PEL	2900 mg/m3	
		500 ppm	

US. ACGIH Threshold Limit Values

Components	Type	Value	Form
2-Methylpentane (CAS 107-83-5)	STEL	1000 ppm	
	TWA	500 ppm	
Dipropylene glycol monomethyl ether (CAS 34590-94-8)	STEL	150 ppm	
	TWA	100 ppm	
n-Hexane (CAS 110-54-3)	TWA	50 ppm	
Petrolatum, Micro Soft Wax (CAS 8009-03-8)	TWA	5 mg/m3	Inhalable fraction.
Stoddard Solvent (CAS 8052-41-3)	TWA	100 ppm	

US. NIOSH: Pocket Guide to Chemical Hazards

Components	Type	Value	Form
2-Methylpentane (CAS 107-83-5)	Ceiling	1800 mg/m3	
		510 ppm	
	TWA	350 mg/m3	
		100 ppm	
Dipropylene glycol monomethyl ether (CAS 34590-94-8)	STEL	900 mg/m3	
		150 ppm	
	TWA	600 mg/m3	
		100 ppm	
Distillates (petroleum), hydrotreated light (CAS 64742-47-8)	TWA	100 mg/m3	
n-Hexane (CAS 110-54-3)	TWA	180 mg/m3	
		50 ppm	
Petrolatum, Micro Soft Wax (CAS 8009-03-8)	STEL	10 mg/m3	Mist.
	TWA	5 mg/m3	Mist.
Stoddard Solvent (CAS 8052-41-3)	Ceiling	1800 mg/m3	
	TWA	350 mg/m3	

Biological limit values

ACGIH Biological Exposure Indices

Components	Value	Determinant	Specimen	Sampling Time
n-Hexane (CAS 110-54-3)	0.4 mg/l	2,5-Hexanedion, without hydrolysis	Urine	*

* - For sampling details, please see the source document.

Exposure guidelines

US - California OELs: Skin designation

Dipropylene glycol monomethyl ether (CAS 34590-94-8)	Can be absorbed through the skin.
n-Hexane (CAS 110-54-3)	Can be absorbed through the skin.

US - Tennessee OELs: Skin designation

Dipropylene glycol monomethyl ether (CAS 34590-94-8)	Can be absorbed through the skin.

US ACGIH Threshold Limit Values: Skin designation

Dipropylene glycol monomethyl ether (CAS 34590-94-8)	Can be absorbed through the skin.
n-Hexane (CAS 110-54-3)	Can be absorbed through the skin.

US NIOSH Pocket Guide to Chemical Hazards: Skin designation

Dipropylene glycol monomethyl ether (CAS 34590-94-8)	Can be absorbed through the skin.

US. OSHA Table Z-1 Limits for Air Contaminants (29 CFR 1910.1000)

Dipropylene glycol monomethyl ether (CAS 34590-94-8)	Can be absorbed through the skin.

Appropriate engineering controls Good general ventilation (typically 10 air changes per hour) should be used. Ventilation rates should be matched to conditions. If applicable, use process enclosures, local exhaust ventilation, or other engineering controls to maintain airborne levels below recommended exposure limits. If exposure limits have not been established, maintain airborne levels to an acceptable level. Eye wash facilities and emergency shower must be available when handling this product.

Individual protection measures, such as personal protective equipment

Eye/face protection Wear safety glasses with side shields (or goggles).

Skin protection

Hand protection Wear protective gloves such as: Neoprene. Nitrile.

Other Wear appropriate chemical resistant clothing. Use of an impervious apron is recommended.

Respiratory protection If engineering controls are not feasible or if exposure exceeds the applicable exposure limits, use a NIOSH-approved cartridge respirator with an organic vapor cartridge. Use a self-contained breathing apparatus in confined spaces and for emergencies. Air monitoring is needed to determine actual employee exposure levels.

Thermal hazards Wear appropriate thermal protective clothing, when necessary.

General hygiene considerations When using, do not eat, drink or smoke. Always observe good personal hygiene measures, such as washing after handling the material and before eating, drinking, and/or smoking. Routinely wash work clothing and protective equipment to remove contaminants.

9. Physical and chemical properties

Appearance	
Physical state	Liquid.
Form	Aerosol.
Color	Dark amber.
Odor	Petroleum.
Odor threshold	Not available.
pH	Not available.
Melting point/freezing point	-244.7 °F (-153.7 °C) estimated
Initial boiling point and boiling range	118.4 °F (48 °C) estimated
Flash point	< 0 °F (< -17.8 °C) Tag Closed Cup
Evaporation rate	Fast.
Flammability (solid, gas)	Not available.
Upper/lower flammability or explosive limits	
Flammability limit - lower (%)	0.7 % estimated
Flammability limit - upper (%)	14 % estimated
Vapor pressure	1437.2 hPa estimated
Vapor density	> 1 (air = 1)
Relative density	0.72 estimated
Solubility (water)	Negligible.

Partition coefficient (n-octanol/water)	Not available.
Auto-ignition temperature	404.6 °F (207 °C) estimated
Decomposition temperature	Not available.
Viscosity (kinematic)	Not available.
Percent volatile	79.2 % estimated

10. Stability and reactivity

Reactivity	The product is stable and non-reactive under normal conditions of use, storage and transport.
Chemical stability	Material is stable under normal conditions.
Possibility of hazardous reactions	No dangerous reaction known under conditions of normal use.
Conditions to avoid	Avoid heat, sparks, open flames and other ignition sources. Contact with incompatible materials.
Incompatible materials	Strong oxidizing agents.
Hazardous decomposition products	Carbon oxides. Aldehydes.

11. Toxicological information

Information on likely routes of exposure

Ingestion	Droplets of the product aspirated into the lungs through ingestion or vomiting may cause a serious chemical pneumonia.
Inhalation	May cause damage to organs through prolonged or repeated exposure by inhalation. May cause drowsiness and dizziness. Headache. Nausea, vomiting.
Skin contact	Causes skin irritation.
Eye contact	Causes serious eye irritation.
Symptoms related to the physical, chemical and toxicological characteristics	May cause drowsiness and dizziness. Headache. Nausea, vomiting. Aspiration may cause pulmonary edema and pneumonitis. Severe eye irritation. Symptoms may include stinging, tearing, redness, swelling, and blurred vision. Skin irritation. May cause redness and pain.

Information on toxicological effects

Acute toxicity May be fatal if swallowed and enters airways. Narcotic effects.

Product	Species	Test Results
Heavy Duty Corrosion Inhibitor		
Acute		
Dermal		
LD50	Rabbit	4361.644 mg/kg estimated
Inhalation		
LC50	Rat	28.2233 mg/l estimated
Oral		
LD50	Rat	4778.8735 mg/kg estimated

* Estimates for product may be based on additional component data not shown.

Skin corrosion/irritation	Causes skin irritation.
Serious eye damage/eye irritation	Causes serious eye irritation.
Respiratory sensitization	Not available.
Skin sensitization	This product is not expected to cause skin sensitization.
Germ cell mutagenicity	No data available to indicate product or any components present at greater than 0.1% are mutagenic or genotoxic.
Carcinogenicity	This product is not considered to be a carcinogen by IARC, ACGIH, NTP, or OSHA.

IARC Monographs. Overall Evaluation of Carcinogenicity

Stoddard Solvent (CAS 8052-41-3) 3 Not classifiable as to carcinogenicity to humans.

Reproductive toxicity	Suspected of damaging fertility.
Specific target organ toxicity - single exposure	May cause drowsiness and dizziness.

Specific target organ toxicity - repeated exposure	May cause damage to organs through prolonged or repeated exposure: Central nervous system. Respiratory system. Skin. Eyes.
Aspiration hazard	May be fatal if swallowed and enters airways. If aspirated into lungs during swallowing or vomiting, may cause chemical pneumonia, pulmonary injury or death.
Chronic effects	May cause damage to organs through prolonged or repeated exposure.

12. Ecological information

Ecotoxicity Toxic to aquatic life with long lasting effects.

Product		**Species**	**Test Results**
Heavy Duty Corrosion Inhibitor			
Aquatic			
Acute			
Fish	LC50	Fish	5189.231 mg/l, 96 hours estimated

Components		**Species**	**Test Results**
Dipropylene glycol monomethyl ether (CAS 34590-94-8)			
Aquatic			
Acute			
Crustacea	EC50	Daphnia	> 5000 mg/l, 48 hours
Fish	LC50	Fathead minnow (Pimephales promelas)	10000 mg/l, 96 hours
Distillates (petroleum), hydrotreated light (CAS 64742-47-8)			
Aquatic			
Acute			
Fish	LC50	Fathead minnow (Pimephales promelas)	45 mg/l, 96 hours
n-Hexane (CAS 110-54-3)			
Aquatic			
Fish	LC50	Fathead minnow (Pimephales promelas)	2.101 - 2.981 mg/l, 96 hours

* Estimates for product may be based on additional component data not shown.

Persistence and degradability No data is available on the degradability of this product.

Bioaccumulative potential No data available.

Partition coefficient n-octanol / water (log Kow)

2-Methylpentane	3.74
n-Hexane	3.9
Stoddard Solvent	3.16 - 7.15

Mobility in soil No data available.

Other adverse effects No other adverse environmental effects (e.g. ozone depletion, photochemical ozone creation potential, endocrine disruption, global warming potential) are expected from this component.

13. Disposal considerations

Disposal of waste from residues / unused products	If discarded, this product is considered a RCRA ignitable waste, D001. Collect and reclaim or dispose in sealed containers at licensed waste disposal site. Contents under pressure. Do not puncture, incinerate or crush. Do not allow this material to drain into sewers/water supplies. Do not contaminate ponds, waterways or ditches with chemical or used container. Dispose in accordance with all applicable regulations.
Hazardous waste code	D001: Waste Flammable material with a flash point <140 F
Contaminated packaging	Empty containers should be taken to an approved waste handling site for recycling or disposal. Since emptied containers may retain product residue, follow label warnings even after container is emptied.

14. Transport information

DOT

UN number	UN1950
UN proper shipping name	Aerosols, flammable, Limited Quantity
Transport hazard class(es)	
Class	2.1
Subsidiary risk	-

Label(s)	2.1
Packing group	Not applicable.
Special precautions for user	Read safety instructions, SDS and emergency procedures before handling.
Special provisions	N82
Packaging exceptions	306
Packaging non bulk	None
Packaging bulk	None

IATA

UN number	UN1950
UN proper shipping name	Aerosols, flammable, Limited Quantity
Transport hazard class(es)	
Class	2.1
Subsidiary risk	-
Packing group	Not applicable.
Environmental hazards	No.
ERG Code	10L
Special precautions for user	Read safety instructions, SDS and emergency procedures before handling.
Other information	
Passenger and cargo aircraft	Allowed.
Cargo aircraft only	Allowed.

IMDG

UN number	UN1950
UN proper shipping name	AEROSOLS, LIMITED QUANTITY, MARINE POLLUTANT
Transport hazard class(es)	
Class	2
Subsidiary risk	-
Packing group	Not applicable.
Environmental hazards	
Marine pollutant	Yes
EmS	Not available.
Special precautions for user	Read safety instructions, SDS and emergency procedures before handling.

15. Regulatory information

US federal regulations This product is a "Hazardous Chemical" as defined by the OSHA Hazard Communication Standard, 29 CFR 1910.1200.
All components are on the U.S. EPA TSCA Inventory List.

TSCA Section 12(b) Export Notification (40 CFR 707, Subpt. D)

Not regulated.

SARA 304 Emergency release notification

Not regulated.

US. OSHA Specifically Regulated Substances (29 CFR 1910.1001-1050)

Not listed.

US EPCRA (SARA Title III) Section 313 - Toxic Chemical: Listed substance

Not listed.

CERCLA Hazardous Substance List (40 CFR 302.4)

Not listed.

CERCLA Hazardous Substances: Reportable quantity

Not listed.

Spills or releases resulting in the loss of any ingredient at or above its RQ require immediate notification to the National Response Center (800-424-8802) and to your Local Emergency Planning Committee.

Clean Air Act (CAA) Section 112 Hazardous Air Pollutants (HAPs) List

Not regulated.

Clean Air Act (CAA) Section 112(r) Accidental Release Prevention (40 CFR 68.130)

Not regulated.

Safe Drinking Water Act (SDWA) Not regulated.

Food and Drug Administration (FDA)	Not regulated.

Superfund Amendments and Reauthorization Act of 1986 (SARA)

Section 311/312 Hazard categories	Immediate Hazard - Yes Delayed Hazard - Yes Fire Hazard - Yes Pressure Hazard - Yes Reactivity Hazard - No
SARA 302 Extremely hazardous substance	No

US state regulations

US. California Controlled Substances. CA Department of Justice (California Health and Safety Code Section 11100)

Not listed.

US. New Jersey Worker and Community Right-to-Know Act

2-Methylpentane (CAS 107-83-5)
Dipropylene glycol monomethyl ether (CAS 34590-94-8)
Stoddard Solvent (CAS 8052-41-3)
n-Hexane (CAS 110-54-3)

US. Massachusetts RTK - Substance List

2-Methylpentane (CAS 107-83-5)
Dipropylene glycol monomethyl ether (CAS 34590-94-8)
Stoddard Solvent (CAS 8052-41-3)

US. Pennsylvania Worker and Community Right-to-Know Law

2-Methylpentane (CAS 107-83-5)
Dipropylene glycol monomethyl ether (CAS 34590-94-8)
Distillates (petroleum), hydrotreated light (CAS 64742-47-8)
n-Hexane (CAS 110-54-3)
Petrolatum, Micro Soft Wax (CAS 8009-03-8)
Stoddard Solvent (CAS 8052-41-3)

US. Rhode Island RTK

n-Hexane (CAS 110-54-3)

US. California Proposition 65

California Safe Drinking Water and Toxic Enforcement Act of 1986 (Proposition 65): This material is not known to contain any chemicals currently listed as carcinogens or reproductive toxins.

Volatile organic compounds (VOC) regulations

EPA

VOC content (40 CFR 51.100(s))	79.2 %
Consumer products (40 CFR 59, Subpt. C)	Not regulated

State

Consumer products	Not regulated
VOC content (CA)	79.2 %
VOC content (OTC)	79.2 %

International Inventories

Country(s) or region	Inventory name	On inventory (yes/no)*
Australia	Australian Inventory of Chemical Substances (AICS)	Yes
Canada	Domestic Substances List (DSL)	Yes
Canada	Non-Domestic Substances List (NDSL)	No
China	Inventory of Existing Chemical Substances in China (IECSC)	Yes
Europe	European Inventory of Existing Commercial Chemical Substances (EINECS)	No
Europe	European List of Notified Chemical Substances (ELINCS)	No
Japan	Inventory of Existing and New Chemical Substances (ENCS)	No
Korea	Existing Chemicals List (ECL)	Yes
New Zealand	New Zealand Inventory	No

Country(s) or region	Inventory name	On inventory (yes/no)*
Philippines	Philippine Inventory of Chemicals and Chemical Substances (PICCS)	No
United States & Puerto Rico	Toxic Substances Control Act (TSCA) Inventory	Yes

*A "Yes" indicates that all components of this product comply with the inventory requirements administered by the governing country(s)
A "No" indicates that one or more components of the product are not listed or exempt from listing on the inventory administered by the governing country(s).

16. Other information, including date of preparation or last revision

Issue date 12-02-2014

Prepared by Allison Cho

Version # 01

Further information Not available.

HMIS® ratings
Health: 2*
Flammability: 4
Physical hazard: 0
Personal protection: B

NFPA ratings
Health: 2
Flammability: 4
Instability: 0

NFPA ratings

Disclaimer CRC cannot anticipate all conditions under which this information and its product, or the products of other manufacturers in combination with its product, may be used. It is the user's responsibility to ensure safe conditions for handling, storage and disposal of the product, and to assume liability for loss, injury, damage or expense due to improper use. The information contained in this document applies to this specific material as supplied. It may not be valid for this material if it is used in combination with any other materials. This information is accurate to the best of CRC Industries' knowledge or obtained from sources believed by CRC to be accurate. Before using any product, read all warnings and directions on the label. For further clarification of any information contained on this (M)SDS consult your supervisor, a health & safety professional, or CRC Industries.

Additional Resources

The following recommended resources can provide additional helpful information related to the requirements for this covered task:

API Recommended Practice 1161, Third Edition, © 2014 American Petroleum Institute. API Publishing Services, 1220 L Street NW, Washington, DC 20005. Distributed by Thompson Reuters (Scientific) LLC, **www.techstreet.com**.

NCCER Module AOCCC-17, *Abnormal Operating Conditions Control Center.*

NCCER Module AOCFG-17, *Abnormal Operating Conditions Field and Gas.*

NCCER Module 62105-14, *Introduction to Pipeline Documents.*

NCCER Module 66101-14, *Introduction to the Pipeline Industry.*

MNL20-2ND, Corrosion Tests and Standards: Application and Interpretation. 2005. Available from ASTM International at **www.astm.org**.

Oil and Gas Pipelines: Integrity and Safety Handbook. R. Winston Revie. 2015. Hoboken, NJ: John Wiley and Sons, Inc.

ANSI/NACE MR0175 / ISO 15116, Part 3. 2015. Houston, TX: NACE International.

Figure Credits

LEWA GmbH, Module Opener, Figure 3
Poly-Mart, Inc., Figure 2
Rohrback Cosasco Systems, Inc., Figure 4 (photo)
CheckPoint Pumps & Systems®, Feature Art Figure 1
Girard Industries, Feature Art Figure 2B
CRC Industries, Appendix

Section Review Answer Key

Answer	Section Reference	Objective
Section One		
1. c	1.1.1	1a
2. a	1.2.1	1b
Section Two		
1. a	2.1.2	2a
2. b	2.2.0	2b

NCCER CURRICULA — USER UPDATE

NCCER makes every effort to keep its textbooks up-to-date and free of technical errors. We appreciate your help in this process. If you find an error, a typographical mistake, or an inaccuracy in NCCER's curricula, please fill out this form (or a photocopy), or complete the online form at **www.nccer.org/olf**. Be sure to include the exact module ID number, page number, a detailed description, and your recommended correction. Your input will be brought to the attention of the Authoring Team. Thank you for your assistance.

Instructors – If you have an idea for improving this textbook, or have found that additional materials were necessary to teach this module effectively, please let us know so that we may present your suggestions to the Authoring Team.

NCCER Product Development and Revision

13614 Progress Blvd., Alachua, FL 32615

Email: curriculum@nccer.org

Online: www.nccer.org/olf

❑ Trainee Guide ❑ Lesson Plans ❑ Exam ❑ PowerPoints Other ____________________

Craft / Level: Copyright Date:

Module ID Number / Title:

Section Number(s):

Description:

Recommended Correction:

Your Name:

Address:

Email: Phone: